T0143256

Huxley's Church and Maxwell's Demon

Huxley's Church and Maxwell's Demon

From Theistic Science to Naturalistic Science

MATTHEW STANLEY

THE UNIVERSITY OF CHICAGO PRESS   CHICAGO AND LONDON

MATTHEW STANLEY is associate professor at New York University's Gallatin School of Individualized Study. He is the author of *Practical Mystic: Religion, Science, and A. S. Eddington.*

The University of Chicago Press, Chicago 60637
The University of Chicago Press, Ltd., London
© 2015 by The University of Chicago
All rights reserved. Published 2015.
Printed in the United States of America
24 23 22 21 20 19 18 17 16 15   1 2 3 4 5

ISBN-13: 978-0-226-16487-8 (cloth)
ISBN-13: 978-0-226-16490-8 (e-book)
DOI: 10.7208/chicago/9780226164908.001.0001

Library of Congress Cataloging-in-Publication Data

Stanley, Matthew, 1975– author.
    Huxley's church and Maxwell's demon : from theistic science to naturalistic science / Matthew Stanley.
        pages ; cm
    Includes bibliographical references and index.
    ISBN 978-0-226-16487-8 (cloth : alk. paper) — ISBN 978-0-226-16490-8 (e-book)
1. Science—Great Britain—History—19th century.   2. Maxwell, James Clerk, 1831–1879.   3. Physicists—Great Britain.   4. Huxley, Thomas Henry, 1825–1895.
5. Naturalists—Great Britain.   6. Naturalism—History—19th century.   7. Religion and science—Great Britain—History—19th century.   I. Title.
    Q127.G4S73 2015
    509.2′241—dc23

                                                                          2014010890

FOR MAYA AND ZOË

Contents

Introduction

E very now and then, enemies agree. The philosopher of biology and anticreationist crusader Michael Ruse describes the practice of science thus: "Inasmuch as one is doing science, one avoid[s] all theological or other religious references. In particular, one denies God a role in creation."[1] Phillip Johnson, the U.C. Berkeley law professor and intellectual leader in the intelligent design (ID) community, says that if considerations of the divine were brought in, science would look "quite different."[2] While clashing over whether this is good or bad, they share a fundamental assumption: the practice of modern science is defined by the absence of religious considerations. Science is only recognizable insofar as God is nowhere to be seen. Adding the divine to science would change it in a profound way—again, for better or worse. The defining characteristic of science is its *naturalism*—broadly meaning the exclusion of supernatural or religious matters.[3] Both naturalists and antinaturalists say that for as long as there has been science, it has been naturalistic.[4] This book is about how we came to believe that.

The intelligent design movement targets naturalism for elimination, explicitly saying that modern science would go with it. The ID adherents' proposed alternatives bear little resemblance to what scientists do today, and are essentially completely different enterprises. Science proponents agree that without naturalism, there is nothing recognizable as science. The 2005 *Kitzmiller v. Dover* court decision in the United States equated naturalism with science to justify keeping intelligent design out of science classrooms, a move consistent with years of judicial precedent.[5] The decision declared that naturalism was "a 'ground rule' of science today which requires scientists to seek explanations in the world around us based upon what we can observe, test, replicate, and verify."[6]

Naturalism's prohibition of religious matters was immediately linked to the very ability to observe the world and test ideas. This is not at all unusual. Ruse and other philosophers such as Robert Pennock assert that basic empiricism and even the use of hypotheses requires a naturalistic approach.[7] Naturalistic science, it seems, is the only way to investigate the universe. Naturalism has provided "fantastic dividends."[8] It works better than anything else.

But it was not always this way. Naturalism has a history. The existential connection of naturalism with science is a relatively recent development. Further, naturalism has a specific birthplace. Despite naturalism's high profile in modern American courts, its roots are in Victorian Britain. It was not until the end of the Victorian period (1837–1901) that naturalism became a common way to think about science, and it was a distinctively British creation. Regardless of this late and local appearance, naturalistic science has come to be seen as universal and eternal. Somehow the long-standing practice of nonnaturalistic science has been forgotten. Science that was developed, considered, and evaluated in deeply religious contexts became sanitized of its divinity. We have come to think of science as *obviously* naturalistic, without even the possibility of an alternative. Examining the history of how we came to this point will help us understand the meaning and role of naturalism in modern America.

The use of the term itself is usually traced back to Thomas Henry Huxley—Charles Darwin's bulldog. In fact, *scientific naturalism* was originally used by American evangelicals in the mid-nineteenth century as a pejorative term to describe science practiced without reference or deference to religious matters. Huxley appropriated the term for his own purposes to indicate expertise rather than scandal. His first public use of it was in the prologue to his 1892 *Essays upon Some Controverted Questions* to indicate the form of science that he and his allies had been practicing for decades.[9] Bernard Lightman and Gowan Dawson suggest that Huxley began using the term after he had lost control of his earlier label—*agnosticism*—to Herbert Spencer. So while *scientific naturalist* was only used for self-identification at the end of the century, it is still a helpful label for values that had been in use since the 1850s by a particular group. These Victorian scientific naturalists desired a science that would be conducted in a "wholly secular temper." There could be no appeals to the supernatural or religious considerations brought into science. Frank Turner described their views as based on the tripod of atomic theory, evolution, and the conservation of energy.[10] These theories, with energy physics in particular, were presented by the scientific naturalists as show-ing that the universe was closed to external forces. Matter and force were posited to be the basic elements of science, not spirit or divine will. From these foundations the scientific naturalists launched attacks against the-ology and religious institutions, battling for power and cultural prestige. The naturalists had to fight for their definition of science.

The history of naturalism in a broader conceptual sense is more con-troversial.[11] It is complicated by a frequently drawn distinction between "methodological" naturalism and "metaphysical" or "ontological" nat-uralism. Methodological naturalism (what ID dismisses as "provisional atheism") states that science should be done *without reference to or con-sideration of* the supernatural or the divine.[12] This is to be distinguished from metaphysical naturalism, "the belief that there is nothing other than nature as we can see and observe it (in other words, that atheism is the right theology for the sound thinker)."[13] That is, that no supernatural or divine forces *exist*.

This distinction is offered by modern scientific naturalists as a way to provide space for religious belief even while doing science without God.[14] One could, it is suggested, be a methodological naturalist while also a metaphysical theist. However, there are difficulties. Even if one accepts the proposal of such a split between belief and practice, it is not always clear how to distinguish them.[15] Indeed, the combatants on both sides of the ID controversies regularly accuse each other of carelessly switching between the methodological and metaphysical varieties of naturalism.[16]

The purpose of the distinction, though, is a valuable one. It draws at-tention to how the *practice* of science distinguishes it from ID, rather than relying on specific claims about the physical world. The method-ological principles that underlie science are essential to its function. And both ID and its opponents agree that those principles as they operate in science today are distinct from religion.

There were, however, alternatives to this view. In the Victorian pe-riod, naturalism was only one possibility of how to practice science. Be-cause science today is naturalistic, it is easy to overestimate the influ-ence of scientific naturalists in the past.[17] Frank Turner has documented a group of British men of science who were hesitant to fully embrace nat-uralistic scientific practice: "They would not tolerate the curtailment of curiosity and the limitation of moral horizon that acceptance of scien-tific naturalism seemed to require."[18] But these figures were somewhat marginal and idiosyncratic. Far more central was the tradition of practicing science in close embrace with Christianity. This had been the standard in Britain since the days of Robert Boyle and Isaac Newton.

I will refer to this tradition as "theistic science." Its practitioners were overwhelmingly Christian (and largely Protestant), but the term *Christian science* would be confusing for obvious reasons. One could easily divide theistic science into many subcategories, such as Catholic, voluntarist, scriptural, and so forth. However, this would rapidly grow confusing. This book will instead focus on the common traits of theistic science, and how that methodological community functioned despite its internal diversity. It is also important to note that theistic science was quite distinct from what the ID community imagines its adherents' version of science to be. Theistic science, like naturalistic science, should be considered in terms of methodology, not in terms of particular theoretical allegiances. Theistic science was a way to *do* science. This is not a story about science versus religion. All parties discussed here, theist and naturalist, cared deeply about science and wanted it to thrive. Rather, the question was about how religious ideas and values should appear in scientific practice, if at all.

The core of this book explores the relationship between the methodological values of theistic and naturalistic science: that is, the foundational principles on which scientific researchers were expected to base their work. As already discussed, our modern expectation is that theism should dramatically change the way science is done. Here I will argue that this was not the case in Victorian Britain, the birthplace of scientific naturalism. Instead, both theistic and naturalistic science held virtually identical methodological values. While this is remarkable on its own, it is even stranger when we see that each group argued that proper scientific methodology could only be justified in their worldview. That is, naturalistic men of science thought science could only be done naturalistically; theistic men of science thought science could only be done theistically. Despite this split, the foundations of their practices were extremely similar, and they were able to function smoothly (if loudly) as members of the same intellectual communities.

This strange pairing is illuminated by the concept of "valence values," which I have developed elsewhere. At times, the same values are found in different groups, disciplines, or traditions. These shared, or valence, values help bond those sources together despite deep-seated differences.[19] Here, the valence values are a number of assumptions and expectations about scientific practice. So a theist and naturalist might disagree on the age of the Earth, or even the results of a particular experiment, but they share the value of (for example) empirical investigation that makes the debate possible in the first place. There were real fractures and splits within the Victorian scientific community; nevertheless, it held together due to powerful shared values regarding the foundations of science. This could also be considered as one of Peter Galison's trading zones, where groups are able to collaborate despite significant differences in training and goals.[20] Such groups retain their distinctness, but "can nonetheless coordinate their approaches around specific practices."[21] In this case those practices were the basic methodological assumptions and goals of science itself.

To explore these issues, this book will focus on one major representative from each of the theist and naturalist camps, with supporting figures appearing as necessary. A close view of this sort allows for deep analysis of specific critical points where naturalism and theism touched scientific practice, and how the details of biography and local context helped shape those intersections. Naturalistic science will be represented by T. H. Huxley (1825–95), pioneering biologist, iconoclastic science educator, and public spokesman for science. He was one of the major figures in creating and propagating naturalistic science, and was closely involved with virtually all the strategies, forces, and social developments that eventually led to the dominance of naturalism. Theistic science will be examined through James Clerk Maxwell (1831–79), the physicist whose work revolutionized electricity, magnetism, thermodynamics, and optics. Maxwell's contributions to science have survived the test of time, and as a conservative evangelical, he cannot be dismissed as someone who was not genuinely religious. He was a product of the long-standing tradition of theistic science that Huxley sought to overthrow, and helps demonstrate the powerful social, cultural, and intellectual forces that made that tradition productive. A surprising absence from this story is Charles Darwin. While Darwin's work became central to the worldview of the scientific naturalists, he rarely participated in the struggles to validate naturalistic science. This is not to say that he was not an important figure in Victorian science, only that he is not the best lens through which to examine the rise of naturalistic science.[22]

Chapter 1 situates us in Victorian Britain. Despite precedents in France, scientific naturalism in the modern sense was a peculiarly Victorian creation, and we need to understand the particulars of religion and science that made it possible. These will be addressed through overviews of Maxwell's and Huxley's careers, particularly with respect to the existing institutions of religion. The established Church of England dominated intellectual and educational life in many ways, both propagating theistic science and providing the impetus for the development of naturalistic science. Despite Anglican power, this was an age of increasing religious diversity, which provided important resources for the growth of naturalistic science. The scientific naturalists sought not only to battle established religion, but also to set up their worldview as an alternative framework for a full intellectual and cultural life. Huxley wanted not only a new science, but a new church.

Chapter 2 examines the concept of the uniformity of nature—the claim that the laws of nature are constant and never violated. Uniformity is a bedrock principle of modern scientific naturalism, and was similarly embraced as the core of what it meant to do Victorian science. In an important sense, all the other issues discussed in this book flow from this basic principle. By the 1850s, uniformity had already been established as a foundation of theistic science. The argument was that uniformity only made sense in a universe with a caretaker deity. This was embraced by both clerics and men of science, with very few protesters in the scientific community. This chapter will investigate Maxwell's electromagnetic theory to show how closely the theistic version of uniformity was tied to his scientific practice. Huxley needed to make the case that his worldview could also support a lawful universe, and he worked hard to reinterpret unity and uniformity as being naturalistic. He posited that one could only assume uniformity if there was *no* active deity able to disrupt natural processes. Uniformity was closely tied to concerns regarding how to think scientifically about miracles, and the chapter closes with a discussion of those concerns.

The correct understanding of what kind of knowledge science provides was a major concern for the Victorians, and remains so today. The crux of the issue is whether science provides certainty, and where the limits of speculation and theory lie. Pennock and Philip Kitcher attack ID supporters for desiring absolute truth and being unable to handle unknowns; the ID philosopher Alvin Plantinga accuses naturalists of being "wildly mistaken" for thinking their ideas are "*certain*."[23] Chapter 3 looks at the Victorian origin of these debates over the nature of provisional knowledge and the limits of scientific explanation. Huxley's agnosticism and Maxwell's models provide useful entry points for understanding the way they dealt with these concerns. Both theists and naturalists agreed that scientific knowledge was uncertain, subject to revision, and that science could only address phenomena insofar as they connected with the unity of nature. This led to their surprising agreement about the inability of science to discuss the beginning of the universe.

The science classroom was as important an issue for scientific naturalists in the nineteenth century as it is today. Chapter 4 shows how theists and naturalists were deeply involved with science education, and how they worked with similar goals and principles in their teaching. The focus of the chapter is on the case study of F. D. Maurice's Working Men's College, a Christian Socialist organization that saw science as a critical part of educating the working classes. Both Maxwell and Huxley taught at the Working Men's College, and their strong views on how science should be taught demonstrates how theists and naturalists saw science education as critical to the propagation of their own views and values. Maxwell followed the tradition of teaching science in a fundamentally theistic way, and Huxley pushed for wholly naturalistic science teaching. However, they agreed on enough of the foundations of science education that they both volunteered to teach at the same institution, in similar ways. In particular, theists and naturalists agreed on the moral and intellectual value of learning about natural laws, which was supposed to be of particular value to workers as they became part of the British polity.

Closely related to science education is the question of intellectual freedom. In the twenty-first century it is common for both naturalists and antinaturalists to accuse the other side of dogmatism and trying to impose their views on others.[24] Naturalism as a concept has been entangled with these issues from the start, and chapter 5 explores how freedom of belief played into Victorian ideals of how science should be done. The scientific naturalists vocally presented themselves as rebels against theological oppression, arguing that only their worldview allowed true freedom of inquiry. But theistic men of science also saw themselves as champions of intellectual freedom. The key overlap was both groups' suspicion of sectarianism and dogma. Everyone agreed that forced belief in any form was a profound threat to the practice of science. Huxley channeled this value into his attacks on the established Church, but Maxwell's embrace of it was shaped by his evangelicalism. This shared value of intellectual freedom was critical for the functioning of a scientific community that included both theists and naturalists.

The tapestry of shared values of theistic and naturalistic science begins to fray in chapter 6. The emerging science of psychology raised the difficult question of whether the processes of the mind were included within the uniformity of nature. The mind seemed to be quite different from other natural phenomena, and it was unclear how to treat it. Huxley appropriated William Carpenter's theistic physiology to argue for a fully naturalistic approach to mind and body—automatism. This chapter focuses on free will, a particularly controversial aspect of automatism. The naturalists insisted on determinism, while theistic scientists such as Maxwell struggled to maintain a view of mind compatible with Christian visions of the soul. Maxwell sought to retain the validity of both his evangelicalism and uniformity of nature, which required deep introspection into the reach of science and the nature of scientific explanation. Many of these issues found their way into his scientific practice through his investigations of the second law of thermodynamics and his eponymous "demon." In the end, theists and naturalists were unable to reconcile their different assumptions about the scientific analysis of the mind: was consciousness a fact upon which one could build science, or an epiphenomenon that needed to be explained away?

Despite this growing split, the victory of naturalistic science did not emerge from any demonstration of its methodological superiority. By the end of the nineteenth century, theistic and naturalistic science had been functioning side by side for decades, and it was only due to deliberate strategic choices that Huxley and his allies came to triumph. Chapter 7 argues that their key strategy was to make naturalistic science seem obvious and unique. Their chief tactics were to gain control of science education in the long term, and work to reframe concepts (such as uniformity) as solely naturalistic despite their theistic roots. The core of the plan was to reinterpret the history of science to erase its theistic past, and make science look as though it had always been naturalistic. These moves all required that naturalistic science largely share the same values as the theistic science it sought to replace, and led to a gradual generational change rather than a sudden revolution. Naturalism was given a long history, to make it seem impossible that science was ever practiced any other way. All of these strategies were critically enabled by large-scale social transitions in Britain, and were successfully brought to America as well. The Victorian scientific naturalists were so successful in telling their new story about science that today it is accepted by both naturalists and their enemies in the ID camp.

At the beginning of the Victorian period, it was expected that men of science take religious considerations into account. By the end, it seemed impossible that they would do so. *Huxley's Church and Maxwell's Demon* argues that this shift was, surprisingly, due largely to the deeply shared assumptions between theists and naturalists about how to do science. Unlike ID's choice to reject the basic elements that make science work, Victorian theists found harmony between scientific methodology and Christian belief. And it was the scientific naturalists' embrace of these same methodologies that led to their eventual victory.

CHAPTER ONE

Religious Lives

There were many ways to experience religion in Victorian Britain. Despite an established Church, religious belief and practice varied tremendously with class and locale. High Church Anglicans had their worship supported by the state, and subscribed to the Thirty-Nine Articles, which were the gateway to full participation in politics and society. Dissenters found their own ways to believe and practice, attending their humble chapels in industrial centers that thrived on commerce rather than links to the religious establishment. Radicals sought to disrupt the whole system.

The practice of science spread across these boundaries. Maxwell and Huxley were contemporaries who worked in the same scientific community, but their religious lives were dramatically different. Their assumptions and conclusions about the relationship of religion to their scientific work were heavily conditioned by these varied experiences. Neither were typical Victorians, though their religious lives bring out the critical themes and factors that set the groundwork for the coexistence of theistic and naturalistic science.

Maxwell

James Clerk Maxwell was born in 1831, an only child, to an aristocratic Scottish family that for generations had been interested in industry, agricultural improvement, mining, and natural philosophy.[1] The compound last name Clerk Maxwell came from the peculiar requirements of the inheritance of the family estate of Middlebie, a stretch of hilly land near Corsock in Dumfries and Galloway. Scotland of these years was full of religious ferment. The evangelical revival had brought great energy to Christianity in both England and Scotland. The country was overwhelmingly Presbyterian at the start of the century, with a strong strain of Calvinism.[2] The Church of Scotland splintered badly in 1843 over issues of religious freedom, evangelicalism, and church-state relationships. This "Great Disruption" saw a third of the church split off under Thomas Chalmers.[3] The struggles around the emergence of the Free Church were a prominent issue in the Maxwell household, and young James's father chose his teachers carefully lest the boy be swept up.[4] They were quite literally in the heart of the religious schism: the Maxwells attended St. Andrew's Church, the very building that sheltered the General Assembly that led to the Disruption.[5] James's mother, who died when he was nine, was an Episcopalian, and his aunt saw that he was educated in that tradition as well—he attended both services each Sunday.[6]

His father, John Clerk Maxwell, was a lawyer, but also deeply interested in science and industrial processes. He joined the Royal Society of Edinburgh and the Society for the Encouragement of the Useful Arts in Scotland, and personally designed and built the estate house (later known as Glenlair) at Middlebie. When the adult James visited Birmingham, John recommended that he inspect the various industries there:

> View, if you can, armourers, gunmaking and gunproving—swordmaking and proving—*Papier-mâchée* and japanning—silver-plating by cementation and rolling—ditto, electrotype—Elkington's works—Brazier's works, by founding and by striking up in dies—turning—spinning teapot bodies in white metal, etc.—making buttons of sorts, steel pens, needles, pins, and any sorts of small articles which are curiously done by subdivision of labour and by ingenious tools—glass of sorts is among the works of the place, and all kinds of foundry works—engine-making—tools and instruments (optical and philosophical) both coarse and fine.[7]

This interest in how things worked was an enormous influence on James as a child, who was notorious for constantly asking "show me how it doos" and investigating the hidden courses of bell wires. And if he was unsatisfied with the answer to his "What's the go o' that?" he would persist with "But what's the *particular* go of it?" Or he would ask "But how d'ye know it's blue?" to the point where his aunt confessed that it was "humiliating to be asked so many questions one could not answer 'by a child like that.'"[8]

After burning through tutors at home, the inquisitive child was sent to Edinburgh Academy. His country manner and homemade jacket and boots gave him a "somewhat rustic and somewhat eccentric" reputation, along with the nickname "Dafty."[9] There he became an expert in Latin and Greek and also met his lifelong friend and ally in physics P. G. Tait. With little formal instruction in mathematics or natural philosophy, at age fifteen Maxwell published his first paper: a treatise on oval curves for the Royal Society of Edinburgh. In 1847 he moved up to the University of Edinburgh, where he found more of an intellectual challenge from his instructors Sir William Hamilton and James Forbes.[10]

It was expected that James would follow his father into the law, but he gradually realized that he felt called to study "another kind of laws."[11] These were the laws of nature. That natural laws were the foundation of science was well accepted by 1850.[12] The previous generation of British natural philosophers (perhaps most influentially John Herschel and William Whewell) had hammered out a robust view of scientific investigation in which goals, methods, and tools were built around the expectation of a lawful universe.[13] Maxwell had familiarized himself with not only the great laws of Newtonian science, but also the cutting-edge speculations of Michael Faraday on electricity, Richard Owen on animal morphology, and the controversial *Vestiges of the Natural History of Creation*.[14] All these laws were seen as the result of a creative and caretaking deity. The laws of nature could not exist apart from divine power. The standard midcentury understanding of laws was described by Susan Cannon: "So the only rational belief is that a good God has been the great organizer of the world, which left to itself would be a corpuscular chaos. Any other belief is not so much irreligious; it is recognizably unscientific."[15]

Cambridge

It was accepted by everyone in Maxwell's life that the only place for him to study natural philosophy was at Cambridge. His entrance there in 1850 was unproblematic. The high fees were of no concern to his wealthy family. Other admission issues—religious tests—were also smooth. Cambridge's special place within the established Church required that students there swear to the Thirty-Nine Articles of the Anglican Church. Maxwell's broad religious upbringing meant this was no obstacle, and he was perfectly comfortable at the overwhelmingly Anglican institution.

After a brief time at Peterhouse College, he transferred to Trinity, where his mathematical skill was better appreciated. He began training with the famous coach William Hopkins, who reported that "it is not possible for [Maxwell] to think incorrectly on physical subjects."[16]

At Cambridge he met George Gabriel Stokes, the Lucasian Professor of Mathematics. He already had a substantial reputation in mathematical physics, and Maxwell attended his lectures regularly. Stokes was a well-known Anglican with an evangelical bent who engaged in substantial theological investigation.[17] Maxwell's experimental work while still a student attracted wide note. William Thomson (later Lord Kelvin) asked Maxwell to make him some "magne-crystallic" preparations that would be passed on to John Tyndall, who would then "discover the secrets of nature, and the origin of the magne-crystallic forces."[18] Thomson had been consulted by Maxwell's father about the boy's education, and was already professor of natural philosophy at Glasgow doing novel investigations of electricity and heat. He would go on to be one of the most influential men of science in Britain, bringing the distinctive values of his latitudinarian Christianity to all his work.[19] Stokes, Thomson, Maxwell, and Tait would form the core of what Crosbie Smith calls the North British group, who practiced theistic science with deep industrial connections and would largely create the new energy physics.[20]

Tait had preceded Maxwell to Cambridge, and their friendship grew. Maxwell's genial sociability made him popular, though his acceptance was not unqualified. His "Gallowegian" accent made him difficult to understand, and his "peculiar mode of speech" gave rise to puzzling rhymes.[21] Further, "his love of speaking in parables, combined with a certain obscurity of intonation, rendered it often difficult to seize his meaning."[22] He was a famous punster and wrote endless comic poems, much to the amusement and suffering of his friends. Consider his Newtonian parody of Robert Burns's "Comin' through the Rye":

> Gin a body meet a body
> Flyin' through the air
> Gin a body hit a body,
> Will it fly? And where?
> Ilka impact has its measure.
> Ne'er a ane hae I,
> Yet a' the lads they measure me,
> Or, at least, they try.[23]

Maxwell also participated actively in physical culture, reporting his achievements in gymnastics, swimming, and boating to his father. After learning to juggle and experimenting on falling cats, he even tried taking his exercise by running through the lodgings at 2:00 a.m., leading to his fellow students pelting him with missiles. Andrew Warwick has documented how intense physical exercise was an important part of the culture of Cambridge mathematics, and Maxwell's outrageous training was not unusual preparation for the rigors of the Tripos.[24]

Maxwell's education in Edinburgh had not fully prepared him for Cantabrigian expectations. He had to read extensively on moral philosophy, particularly Immanuel Kant and Thomas Hobbes. He also spent time with the work of William Paley, whose writings were enormously influential at the time. Paley's *Evidences* (required for the BA exam) articulated the contemporary Anglican approach to the historical and scriptural foundations of Christianity, but it was his *Natural Theology* that was more significant for a student at Cambridge interested in science.[25] This classic text articulated the venerable practice of using the study of the natural world to establish the existence and characteristics of a creative deity. Natural theology was overwhelmingly the standard context for the practice of science in the early nineteenth century. Robert Young and Susan Cannon have described this as creating a view of truth in which science and religion were seen as a single intellectual project.[26] John Hedley Brooke has shown how natural theology had an important social function, particularly in Britain. Its tactic of talking about the divine through the lens of nature, rather than scripture or sectarian doctrine, made it a useful religious glue across the doctrinal and confessional divisions that had caused so much trouble in the seventeenth century.[27] It provided inclusiveness without contention: "Belief in a beneficent designer could serve as a lowest common denominator for men whose primary interest was the pursuit of science."[28]

By the time Maxwell arrived at Cambridge, natural theology had become virtually synonymous with Paley's design argument. His simple suggestion that a watch implied a watchmaker relied on a combination of awe, complexity, and preexisting belief that was extremely popular. Proof per se was not the major intention or effect of natural theology; rather, it was useful for educational or apologetic purposes (and perhaps occasionally to win over a religious waverer).[29] However, there were many varieties of the argument, and the persuasiveness of various forms of natural theology was highly conditioned by time and place.[30] Even within Britain, natural theological arguments multiplied in form, style, and strategy. The *Bridgewater Treatises on the Power, Wisdom, and Goodness of God, as Manifested in the Creation* appeared in the 1830s, recruiting authors from across the political and theological spectrum to publish on how a wide variety of natural phenomena (from geology to astronomy to the human hand) could provide knowledge of God.[31] Few of the Bridgewater authors were willing to endorse Paley's Enlightenment confidence, and instead made it clear that their arguments were inductive, with no certainty. Nonetheless, natural theology was a standard part of science education for Maxwell's generation.[32] He read Bishop Butler's *The Analogy of Religion* at least twice as a young man, becoming familiar with that divine's arguments for the rationality of Christianity based on the uniformity and lawfulness of the natural world.[33] At Trinity he was surely exposed to William Whewell's own Bridgewater treatise on astronomy and his arguments for seeing design in the simplicity and perfection of the laws of physics.[34] Whewell was deeply concerned that natural theology be compatible with the progress of science, arguing that scripture needed to be reinterpreted in light of new discoveries and that biblical statements had been accommodated for the common man and were therefore inadmissible to scientific discussion.[35] For a student at Cambridge, religious considerations were a basic and obvious part of learning science. Theistic science was the norm.

Cambridge was a religiously stimulating place in the early 1850s. Maxwell joined the famous discussion club known as the Apostles, where he was both introduced to new ideas and developed his own. He also became familiar with the preaching of Harvey Goodwin and evangelical thought through his friends such as G. W. H. Tayler. Evangelical Christianity was not a separate sect, but rather a cross-denominational movement to reconceptualize the relationship of God and man through individual reflection and action.[36] In the evangelical framework, man was naturally depraved via original sin and was wholly other from the divine. Life was the opportunity to prove one's morality through the exercise of free will to choose a godly life over a worldly one. The individual conscience was the critical element in this scheme: evangelicalism discarded Calvinist predestination in favor of an emphasis on man's free ability to accept God's freely offered grace. Evangelicalism was driving the revival of the Victorian church, and Maxwell could not have avoided it at Cambridge.

These issues became critical for Maxwell in summer 1853 when he spent his vacation with Tayler's uncle, an evangelical rector in Suffolk.

He was deep into studying for the rigorous Tripos exam at this time, working under "high pressure."[37] In addition to his mathematics, Maxwell was reading one of his favorite books: Sir Thomas Browne's *Religio Medici*, a seventeenth-century meditation on religion in an increasingly philosophical age. He fell ill and collapsed while studying, resulting in an intense conversion experience. Maxwell emerged from his conversion with a fierce evangelical faith. Writing to his summer host after his return to Cambridge, he described his new religious outlook:

> All the evil influences that I can trace have been internal and not external, you know what I mean—that I have the capacity of being more wicked than any example that man could set me, and that if I escape, it is only by God's grace helping me to get rid of myself, partially in science, more completely in society,—but not perfectly except by committing myself to God as the instrument of His will, not doubtfully, but in the certain hope that that Will will be plain enough at the proper time.[38]

Maxwell's newfound evangelical stance was quite clear: humans had an essentially depraved nature that could only be reformed through a complete reliance on divine grace. It is interesting to note his intent to use his work in science as part of his religious duty, but the dominant thought of this passage was certainly a statement of Maxwell's acceptance of the reality and overwhelming importance of a correct understanding of God's will. The evangelical outlook required a God who provided grace solely as a free choice and human beings who acknowledged that free choice through exercise of their own will.

That summer Maxwell threw himself into reading to better understand his new faith. He devoured sermons of all kinds, including F. D. Maurice's *Theological Essays*, which were soon accused of heresy. He did not, apparently, agree with all of Maurice's ideas, but the controversial theologian's emphasis on social activism resonated with Maxwell's own paternalist attitudes.[39] He followed Church of England politics closely enough to baffle his father with the details, though it does not seem that he sided with any one group.[40] Maxwell never subscribed to any one sect or denomination, moving comfortably among many, and he maintained that he was never in "bondage to any set of opinions."[41]

Into the World

Two years after Maxwell graduated in 1854 as Second Wrangler and Smith's Prizeman, he became a professor of natural philosophy at Marischal College in Aberdeen. He was not a particularly skilled teacher, and his biographers noted that he was "not on the whole successful in oral communication" and that "between his students' ignorance and his vast knowledge, it was difficult to find a common measure."[42] He did have a close relationship with the students, however, even checking books out of the library for them. After just four years he was unemployed— the consolidation of the two universities in Aberdeen made his position redundant.

His time in Aberdeen produced both his Adam's Prize–winning work on Saturn's rings and a marriage. He was deeply devoted to his wife, and in a typically Victorian way, their relationship was intertwined with Christianity. Although they never had children, they often discussed the religious significance of family. Maxwell to his wife:

> Here is more about family relations. There are things which have meanings so deep that if we follow on to know them we shall be led into great mysteries of divinity. If we despise these relations of marriage, of parents and children, of master and servant, everything will go wrong. . . . But if we reverence them, we shall even see beyond their first aspect a spiritual meaning, for God speaks to us more plainly in these bonds of our life than in anything that we can understand. So we find a great deal of Divine Truth is spoken of in the Bible with reference to these three relations and others.[43]

Their knowledge of scripture was deep and wide.[44] When together, they would read a section of scripture together every night. When apart, which was rare, they would read the same passage and discuss it by letter:

> Now let us read (2 Cor.) chapter xii., about the organisation of the Church, and the different gifts of different Christians, and the reason of these differences that Christ's body may be more complete in all its parts. If we felt more distinctly our union to Christ, we would know our position as members of His body, and work more willingly and intelligently along with all the rest in promoting the health and growth of the body, by the use of every power which the spirit has distributed to us.[45]

These letters reveal the Maxwells' profound concern with scripture, sin, redemption, and living in consonance with God's will. The correspondence was deeply Christian, but also ecumenical. Maxwell often reported his experiences visiting churches of many denominations, orthodox and Nonconformist, during his travels.[46]

In 1860 he took a position at King's College London. His theoretical and experimental research varied widely, across electricity, magnetism, kinetic theory, and color blindness. While there, he and his wife attended a Baptist church run by a Mr. Offord. Maxwell described him as someone "who knows his Bible, and preaches as near it as he can, and does what he can to let the statements in the Bible be understood by his hearers. We generally go to him when in London, though we believe ourselves baptized already."[47] Maxwell's ability to move between orthodox Cambridge and Dissenting London was remarkable.

After resigning from King's, Maxwell returned to his country house at Glenlair to write his *Treatise on Electricity and Magnetism*. The life of a country laird agreed with him, and he greatly enjoyed horse riding and conducting daily prayers for his servants. One of those prayers, authored by Maxwell, has survived:

> Almighty God, who hast created man in Thine own image, and made him a living soul that he might seek after Thee and have dominion over Thy creatures, teach us to study the works of Thy hands that we may subdue the earth to our use, and strengthen our reason for Thy service; and so to receive Thy blessed Word, that we may believe on Him whom Thou hast sent to give us the knowledge of salvation and the remission of our sins. All which we ask in the name of the same Jesus Christ our Lord.[48]

He raised significant money for and helped organize renovations at the local kirk, where he is remembered with a stained-glass window today. He was an elder at Parton Kirk and always made sure he was present to officiate at the midsummer communion. While in the north, he maintained ties with Cambridge, acting as moderator or examiner for the Mathematics Tripos.

It was only with difficulty that he was drawn away from Glenlair to become the first Cavendish Professor of Experimental Physics at Cambridge. The Duke of Devonshire put up substantial funds in 1871 to create the position and endow the laboratory. Maxwell oversaw the construction of the lab, drawing up detailed plans. He was a fairly distant instructor, spending most of his time at the lab on his own projects such as electrical standards work.[49]

Maxwell died suddenly in 1879 at the height of his powers. He left a legacy of brilliant but often difficult-to-understand theory. His life as a conservative aristocrat and deeply religious man was well remembered by his friends and contemporaries. In particular, memoirs remarked on his broad reading and willingness to engage with different ideas. Always, though, he returned to Christianity. As he remarked to a friend: "I have looked into most philosophical systems, and I have seen that none will work without a God."[50]

Huxley

Like Maxwell, Thomas Henry Huxley was born into a religious family. From that point, their trajectories diverged dramatically. Huxley's father was a mathematics teacher whose school in Ealing failed, driving him to Coventry to try a career in banking. That change of career meant that young Tom's free education was at an end, and he was left to read on his own.[51] Fortunately he had access to his father's library, where he read James Hutton's *Theory of the Earth*, and somewhere found Thomas Carlyle's work. He also had informal mentoring from the merchants and mechanicals industrializing Coventry, which lasted him until he was apprenticed to a drug grinder at age thirteen. His only noted talent was skill in drawing. As the youngest of seven children, he did not receive much in the way of attention or guidance. We know little about the details of his religious upbringing, other than a somewhat clerical tendency: he recalled "preaching to my mother's maids in the kitchen as nearly as possible in [the minister's] manner one Sunday morning when the rest of the family were at church" and practicing delivering sermons from tree stumps.[52]

The youthful Huxley had inherited his mother's sharp eyes, quick wit, and a deep ferocity. This last he put to use in the highly sectarian environment of industrial Coventry. Radicals and Dissenters railed against the privileges of the Anglican Church, and Huxley listened with a sympathetic ear. In his journal he wrote:

> Tried an electrolyte experiment. Had a long talk with my mother and father about the right to make Dissenters pay church rates—and whether there

ought to be any establishment. I maintained that there ought not to be in both cases. I wonder what will be my opinions ten years hence? I think now that it is against all laws of justice to force men to support a church with whose opinions they cannot conscientiously agree. The argument that the rate is so small is very fallacious. It is as much a sacrifice of principle to do a little wrong as to do a great one.[53]

Questions of the power of the Church literally followed on his science experiments. The controversy at hand was the voluntarism issue (a central argument in the relationship between Dissenters and the Church)— could people be made to support a church of which they were not willing members? And in particular, could they be *forced* to support a church? He could not accept the argument that it was only a small amount of damage; it was the principle of coercion that mattered. Fighting for justice was not something that could be compromised.

The first page of his journal was emblazoned with a quote from Novalis: "Philosophy can bake no bread; but it can provide for us God, freedom, and immortality. Which, now, is more practical, Philosophy or Economy?" Huxley's search to turn philosophy into a livelihood would become one of the defining characteristics of his career. At this point, his journal reveals his interest in metaphysics, morality, and endless experiments. He recalled later that at this age what he really wanted to be was a mechanical engineer:

> The only part of my professional course which really and deeply interested me was physiology, which is the mechanical engineering of living machines . . . what I cared for was the architectural and engineering part of the business, the working out the wonderful unity of plan in the thousands and thousands of diverse living constructions, and the modifications of similar apparatuses to serve diverse ends.[54]

Living things were simply a useful way to investigate the issues he was truly interested in.

In 1841 Huxley went to London as an apprentice to Thomas Chandler, a doctor of low repute in the East End. He saw the worst of the suffering that poor and filthy neighborhood had to offer and was deeply moved by the terrible conditions there. He planned to study himself out of poverty. With incredible focus he spent every night studying chemistry, history, physics, and geometry. Adrian Desmond points out that Huxley was the ideal of what the cult of Victorian self-improvement imagined—tireless, self-driven effort, unsupported by a penny from anyone else.[55] His attempts at formal education trained him in radical, materialist science.[56]

It is unclear precisely how, but it was in this period that Huxley developed his basic ideas of science as a naturalistic enterprise. That is, that scientific investigation should never refer to divine powers, supernatural forces, or a priori categories. Similar ideas had been in play in France for some time, and Huxley clearly owed Continental thinkers some debts. Like figures such as Pierre Simon de Laplace, Huxley's most important tool for naturalism was an emphasis on natural laws, and particularly their universal character. He thought that strict adherence to this rule meant that the naturalistic universe was a closed system functioning only according to its own principles, without teleology or guidance from on high. A crucial corollary was that there was no need to give authority to clerics or institutions who claimed to speak for those external forces. This was a radical view of science well outside the Victorian mainstream (often associated with French radicalism), and Huxley would be hard pressed to make a career with it.

In 1842 he and his brother started as students at Charing Cross Hospital on exam-qualified free scholarships. The thin, lanky Huxley, sporting a terrible haircut, learned by doing at the hospital, treating all the misfortunes of Victorian London. He continued his nighttime studies, encouraged by his mentor Thomas Wharton Jones to look for wider principles beyond the facts of the living body.[57]

By 1845 his scholarship ended. Too young to get a license to practice, but too far in debt to continue his studies, Huxley looked to the Royal Navy. As an assistant surgeon on the HMS *Rattlesnake*, he hoped that he could do some real field science while paying down his debts. The ship headed to northern Australia to make the area safe for merchants and settlement. He slept in an alcove off the gun room and complained of the class divisions on the ship.[58]

On the journey he began studying jellyfish. A paper he wrote on board was published back home at the Linnean Society, whose president was the father of the *Rattlesnake*'s captain. While pleased at the recognition, it burned Huxley to have to rely on nepotism.[59] In addition to finding valuable research material in Australia, he also discovered his future wife, Henrietta Anne Heathorn, who would join him in England years later. Despite her rather conventional Anglicanism, they were a lively and loving match.

Upon his return to Britain in 1850, Huxley was again adrift. His longing for a career in science was hampered by his unimpressive educational background and lack of connections. Barred from Cambridge by both money and religious test, he had little to recommend him. One contact he did have was Edward Forbes, paleontologist and oceanographer at Edinburgh and later the Royal School of Mines, whom he had met at the British Association for the Advancement of Science (BAAS) before leaving on the *Rattlesnake*. Forbes took the young Huxley under his wing and introduced him to the London scientific scene. The elder man of science helped Huxley petition the Admiralty for support while he wrote up the scientific results of the expedition. He ended up receiving six months at half pay to accomplish the same task for which Darwin, with his university connections, had received one thousand pounds after the *Beagle*'s voyage.[60]

Forbes's Red Lion Club provided Huxley's social life. Through the club he met many who would be lifelong friends. He found kindred spirits in Edwin Lankester, a doctor with a deep interest in natural history and public health, and the fellow frustrated naval surgeon George Busk. The secularist writer (and future editor of the *Fortnightly Review*) G. H. Lewes introduced him to literary London. And the fiery former railway engineer Herbert Spencer would become a close, if contentious, friend for most of Huxley's life. Huxley cheered the deep antagonism to the established Church that grew out of Spencer's Nonconformist background, but was always unsure of his friend's grand evolutionary vision for the whole of the universe. Through these colleagues he increasingly rubbed elbows with secularists, political radicals, and others who had similarly seen the back side of Anglican privilege.

While living on the navy's pittance, Huxley continued to look for work in science, to no avail. He wrote to Henrietta:

> I am sick of writing, weary of longing. The difficulties of obtaining a decent position in England . . . seem to me greater than ever they were. . . . To attempt to live by any scientific pursuit is a farce. . . . A man of science may earn great distinction—great reputation—but not bread. He will get invitations to all sorts of dinners & conversaziones, but not enough income to pay his cab hire.[61]

There were so few paid professorships in science, and most of those were on the other side of the Oxbridge wall and occupied for life by clerics. Huxley accumulated friends in similar straits. Most important was the Irish physicist John Tyndall. Tyndall had worked as a surveyor during the railway boom of the 1840s, and then became a mathematics teacher at Queenwood College. After developing a deeper interest in science, he went to Germany to study physics. Upon returning to England with a PhD in hand, he was frustrated while searching for employment in science.[62] Both were constantly on the lookout for jobs, including the appealing Toronto professorship, which was snatched up by a candidate with local family connections. Along with Tyndall came the chemist Edward Frankland, who had also studied with Robert Wilhelm Bunsen in Marburg. Frankland had moved through Anglicanism and evangelicalism until his encounter with German skepticism and Thomas Paine, which left him with a strong suspicion of theological systems and organized religion.[63] The only member of the group with a secure professional future was the botanist Joseph Hooker, whose father was in charge of Kew Gardens. While in London, Huxley also made the acquaintance of the famous paleontologist Richard Owen. Owen was of a rough background not unlike Huxley and Tyndall, but had quite successfully found patrons in the Anglican establishment and moved into powerful positions.[64]

Continuing with his zoological research, Huxley began accumulating honors and awards. But even being elected Fellow of the Royal Society in 1851 was bittersweet when he could barely afford the attendant cost.[65] He scrounged to make ends meet with anything related to science: he wrote book reviews, translated anatomical books from German, wrote entries for encyclopedias, graded exams, or anything else that would pay. He quickly became known for his sharp pen, and his economic efforts began earning him a literary reputation.[66] Huxley was deeply concerned that this scientific piecework would make him a "crass journalist" and damage his reputation as a man of science.[67] To ameliorate this, his writings began to enforce a strict demarcation between mere writers and real researchers (meaning himself). Amateurs were no longer to be allowed to speak on science.

One small editing project particularly piqued his interest. While being paid poorly to correct the proofs of William Benjamin Carpenter's *Principles of Comparative Physiology*, he realized that he had found a kindred soul. Carpenter, the son of a Unitarian minister, had struggled against the Anglican hold on science professorships just as Huxley did, eventually finding a position as registrar at London University. His voluminous works on physiology continually emphasized that living processes were as lawful and uniform as phenomena in physics or astronomy. Carpenter, despite his clear theism, seemed a perfect ally for Huxley's interest in moving the world of life from divine fiat to scientific regularity.

In the end, the excellence of Huxley's science did bring him a steady income. The Royal School of Mines on Jermyn Street welcomed him to its cramped quarters, giving him a professorship that paid him not quite enough to thrive. But it was work in science, and was enough for him to begin a family. Once Henrietta arrived in England, they started what would end up being a large, loving clan. Huxley strove for his family to be a model of that most Victorian of values: respectability. Regardless of his heterodoxy, the family was indistinguishable from any other middle-class home. Henrietta insisted on baptizing their children, and many of the X-Club served as godparents.[68] Despite his affection for his children, Huxley could never stop working, probably due to habits set when every hour needed to bring in extra money. Even on vacation in South Wales he went collecting for specimens. It was there that he met Frederick Dyster, a local doctor who became a lifelong confidant. His tendency to overwork brought him to illness and the edge of breakdown multiple times. In one particularly bad case, his friends took up a collection to force him onto a holiday.[69]

Huxley and his friends' difficulties in finding work became one of their defining characteristics. Even further, their decision about who was responsible for their difficulties helped shape their identities for the rest of their lives: the Church of England. Or more generally, the protest that orthodoxy and the clergy's control over the educational and intellectual institutions of Britain had made it impossible for self-made men such as themselves to flourish.[70]

Huxley's Church and New Reformation

Huxley and friends began an assault against the Anglican Church's control that would last decades. They wanted to ensure, however, that their attacks were not misconstrued. They emphasized that they were not against *religion*, they were against *theology*. Most of the scientific naturalists had religious upbringings and maintained their Christian values of sincerity, honesty, moral earnestness, and respect for the Bible.[71] Tyndall reminisced about his rigorous Irish Protestant education:

> I was well versed in Scripture: for I loved the Bible, and was prompted by that love to commit large portions of it to memory. Later on I became adroit in turning my Scriptural knowledge against the Church of Rome, but the characteristic doctrines of that Church marked only for a time the limits of enquiry. The eternal Sonship of Christ, for example, as enunciated in the Athanasian Creed, perplexed me.[72]

This was generally the position of the scientific naturalists: true religion was worthy of love and imitation, but doctrines and theologies were wrong at best and corrupt at worst. That distinction was critical for the way the naturalists presented their own ideas and goals. It allowed them to maintain their respectability and to claim that they provided spiritual leadership for British society.

A foundational resource for this distinction was the work of Thomas Carlyle. As Frank Turner noted, most of the naturalists read and admired Carlyle when they were young.[73] Carlyle persuaded them that it was possible to be rational and scientific without being materialistic or atheistic. His critique of the Anglican clergy was inspirational, and his call for the replacement of the old aristocracy with a new meritocracy was exactly what those young men of science hoped for. Huxley said Carlyle had taught him "that a deep sense of religion was compatible with the entire absence of theology."[74]

This sense of religion as apart from theology was usually characterized in a Romantic fashion, emphasizing emotion and feeling instead of reason. Tyndall declared that "religion is not a *persuasion*, it is a *life* . . . *it* must come from the heart . . . finds a root in human nature which is deeper than all sensuous experience and lies below our modern science of logic."[75] Turner argued that this move allowed naturalism to sit alongside experiences and ideals usually called religious: "In essence it attributed to the material world and experiences therein the characteristics previously associated with spiritual realms of aspiration."[76] A reverence for nature replaced orthodox ritual. While "religion" was emotional, "theology" was intellectual and made claims that were subject to rational analysis and disproof. Theology, then, was indeed in constant and fruitless competition with science. As Huxley famously wrote: "Ex-

tinguished theologians lie about the cradle of every science as the stran-
gled snakes beside that of Hercules; and history records that whenever
science and orthodoxy have been fairly opposed, the latter has been
forced to retire from the lists, bleeding and crushed if not annihilated;
scotched, if not slain."[77] Real religion, based on emotion, could remain
untouched by science.

The naturalists made the case that this kind of religion was not their
invention, but was rather the *true* religion that had been disguised by
centuries of clergy. Huxley, Tyndall, and even William Kingdon Clifford
expressed reverence for Jesus and the teachings of the primitive church.[78]
In particular, their religion's purity was opposed to the corruption of the
Anglican Church.[79] They saw their attacks as part of a "New Reformation" intended to restore that pure Christianity. James Moore has argued that the naturalists thought of themselves as religious reformers
and saw their efforts as a transformation just like the one that liberated
England from Rome.[80] Huxley used religious symbols and language as a
strategy to "sanctify his own reformism" despite his attacks on theological tradition.[81]

Science, he said, was the key to this reformation. Even the reforma-
tions of the past had been due to scientific thinking:

> And all the reformations in religion—all the steps by which the creeds you
> hold have been brought to that comparative purity and truth in which you
> justly glory—have been due essentially to the growth of the scientific spirit,
> to the ever-increasing confidence of the intellect in itself—and its incessantly
> repeated refusals to bow down blindly to what it had discovered to be mere
> idols, any more.[82]

All the issues that motivated the first Protestant Reformation—corruption, dogma, blind obedience to tradition—were again at work in the
nineteenth century.[83] This reformation, once complete, would create a
new church—Huxley's church.

Some of Huxley's use of this language was clearly playful, such as his
embrace of the satirical title "Pope Huxley" or his doodle showing Darwin blessing a supplicant to the church of naturalism.[84] And while he denied that agnosticism was a creed, his sense of creating a new religion
for a scientific age was often genuine. Titling his essays *Lay Sermons*
was not an accident, and he wrote meaningfully about joining scientific

to religious education: "I should like to see a scientific Sunday-school in every parish, not for the purpose of superseding any existing means of teaching the people the things that are for their good, but side by side with them."[85]

Even his great enemy, the Church of England, could provide a model. It was not even critical that science per se be taught.

> For example, I can conceive the existence of an Established Church which should be a blessing to the community. A Church in which, week by week, services should be devoted, not to the iteration of abstract propositions in theology, but to the setting before men's minds of an ideal of true, just, and pure living; a place in which those who are weary of the burden of daily cares, should find a moment's rest in the contemplation of the higher life which is possible for all, though attained by so few; a place in which the man of strife and of business should have time to think how small, after all, are the rewards he covets compared with peace and charity. Depend upon it, if such a Church existed, no one would seek to disestablish it.[86]

It was not scientific facts about the natural world that were important; it was the deeper spiritual and moral benefits that needed to be taught. The doctrines of his imagined church were not intended to be novel; they were all the values that Huxley saw as shared between science and true religion. This was in an important sense the underlying goal of all his work to promote and shape science. In his waning years he commented that if he could give a "little shove to the 'New Reformation,'" then "I shall think the fag-end of my life well spent."[87]

Darwin

The scientific naturalists' reverence for a pure religion and a new reformation was largely obscured in the public eye by their vitriolic attacks on Christian theology and the established Church. Huxley's diction was unsparing and militaristic:

> Science exhibits no immediate intention of signing a treaty of peace with her old opponent, nor of being content with anything short of absolute victory and uncontrolled domination over the whole realm of the intellect. Her

champions ask why they should falter? Which of the memorable battles that
have been fought have they lost? When have they ever retreated from ground
they have once occupied, or surrendered a duly fortified stronghold?[88]

This was the birth of the "warfare" metaphor of science and theology,
easily shorthanded into the war between science and religion by careless
observers.[89] Science was declared to be the inevitable victor. In one notebook Huxley charted the complex history of the Abrahamic religions
(fig. 1), only to literally wipe them out with the exponential growth of
science. Huxley and Tyndall's blistering pens brought sharp and cease-
less attacks against the clerics and theologians they saw as holding them
back. Richard Owen was one of Huxley's first targets. Huxley savaged
the elder paleontologist's theistic work and decried him as a puppet of
the Anglican establishment.[90]

Their assault on theistic science gained new vigor with the 1859 publication of Darwin's *The Origin of Species*. As Darwin considered breaking his many-year silence on his ideas, he spoke with his confidant John
Lubbock about which men of science might be sympathetic. Lubbock, a
naturalistic entomologist, suggested Huxley. Darwin gently probed Huxley's views on species change, only to be rebuffed by Huxley's strong feelings about the fixity of species.[91] But when the *Origin* appeared, Huxley
swore his loyalty despite his reservations about the truth of natural selection. What he was fascinated by was not so much the content of the book
as its purely naturalistic approach to the living world. As Ruth Barton
has argued, "Huxley's defence of Darwin was less a defence of Darwin
than the kind of theory he propounded."[92] The naturalists finally had
their Newton: a deep, profound contribution to science that showed the
potential of their worldview. Whether Darwin himself agreed with them
was of less significance. What was important was how he could be used
as a figurehead for their New Reformation. As Paul White has pointed
out, Darwin was an especially good symbol for Huxley because he had
no institutional status—despite his enormous network of connections, he
could be held up as a disinterested outsider.[93]

Huxley quickly became "Darwin's bulldog," far and away the most
aggressive and effective advocate for that hermetic evolutionist. He
stressed the antiorthodox elements of the *Origin*, calling it "a veritable
Whitworth gun in the armoury of liberalism."[94] Rather frustratingly for
Darwin, however, Huxley spent little time discussing the actual details
of natural selection.[95] He was excited about this new kind of explanation

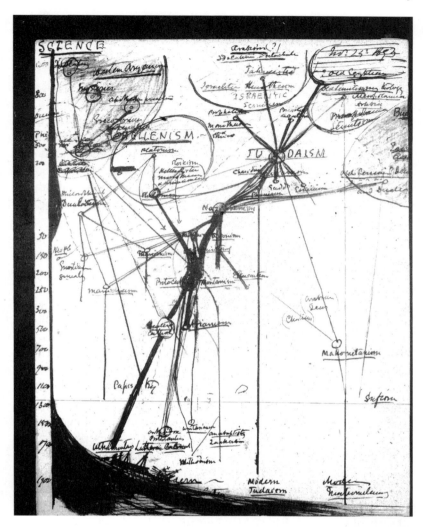

FIGURE I. Huxley's chart on the historical development of religions. Huxley Papers, HP 66.1.

of life, not the mechanics of the theory. His increasingly powerful po-
lemics got everyone reading the book, even if he did not help them understand it.

The debates over Darwin became set in the public imagination as an
epic battle between science and theology. Huxley's clash with Bishop

Wilberforce at Oxford seemed to encapsulate the situation: a young, sharp upstart verbally trouncing the pompous, overconfident cleric.[96] Huxley relished his role as controversialist, happily skewering anyone unfortunate enough to cross swords with him. When Owen tried to show that the primate brain made Darwinian evolution untenable, Huxley became an expert overnight to crush him. Endless debates followed. He did not shy away from confronting the prime minister on the territory of biblical scholarship, nor challenging Matthew Arnold on education and literature.[97] A seemingly continuous parade of underprepared clerics and amateur scientific investigators were buried under the products of his pen. Fighting was an indelible part of his character. Indeed, the opportunity for one more battle with Gladstone appeared to bring him back from death's door.[98]

The X-Club

It may be too strong to say that the scientific naturalists engaged in a conspiracy to take over science—but it is close. They were ambitious, bound by tight personal ties, and worked hard to achieve their joint goals. The clearest manifestation of this was the 1864 founding of the X-Club.[99] Originally simply a group of friends dining together regularly at the St. George's Hotel, the X-Club became synonymous with naturalistic, research-oriented, professional science. The group consisted of Huxley, Hooker, Lubbock, Tyndall, Frankland, Busk, Spencer, the mathematician William Spottiswoode, and Thomas Hirst. Hirst described his own view of their origin: "Beside personal friendship, the bond that united us was devotion to science, pure and free, untrammeled by religious dogmas."[100] It was clear to all observers that this closed club contained many of the most active and vocal proponents of the new science, giving rise to all sorts of vague suspicions. Members of the club dominated the council of the Royal Society for the rest of the century and were well known for maneuvering to control scientific appointments, elections, and grants across Britain. Huxley's personal success in achieving leadership positions, as well as his ability to place students and allies, made him one of the most influential men of science of the time.

With his increasing stature, a new generation of men of science flocked to Huxley's flag of naturalistic science. One young, ferocious addition to the group was William Kingdon Clifford, a talented mathematician who delighted in controversy.[101] His skill with non-Euclidean geometry was brought to bear on naturalistic epistemology, but also led to a friendship with Maxwell. Another promising disciple was St. George Jackson Mivart, trained originally as a lawyer and then in biology by both Owen and Huxley. Mivart called himself a "hearty and thoroughgoing disciple of Mr. Darwin."[102] His excellence in biology was rewarded with a professorship of comparative anatomy at St. Mary's Hospital Medical School in London. However, Mivart was a reverent Catholic whose attempts to reconcile natural selection with Rome's doctrines eventually led to his expulsion from the Darwinian circle.

Huxley's ideas were spread across the Anglophone world through his popular writings. While his legacy here was tremendous, he was originally reluctant to enter this arena. Lightman has shown that Huxley did not originally think there would be any interest in published versions of his lectures and was only pushed into it by Darwin and Charles Lyell.[103] His rhetorical skill was extraordinary, leading Darwin to write in a letter that it was an "injustice that any one man [should] have the power to write so many brilliant essays. . . . There is no one who writes like you."[104] Once his employment was secure, Huxley entered more into lecturing and writing, particularly for workers. The lectures grew into full syllabi and textbooks that Huxley used to create a whole new kind of science education in South Kensington: "biology." His efforts to restructure the way science was taught ranged from exam standardization to running a school for workingmen to sitting on the first London School Board. In both popularization and education, Huxley struggled to enforce the authority of professionals such as himself—ill-informed amateurs with an attraction to religiously tinged natural history, particularly clergy, were to be excluded.

He also took the debate straight to his enemies. At the Metaphysical Society he engaged with representatives of the major intellectual traditions and groups of Victorian Britain.[105] That debating group, organized by James Knowles, saw fierce contests over the nature of science. Huxley (a founding member) argued for scientific naturalism, but Christian intellectuals were not prepared to cede control of science. They defended theistic science as the proper and traditional way to investigate the natural world. The debaters did not see the struggle as between science and religion, but rather between theistic and naturalistic views of science. Huxley's irreverence was not lessened by the impressive company, however—he continually doodled and drew caricatures during the meetings.

Huxley ended his life as a grand old man of British science. By the time of his death in 1895, his goals to reshape science had been essentially accomplished. The repeal of the Test Acts meant that Oxford and Cambridge were open to all, the number of science professorships was vastly increased, and his vision of a professional, naturalistic science was well on its way to reality. He had become an icon of the changes in science in a remarkable way. Andrew Lang was only half joking when he commented that "in England, when people say 'Science' they commonly mean an article by Professor Huxley in the *Nineteenth Century*."[106]

Conclusion

Maxwell's and Huxley's views on the relationship between religion and the methodological values of science were not accidental, and they were not trivial choices. They were deeply bound to the social and cultural forces in which they grew, practiced, and prospered. Both had their personal idiosyncrasies, but their positions on theistic and naturalistic science were indicative of significant groups of Victorian practitioners (certainly not limited to the figures mentioned in this chapter).

Maxwell's theistic science was supported by centuries of tradition and the demonstrated success of all the great names of British science. It made science an ally of mainstream religion, particularly the established Church and its close relationship to the political power structure. The universities were set up to teach science theistically, with natural theology, in a way that brought together the diverse branches of Christianity and encouraged the investigation of the natural world. It was the standard.

Huxley's naturalistic science was an upstart, associated with political radicals and dangerous French thinkers. In the middle of the century it had no institutional home or established methods of teaching or investigation. Whether it even formed a coherent worldview was not at all clear. Without an established track record or triumphs to claim, naturalistic science had to constantly defend its relevance and validity. It would eventually claw its way into respectability and power, but only with deliberate effort by committed actors. It was the challenger.

Both Huxley and Maxwell saw science as being inextricably linked to their worldview—perhaps as the investigation of God's creation, perhaps as the only practice that could challenge the Church. Neither thought science could be justified outside the contexts in which they practiced it, and indeed thought such a claim was absurd. The following chapters will show in detail how both the theists and the naturalists argued for these unbreakable links, and how those individual issues were seen to be essential for science. Underlying all these issues was the tension of two strongly opposed groups arguing for virtually the same foundational elements of science. Victorian science somehow functioned with theists and naturalists working alongside each other; this is the story of how they did so.

CHAPTER TWO

The Uniformity of Natural Laws

John Herschel's *Preliminary Discourse* (1830) shaped a generation of science.[1] The attention of natural philosophers, he said, should be focused on the principles of nature—its laws.[2] He wrote that it is clear to "us from our earliest infancy . . . that events do not succeed one another at random, but with a certain degree of order, regularity, and connection." Because this has gone on "uninterruptedly, for ages beyond all memory, [it] impresses us with a strong expectation that it will continue to do so in the same manner; and thus our notion of an *order of nature* is originated and confirmed."[3] One might be ignorant of the laws of nature, they might yet be hidden, but they govern all things. They function "uniformly and invariably."[4]

This proposal, that the rules of the natural world were never violated or paused, and apply to all things in all circumstances, is often called the *uniformity of nature*. It has been articulated in many different ways, but by the time Herschel wrote his formative text, the assumption that the universe was governed by uninterrupted laws was a fundamental part of natural philosophy.[5] By the end of the nineteenth century, Huxley and his allies were using this concept as a bludgeon to drive theism out of science, and it continues to be used so today under the rubric of scientific naturalism. It is impossible, say the naturalists, for divine action or intervention to have any role in a world that runs by uniform natural laws.

And yet, Herschel wrote that natural laws had their origin in the "Divine Author of the universe" and that the uniformity of those laws came from "the constant exercise of his direct power in maintaining the system of nature, or the ultimate emanation of every energy which material agents exert from his immediate will, acting in conformity with his own laws."[6] How can it be that uniformity was seen as rooted in theism in the early Victorian period, when it was presented as an enemy of theism by the end? The critical point is to understand that uniformity was a core principle of science throughout the nineteenth century for *both* theists and naturalists, despite each group claiming it as uniquely theirs. Uniformity seemed to be an obviously theistic category for those of Herschel's generation, and it was only through the deliberate efforts of the scientific naturalists that it came to be seen otherwise.

Theistic Laws

The Victorian case for theistic laws of nature was laid out most clearly and influentially by the Reverend Baden Powell in his 1838 *The Connexion of Natural and Divine Truth.*[7] Powell was Savilian Professor of Geometry at Oxford, an active member of the BAAS, and worked tirelessly for university reform and to persuade the Anglican Church to adapt to new ideas. His book was a virtual tribute to the importance of natural laws, and in particular, their uniformity. His first chapter was built around a section titled "Belief in the Uniformity of Nature," in which he set out the fundamental idea without which science could not function:

> Now there is one grand, fundamental principle, without which no induction of laws from particular instances, no generalizations of individual truths, no regular or systematic study of nature, could ever proceed: and this is our conviction of a permanence and uniformity in the order of natural things: our belief that that which has happened in succession for days and years past, will, under the same circumstances, continue to happen for time to come: our persuasion that what so takes place in one instance, in one place, will and does take place, under the same conditions, in all other instances, and in all other places. We suppose, that is, that nature *is so constituted*, that there exists *some* principle of undeviating regularity in the connexion of qualities and properties, of causes and effects, even though we should fail in always tracing it.[8]

Powell asserted that this fundamental axiom could not be rigorously justified, but that it seemed to be an inherent principle in human nature (in addition to being reinforced by the cultivation of the mind). In addition to being necessary for proper inductive science, uniformity also provided an impulse to philosophical investigation and allowed one to search beyond the limits of sensible experience.[9] Lyell's geology and Laplace's nebular theory were wholeheartedly embraced. Despite their implications of a very old Earth created only by physical processes, they were held up as paramount examples of how using uniformity could drive science forward. The human sense that nature was uniform was so deep "that we cannot bring ourselves to believe in the capricious violation even of *one* of her laws."[10]

The critical point of Powell's argument came when he discussed an implication of this uniformity:

> It is when results are reducible to regular rules, when observed actions are found to be consistent with some fixed and constant system; when phenomena can be traced up to their determinate laws, or in other words (agreeably to what has been above maintained), to their *physical causes*, then, and then alone, it is, that we can ascend to the idea of a regulating *moral cause*; and deduce the conclusion of super-intending volition and designing intelligence.[11]

That is, the uniformity of nature itself was evidence for a divine creator. A universe in which natural laws were uniform and regular was one that was clearly designed to be that way. Powell, following in the natural theology tradition discussed in chapter 1, argued that the characteristics of natural laws could be used as windows into the divine. More scientific knowledge, particularly of laws, was more knowledge of God: "The evidences of the Divine operation seem to me manifested precisely in proportion as *we can* trace material laws and physical laws."[12] If phenomena were not constrained and regulated by these uniform laws, Powell argued, there would be no evidence for a creator at all. In a world without uniformity, "We are involved in the anarchy of chaos and the darkness of atheism."[13]

Baden Powell warned that asserting divine causation for apparent gaps in uniformity would be dangerous for religion, because "enlarged discovery shall disclose the connection and explanation of these appearances by regular laws, [and] their argument for a Deity will fall to the ground!" Reducing God to only a "confession of ignorance," as in the case of Newton's arguments for planetary stability, was bad science and bad religion. Why? Because it allowed for violations of uniformity, which was critical for both: "Law and order, physical causation and uniformity of action, are the elevated manifestations of Divinity, creation and providence."[14] Some critics of this position claimed it restricted God's action, saying that a God who could not intervene in special circumstances was no God at all. But, again, it was uniformity, not interruptions of it, that truly showed us the nature of things: "To speak of apparent anomalies and interruptions as *special* indications of the Deity, is altogether a mistake. In truth, so far as the *anomalous* character of any phenomenon can affect the inference of presiding Intelligence at all, it would rather tend to *diminish* and detract from that evidence."[15]

Powell approvingly quoted his contemporaries Charles Babbage and William Whewell, whose natural theological writings similarly stressed that natural laws were directly due to God's action.[16] Such theistic laws were a commonplace assumption among natural philosophers of the period, and they were certainly axiomatic at Cambridge by the time a young, thick-accented Scot came to study there in 1850.

Maxwell and Uniformity

As discussed in chapter 1, Maxwell's training at Edinburgh and Cambridge placed him squarely in the mid-Victorian approach to laws of nature. Newton's laws were the model for which a man of science should strive. We can see Maxwell embracing, even in his early work, the assumption of the uniformity of these natural laws in both time and space. Spatially, his award-winning investigation of Saturn's rings began with a statement that it was inconceivable that terrestrial mechanics might not apply to distant planets.[17] Temporally, he explained to his students that once a natural law was grasped, one could ponder its effects on events that have not yet occurred—assuming, of course, that that law will hold good in the future.[18] A basic principle of science was that "place and time are not among the conditions which determine natural processes."[19] Maxwell did not welcome violations of uniformity. He rejected otherwise impressive hypotheses for the cause of gravity because they suggested interruptions of conservation of energy.[20]

Maxwell was a fairly conservative Victorian evangelical Christian, who took scripture quite seriously. As with Powell, Maxwell thought that the uniformity of the cosmic order had greater significance:

> I think that each individual man should do all he can to impress his own mind with the extent, the order, and the unity of the universe, and should carry these ideas with him as he reads such passages as the 1st Chap. of the Ep. to

Colossians . . . , just as enlarged conceptions of the extent and unity of the world of life may be of service to us in reading Psalm viii.; Heb. ii. 6, etc.[21]

The unity of nature was a theological concept as well as a scientific one. The scriptural passages Maxwell referred to here emphasized God's role as creator of the natural world ("For by him were all things created, that are in heaven, and that are in earth, visible and invisible") and the awe that God designed his creation for man ("What is man, that thou art mindful of him?").[22] Thus, Maxwell was powerfully linking the unity and order of nature not just with divine creation itself, but also with the role of man in that creation. In the same letter he argued that we can see "wisdom and power" in the uniformity of natural laws just as effectively as in the beneficial adaptations of living creatures: "Uniformity, accuracy, symmetry, consistency, and continuity of plan are as important attributes as the contrivance of the special utility of each individual thing."[23]

In addition to uniformity being another premise in the argument from design, it was a tool given to man by God to fulfill the commandment to subdue nature. Maxwell instructed his students that once they understood the constancy and universality of natural laws, they would

begin to understand the position of man as the appointed lord over the works of Creation and to comprehend the fundamental principles on which his dominion depends which are these—To know, to submit to, and to fulfil, the laws which the Author of the Universe has appointed. Attend to these laws and keep them, you succeed, break them, you fail and can do nothing.[24]

Natural laws were a manifestation of the divine will, just like biblical commandments. The principles that ran the steam engine and created the wealth of industrial Britain were gifts from God. Conscientious people had an obligation to be mindful of the immutability of both natural and moral law.

Religious defenses of uniformity were found in the life sciences as well. William Carpenter, the pioneering physiologist and liberal Unitarian, argued that "orderly uniformity" was the distinguishing feature of a law of nature, and that this uniformity revealed the law's divine origins:

It is thus that when we pass from the sphere of human government to that of the Divine, and speak of the universe as "governed" by the "laws" of a supreme Ruler, we mean that his power is exerted, not like that of an arbitrary potentate who changes his course of action as his own caprice or passion may direct, but like that of a benevolent sovereign whose rule is in uniform and orderly conformity with certain fixed principles, originally determined as conducive to the welfare and happiness of his people.[25]

In addition to being dependent on the Creator, uniformity also revealed his nature. God's benevolence and consistency could be seen in all realms: "We see that the hypothesis [of uniformity] coincides with all which Science and Religion alike teach, respecting the invariability of His mode of working."[26]

Maxwell's adherence to uniformity impacted his scientific work dramatically in his unification of electricity, magnetism, and light. In 1861 Maxwell was developing further his first forays into electricity and magnetism found in his "On Faraday's Lines of Force." Largely inspired by William Thomson's work on vortex molecules and magneto-optic rotation, Maxwell was attempting to systematize all that was known about electrical and magnetic phenomena. This effort would become his famous series of papers "On Physical Lines of Force."[27]

One of the most distinctive features of the theory Maxwell was developing was a mechanical model wherein electrical and magnetic effects could be explained by the rotation of vortex molecules and the movement of interspersed "idle wheel" particles.[28] Electrostatic effects were accounted for by the elasticity of the medium made up of these molecules and particles.[29] An unexpected effect of postulating such elasticity was the possibility that the medium could support transverse waves of electromagnetic effects. On his theory, the speed of such waves would be equal to the ratio of electrostatic to electromagnetic units (represented by v).[30] Maxwell scoured the physics literature to see if anyone had made the measurements necessary to calculate this ratio, and found that Rudolf Kohlrausch and Wilhelm Weber had done so in 1857.[31] The result of 193,088 miles per second was remarkably close to 193,118 miles per second, the most recent measurement of the speed of light, c.[32] This suggested a profound conclusion, "that the luminiferous and the electromagnetic medium are one."[33]

Maxwell embraced the unification of natural laws suggested by his mechanical model. He was convinced that there was a true connection between optics and electromagnetism—that there was a fundamental principle hidden in the chaos of observable phenomena. For him, such unification was essential to the very notion of physical explanation: "When any physical phenomenon can be described as an example of a general principle which is applicable to other phenomena the phenomenon is said to be explained."[34] The primary conceptual tool Maxwell used for searching for such underlying principles was that of the "analogy."

By analogy he meant "that partial similarity between the laws of one science and those of another which makes each of them illustrate the other."[35] More informally, and reflecting Maxwell's lifelong interest in comic poetry: "Now, as in a pun two truths lie hid under one expression, so in an analogy one truth is discovered under two expressions."[36] In simplest terms, a physical analogy was a claim of an underlying similarity between two apparently different phenomena or laws. For Maxwell, analogies were a powerful way to approach scientific problems. They functioned as a happy medium between unwarranted hypotheses and highly abstract mathematical theories.[37] They allowed one to describe something unfamiliar or complex (such as electromagnetism) in terms of something familiar or simple (such as idle wheels) without postulating completely speculative entities or discarding physical reasoning. His reliance on and confidence in analogies was certainly encouraged by his training in the Common Sense school of philosophy, which also emphasized the need for constant testing of analogies.[38]

Maxwell discussed several different kinds of analogies, but relevant here is his approach to analogous laws; that is, his interest in how different laws of nature appear to be analogous to one another, and might therefore be linked. A real analogy between disparate laws, such as electromagnetism and optics, would suggest a real unification. The scientific role of analogies was in

> its application to the opinion, that all the phenomena of nature, being varieties of motion, can only differ in complexity, and therefore the only way of studying nature, is to master the fundamental laws of motion first, and then examine what kinds of complication of these laws must be studied in order to obtain true views of the universe. If this theory be true, we must look for indications of these fundamental laws throughout the whole range of science, and not least among those remarkable products of organic life, the results of cerebration (commonly called "thinking"). In this case, of course, the resemblances between the laws of different classes of phenomena should hardly be called analogies, as they are only transformed identities.[39]

Analogy was a route to uncovering the "fundamental laws" that under-pinned all the varied phenomena of the universe, in this case suggested to be the laws of motion. Maxwell was here stating a methodological value, not necessarily a claim about nature. He was arguing that if there were true connections between the various laws of nature, then anal-ogy could be a useful tool. But were there such true connections? He acknowledged that humans seemed to feel a compulsive need to search out such fundamental laws, regardless of whether they existed: "The hu-man mind cannot rest satisfied with the mere *phenomena* which it con-templates, but is constrained to seek for the principles embodied in the phenomena."[40]

Maxwell was well aware of the psychological quirks of the human mind, however, and seriously considered the possibility that the unifica-tion of laws was only a feature of the mind and not the physical world. "Are we to conclude that these various departments of nature in which analogous laws exist, have a real interdependence; or that their rela-tion is only apparent and owing to the necessary conditions of human thought?"[41] It was entirely possible that the concept of an orderly, uni-fied, well-organized universe was simply a human construction that we projected onto the world. To illustrate this, Maxwell presented two pos-sible metaphors for the laws of nature:

> Perhaps the "book," as it has been called, of nature is regularly paged; if so, no doubt the introductory parts will explain those that follow, and the meth-ods taught in the first chapters will be taken for granted and used as illustra-tions in the more advanced parts of the course; but if it is not a "book" at all, but a *magazine*, nothing is more foolish to suppose that one part can throw light on another.[42]

If nature were like a book, then there was a single argument—a com-mon thread holding together the text that could be used to interpret and understand the whole. If so, then in physics, electricity could help you understand magnetism because they were both part of a single "docu-ment." But if nature were like a magazine, where the separate articles had nothing to do with one another and could even be written by differ-ent authors, there was no such assurance. A magazine has no single ar-gument. Rather, it is a collection whose elements may or may not have a connection to each other. There was no guarantee that any one article could help the reader understand any other. If this were the case, there would be no reason to think that electricity could help us understand magnetism. It was not immediately clear why scientists should choose one metaphor over the other. Whether nature was like a book or a magazine was of the highest importance for understanding it. But how was one to decide?

The Design of Nature

Maxwell was convinced that nature *was* like a book, and that its individual elements should be seen as manifestations of deeper unified principles. He thought that the connections between the laws of nature were, literally, a sign from above. Someone investigating physical laws "will see as he advances that the laws of nature are not mere arbitrary and unconnected decisions of Supreme Power, but that they form essential parts of one universal system, in which infinite Power serves only to reveal unsearchable Wisdom and eternal Truth."[43] An earlier draft of this passage (which also appears in Maxwell's Aberdeen inaugural lecture) made the role of laws as divine messenger even clearer, saying physics revealed a world "in which Wisdom and Truth are supreme, and Power is their minister."[44]

Maxwell's claim that laws were "parts of one universal system" was arguing that there was a plan to the interrelationship of natural laws. This interrelationship was a way that God communicated his existence, and it was the unity of laws that revealed this communication. An "arbitrary" distribution of individual laws (like the articles of a magazine) would not suggest anything about a divine plan, but unification (like the chapters of a book) would be highly improbable and therefore was a kind of divine communication. God had a plan for the world, and part of that plan was designing natural laws to fit together like the pieces of a puzzle.

Maxwell's best-known statements on the uniformity of nature appeared in two lectures he gave at the British Association in 1870 and 1873. He told his audience that new techniques of spectroscopy showed that in the Sun

> there are molecules vibrating in as exact unison with the molecules of terrestrial hydrogen as two tuning-forks tuned to concert pitch, or two watches reg-

ulated to solar time. . . . Now this absolute equality in the magnitude of quantities, occurring in all parts of the universe, is worth our consideration.[45]

In addition to the spectroscopic evidence that molecules everywhere in space were identical, he pointed to simple chemical evidence of uniformity through time. Hydrogen liberated from rocks buried since time immemorial was identical to that manipulated in the Cavendish Laboratory. Warning off the scientific naturalists, he declared that "no theory of evolution can be formed to account for the similarity of molecules, for evolution necessarily implies continuous change, and the molecule is incapable of growth or decay, of generation or destruction."[46]

Maxwell used this premise to build on John Herschel's metaphor of molecules as manufactured articles. The incredible uniformity of molecules scattered in space and time showed that they could not have been formed in natural processes, only by deliberate intent.

> They continue this day as they were created—perfect in number and measure and weight, and from the ineffaceable characters impressed on them we may learn that those aspirations after accuracy in measurement, truth in statement, and justice in action, which we reckon among our noblest attributes as men, are ours because they are essential constituents of the image of Him who in the beginning created, not only the heaven and the earth, but the materials of which heaven and earth consist.[47]

His claim was not only that the uniformity of molecules showed the hand of a divine manufacturer, but that the uniformity reflected divine values that humans could see in themselves. A man of science's drive to measure, think, and speak accurately was profoundly connected to the uniformity of nature.

This natural theological argument might, at first glance, seem to be identical to Paley's. An object in nature appears to be the result of manufacture; therefore, a manufacturer must exist. However, the core of Maxwell's argument was actually quite different. Paley emphasized *complexity* as the indicator of God's hand; Maxwell emphasized *unity*.[48] Whereas Paley's argument said we can see design through our inability to understand complexity, Maxwell said we see design through the unity revealed by our scientific efforts. More science revealed more design, not less. This is a strategy we will see Maxwell use again.

Maxwell did not invoke the design argument naïvely. His approach was sophisticated, as he was well read in David Hume and other philosophers and was quite aware of the pitfalls of relying on divine action. An essay fragment of his on evidence for design had nearly half its length devoted to the dangers of such assertions. He listed three fallacies commonly associated with the design argument: putting a final cause where there should be a physical one, asserting incorrect physical connections, and reasoning from facts that are simply wrong.[49] He saw the invoking of final causes (the purposes of things) as the most dangerous. An overreliance on final causes could blind one from seeing important facts and if used poorly could call a halt to natural investigations far too early. It was helpful as a guideline but required significant assistance:

> The doctrine of final causes, although productive of barrenness in its exclusive form, has certainly been a great help to enquirers into nature; and if we only maintain the existence of the analogy, and allow observation to determine its form, we cannot be led far from the truth.[50]

He did not want design to replace empiricism, observation, or analysis. He wanted design to be something that would help science work better. His methodological prescript was that final causes were dangerous if used in isolation (one should not replace physical causes with divine ones); but if you are aware beforehand that there are final causes of some sort at work in nature, that can help guide you to observational truths about nature that you would have otherwise missed. Understanding the reality and character of God's actions could provide assurance that certain kinds of investigation would be fruitful, and keep one from going astray.

Maxwell did not think that any hypothesis proved, or was proved by, any specific religious statements. He was willing to make broad, general claims about the theology of nature (such as in his molecules lecture), but was quite hesitant to directly link particular scientific ideas to any part of religious doctrine:

> But I should be very sorry if an interpretation founded on a most conjectural scientific hypothesis were to get fastened to the text in Genesis, even if by so doing it got rid of the old statement of the commentators which has long ceased to be intelligible. The rate of change of scientific hypothesis is natu-

rally much more rapid than that of Biblical interpretations, so that if an interpretation is founded on such an hypothesis, it may help to keep the hypothesis above ground long after it ought to be buried and forgotten.[51]

Similarly, he warned that "it is of the nature of Science" to be continually "spreading into unknown regions," and that one should be quite wary of attaching permanent religious doctrine to changing scientific ideas.[52] Science was expected to advance and change rapidly in a way religion (specifically Christianity) was not. Maxwell was concerned to protect both science and religion from this mismatch. He did not think religion could change rapidly because it was to be anchored to scripture, which was not subject to significant revision. Interpretations could change, but since the subject of interpretation was fixed, he did not expect such changes to be rapid or dramatic. Science, on the other hand, had no such firm anchor, and could change quickly with new discoveries or formulations. Neither of these rates of change was a problem unto itself—difficulties only arose when one expected the rates of change to match. The danger was to tie a particular scientific hypothesis to a particular point of scriptural interpretation (e.g., God's creation of light in Genesis was the formation of the idle wheel ether) and thus either retard science to the speed of religion or dangerously accelerate religion to a scientific pace. Harmonization of science and religion should instead rely on broad theological truths (e.g., God as benevolent creator) and broad scientific guidelines (e.g., physical forces are fundamentally unified). At this high level, the rates of change of both science and religion could be expected to be glacial and therefore safe.[53] Note that he did expect even these sorts of harmonizations to change—the progress of science was axiomatic for Maxwell.

This continuous progress of scientific thought also played an important role in the way Maxwell thought God had designed the laws of nature. The divine uniformity of natural laws had implications even beyond the simple recognition of a creator. Natural laws were designed with the special feature that they were meant to be discovered:

[Once we understand some science,] we are prepared to see in Nature not a mere assemblage of wonders to excite our curiosity but a systematic museum designed to introduce us step by step into the fundamental principles which are displayed in the works of Creation.[54]

In particular, the unification of laws was intended for discovery. The connections of natural laws were "systematic" in that they were carefully designed to attract the attention of humans and lead them to deeper and deeper principles. Laws were laid out like a trail of bread crumbs to guide the attentive from diverse phenomena to unification via strategic connections: "The Book of Nature, in fact, contains elementary chapters, and, to those who know where to look for them, the mastery of one chapter is a preparation for the study of the next."[55] Maxwell used very similar language in his 1870 BAAS address, but without the explicit theological references: "What we set ourselves to do is to unravel these conditions, and by viewing the phenomenon in a way which is itself partial and imperfect, to piece out its features one by one, beginning with that which strikes us first, and thus gradually learning how to look at the whole phenomenon so as to obtain a continually greater degree of clearness and distinctness."[56] Someone in the audience who was already familiar with these ways of thinking would recognize the religious context immediately; someone who was not would simply read this as a naturalistic methodological precept. Seeing the religious meaning in scientific statements was heavily dependent on the outlook of the reader.

Scientific investigation, then, and particularly the search for unified physical laws, was a task given divine assent and even encouragement. Maxwell's God wanted us to understand the world in deeper and deeper terms. His theology gave him a powerful set of tools for understanding the natural world, and for guiding his investigations in physics. He argued that God made the universe out of laws that were fundamentally unified and that he wanted humans to discover that unity.

Evangelical Attitudes

In examining such issues, it is important to be specific about what one means by "religion" or even by "Christianity." Design arguments and natural theology varied in location, time, and doctrinal context.[57] Maxwell was a conservative evangelical Christian who had a specific understanding of the nature of God. His thinking about the relationship between God and nature was shaped heavily by his evangelical outlook.

Maxwell was raised in Scotland in both the Presbyterian and Episcopal traditions, and only became an evangelical after 1853. As discussed in chapter 1, in that summer he was studying for the Tripos exam at Cambridge while staying with his friend Tayler's uncle, who was an evangelical rector in Suffolk. While studying, he collapsed with a dangerous fever, which resulted in an intense conversion experience under the ministrations of his host. His newfound evangelical faith emphasized sin, grace, and leading a life governed by God's will. Religious truth was beyond human understanding, except insofar as God chose to reveal portions of it through scripture and personal contact with the faithful. Evangelicalism focused strongly on the importance of scripture, particularly against those who asserted the sufficiency of human reason or church authority.

Maxwell's correspondence with his wife gives us a particularly valuable window into his evangelicalism. Here he describes a visit to see his friend and future biographer Lewis Campbell deliver a sermon: "He showed up sin as the universal poison. . . . Lewis preached on 'Ye must be born again.' . . . Then he described the changes on a man new-born, and his state and privileges. I think he has got a good hold of the people, and will do them good and great good."[58] Divine grace, submission to God's will, and Christology were constant themes:

> Think what God has determined to do to all those who submit themselves to His righteousness and are willing to receive His gift. They are to be conformed to the image of His Son, and when that is fulfilled, and God sees that they are conformed to the image of Christ, there can be no more condemnation, for this is the praise which God Himself gives, whose judgment is just. So we ought always to hope in Christ, for as sure as we receive Him now, so sure will we be made conformable to His image.[59]

Almost every surviving letter between Maxwell and his wife makes reference to scripture:

> I have been back at 1 Cor. xiii. I think the description of charity or divine love is another loadstone for our life—to show us that this is one thing which is not in parts, but perfect in its own nature, and so it shall never be done away. It is nothing negative, but a well-defined, living, almost acting picture of goodness; that kind of it which is human, but also divine. Read along with it 1 John iv., from verse 7 to end; or, if you like, the whole epistle of John and Mark xii. 28.[60]

Maxwell was an amateur, but serious, biblical scholar, often inspecting passages in their original language. His frequent use of the metaphor of the book of nature is given deeper meaning by his placement among the evangelicals, who attributed tremendous power to the written word. This provides a small taste of the evangelical religious outlook that heavily shaped Maxwell's adult life, but it is sufficient to give a sense of the values and beliefs in play.

Victorian evangelicals had mixed feelings about science.[61] The standard story has been that evangelicals were suspicious of Enlightenment claims that natural theology and rational religion by themselves could suffice as the foundations of Christianity. They, in contrast, wanted to reserve that role for scripture. Further, humans, steeped in sin and error as they were, could not hope to penetrate the mysteries of the universe with their own abilities. These seemed to combine into a strong antiscience position. Recently historians have warned that this story, while containing elements of truth, should not be taken too far.[62] We now have several studies that show some of the fruitful interactions between science and Victorian evangelicalism.[63] A more complete picture might be this: evangelicals were often skeptical that the natural world could provide significant religious insight vis-à-vis scripture, but they were not on the whole hostile to science.

And those evangelicals who did embrace science had a distinctive approach to thinking about the natural world. John Brooke has argued that an "archetypical evangelical scientist" would probably have a biblically informed philosophy of nature, be averse to speculative hypothesis due to humility before God, be sensitive to the uses and limitations of natural theology, and insist on harmony between true science and true religion.[64] An important distinction to be made when thinking about evangelical science is between *natural theology* (in the sense of grounding religious truths in the natural world) and a *theology of nature* (recognition of the role of God in nature).[65] The evangelicals preferred the latter, meaning that even if an individual held scripture to be primary, there could still be a robust devotional role for science. Natural reason and revelation could work together.

Thomas Chalmers, the massively influential Scottish theologian, provides an important example of how nineteenth-century evangelicals could approach science. Early in his career he was skeptical of natural theology for all the reasons discussed above, but eventually accepted the

utility of scientific knowledge for specific religious purposes.[66] His famous 1817 lectures on astronomy built on Scottish Calvinism and the
Common Sense philosophy to assert both the value and limitations of
natural knowledge. Roughly speaking, his purpose was to defend Christian revelation from objections that had been brought based on the vast-
ness of the universe, but not necessarily to convert unbelievers.[67] Such
evangelical theologies of nature often critiqued Paley for not accepting the "dysfunctional aspect of creation," and integrated visions of sin
and evil.[68] Crosbie Smith points out that while Chalmers argued for a
universe governed by divinely chosen laws, he emphasized both the beginning and ending of the world and its associated utter dependence
on divine will.[69] In particular, the creator arranged the material of the
universe in particular "collocations" that would lead to his desired outcome.[70] As with salvation, God had bestowed certain gifts upon humanity, which could be either embraced or ignored. The action and role of
God could be found in nature, but only through a lens of human sin, fallibility, and complete separation from the divine.

To place Maxwell within this tradition, it is helpful to refer to Boyd
Hilton's reading of nineteenth-century evangelicalism as representing "a
shift in natural religion from *evidences* to *paradoxes*."[71] That is, a major claim of Enlightenment natural religion was that God's benevolence
and foresight could be immediately seen in the harmonies of both nature
and human affairs. However, the wars, economic upheavals, and political unrest in the Britain of the early decades of the nineteenth century
drove many to doubt that divine governance was really so easily seen.
So, instead, many evangelicals began to emphasize that the harmonies
of nature were still present, but were hidden from view behind war and
famine.[72] These were the sorts of paradoxes that Hilton describes early
Victorian evangelicals as grappling with: the defense of a just, beneficent
God who plans ahead despite a world that seemed to be depraved, un-
godly, and chaotic. This emphasis on the hidden aspect of divine action
was quite compatible with the general assumptions of evangelical theology. Humans were fallen, sinful, and fallible and thus could not be relied
upon to find the truths of the world with their own powers. God gave humans the ability to see his actions, but only if they embraced him fully.
A religious perspective was seen as necessary to properly interpret the
world's disorder, just as a religious perspective was necessary to properly
interpret individual human sinful behavior.

Crucial to this theological reasoning, and also to Maxwell's interpretation of nature, was the notion of the revelation of the mysterious. The evangelical God was wholly other, and the greater truths of the world only became known to humanity through his choice in revealing them.[73] This required both an embrace of the unknown and a faith that God would eventually provide what was necessary for comprehending the unknown. Interestingly, Maxwell talked about science in precisely this way: "I have endeavored to show that it is the peculiar function of physical science to lead us to the confines of the incomprehensible, and to bid us behold and receive it in faith, till such time as the mystery shall open."[74] This kind of language, from Maxwell's inaugural lecture at Aberdeen in 1856, seems somewhat awkward from a purely naturalistic view of science but was quite recognizable for evangelical Christians. Indeed, if "physical science" were replaced with "scripture" or "revelation," this passage would seem perfectly at home in an evangelical tract. He continually emphasized that the world "conceals far more than it displays" and that one must be patient with "mysteries within mysteries."[75]

Maxwell's Aberdeen lecture is a rich source for seeing how this theology of nature manifested in the realm of physics:

> Is it not wonderful that man's reason should be made a judge over God's works, and should measure, and weigh, and calculate, and say at last "I understand I have discovered—It is right and true" . . . we see before us distinct physical truths to be discovered, and we are confident that these mysteries are an inheritance of knowledge, not revealed at once, lest we should become proud in knowledge, and despise patient inquiry, but so arranged that, as each new truth is unraveled it becomes a clear, well-established addition to science, quite free from the mystery which must still remain, to show that every atom of creation is *unfathomable* in its perfection. While we look down with awe into these unsearchable depths and treasure up with care what with our little line and plummet we can reach, we ought to admire the wisdom of Him who has arranged these mysteries that we find first that which we can *understand* at first and the rest in order so that it is possible for us to have an ever increasing stock of *known* truth concerning things whose nature is absolutely incomprehensible.[76]

Note the repeated warnings against human pride and arrogance, and the evocative image of man's limited powers represented by "our little line and plummet." The deepest truths of nature were simply beyond our understanding, except where God allowed us to explore. As with the evangelical position on sin and redemption, our ability to know anything about the universe was due only to God's grace in making those things known and available. Comprehension of nature was the result not just of human efforts but of God's free choice to set up the laws of nature such that they could be understood.

Thus, Maxwell's theology of nature gave him both an assurance that natural laws were uniform and also an explanation for why the world did not always look that way: seeing uniformity was divine providence; seeing complexity was intrinsic human failing. This peculiarly evangelical twist on theology of nature had the additional benefit of avoiding a common difficulty of invoking divine action in science. Claiming divine action or design usually calls a sudden halt to scientific inquiry by declaring a phenomenon unexplainable by science and understandable only by religion, usually scripture. But Maxwell's evangelical God had planned the universe very carefully to always have lingering mysteries that humans were supposed to investigate. Mysteries were there to be uncovered and would lead to even more mysteries to solve. There was no level of explanation at which science was instructed to stop, and there was always something more to be found. Further, there was an assurance that the results of these investigations could never be dangerous to religion, because God had planned out exactly what could be comprehended and what would remain opaque. Again from the Aberdeen lecture:

> I have also thought it unnecessary to tell you that the study of the world in which we live is our obvious duty as a condition of our fulfilling the original command "to subdue the earth and have dominion over the creatures." Those who have raised objections to the engrossing pursuit of physical science had done so on the ground of the supposed effects of exact science in making the mind unfitted to receive truths which it cannot comprehend. I have endeavored to show that it is the peculiar function of physical science to lead us to the confines of the incomprehensible, and to bid us behold and receive it in faith, till such time as the mystery shall open.[77]

For Maxwell, physics was an echo of biblical commandments. Pursuing deeper and deeper levels of unification was a religious task in that he was following the path God had laid down for him. God revealed truth to man in scripture and the natural world, and it was an obligation to pursue both: "We all belong to a race endowed with faculties which urge us on to search deep and ever deeper into the nature of things."[78]

In sum: far from uniformity being antithetical to religious thinking, many scientists and philosophers concluded that uniformity only made sense in a theistic world. Without an ordering force (i.e., God), one would expect the universe to be a mishmash of chaotic events. The only guarantee for constancy of the laws of nature was the intent of the lawgiver. Frederick Temple, the future archbishop of Canterbury, and George Campbell, Duke of Argyll, were two important spokesmen for this position in the second half of the century. They acknowledged that the uniformity assumption was critical for science ("on no other assumption can Science proceed at all"), that it was justified both by the results of science ("This idea is a product of that immense development of the physical sciences which is characteristic of our time") and by simple experience ("Millions on millions of observations concur in exhibiting this uniformity").[79] The theists did not reject empiricism, reasoning, testing, or theorizing. Rather, they said that all of those things were dependent on God, and pointed to his role in the universe.[80] Baden Powell chose a quote from the scientist Hans Christian Oersted to describe elegantly this framework:

> The progress of discovery continually produces fresh evidence that Nature acts according to eternal laws, and that these laws are constituted as the mandates of an infinite perfect reason; so that the friend of Nature lives in a constant rational contemplation of the Omnipresent Divinity. . . . The laws of Nature are the thoughts of Nature; and these are the thoughts of God.[81]

Naturalistic Laws

The young Thomas Huxley first publicly confronted this sort of theistic laws in a famously vicious 1854 review of the tenth edition of *Vestiges of the Natural History of Creation*.[82] Robert Chambers's anonymous work describing a world emerging from natural processes was enormously controversial and launched a thousand essays.[83] Many of the attacks on *Vestiges* were due to its reliance on natural laws for creative acts, but Huxley savaged it for the particular *kind* of natural laws it invoked— those designed and governed by a deity. This review was the beginning of a lifetime of Huxley trying to claim the uniformity of natural law for his own purposes.

Huxley was far from intimidated by the book's success to date and declared it on the first page to be "a mass of pretentious nonsense."[84] He quoted the author as saying that the basic proposition of the book is "'creation in the manner of law,' that is, the Creator working in a natural course, or by natural means." This was the key point of Huxley's review, and laid out an agenda that he would return to again and again: "natural course" and "Creator" were as oil and water. The combination of laws and divine action had no "intelligible meaning."[85]

Huxley recast Chambers's central point: "Stripped of unnecessary verbiage, it comes to this, that 'Creation took place in an orderly manner, by the direct agency of the Deity.' A proposition which is as old as the Book of Genesis." He thus redefined theistic natural laws as not really natural laws at all, but rather a simple cover for a slavish biblical literalism.[86] Divine action, he said, was always disruptive.

> If the Deity be ever present, and phenomena are the manifestation of his will—law being simply a name for the order in which these occur—what is every phenomenon but the effect of a "creative fiat," an "interference," an "interposition of creative energy"?[87]

He said such disruptions and interference violated uniformity, that basic premise which made it possible to understand the natural world. Huxley worked from the axiom that any role for the divine in the natural world made uniformity impossible. His entire review was a grab of conceptual territory. *Vestiges* was a dangerous failure because it mistook its theological ideas for the proper claim

> that the past may be interpreted by the present; and that the succession of phenomena in past times, took place in a manner analogous to that which occurs at the *present day*. Such a proposition is the base of the modern science of history, whether natural or civil . . . "natural laws" are nothing but an epitome of the observed history of the phenomena of the universe.[88]

Here Huxley laid out the principles that would guide his use of the laws of nature: they could have no connection to divine forces; they were observational truths; they were uniform. These points (particularly the first) were far from broadly accepted, and he would battle for them over the course of decades.

Unfortunately, a review, however striking, would not change the way men of science thought about laws. Huxley would need to show how scientific *practice* could be based on and benefit from this naturalistic approach. The young naturalist found the opportunity to do this when he challenged one of the great paleontologists of the age: Richard Owen. Owen's heroic work—he famously reconstructed an entire extinct animal from one small bone—was based firmly in the tradition of theistic science. He saw natural laws as divine rule, and this appeared in his paleontological practice through his use of the archetype, the "primal pattern" on which God created animals.[89] These archetypes remained static while the actualized form could become more specialized over time, as seen in the fossil record. His favorite example was the structure of the vertebrate skull, which was held to show divine forethought and benevolence.[90] Owen accepted that the emergence of new species was due to secondary causes in the form of natural laws, and not to direct divine intervention—as was quite typical. He gave the continual, progressive creation of species scriptural support.[91] His goal was "to prove that Man was in the Divine Mind at the time of Creation."[92]

Owen's archetypes were explicit manifestations of theistic thinking, and Huxley set out to show how the work of the "British Cuvier" had been corrupted. Huxley was still a young naturalist of far less authority than Owen, and these early offensives showed the argumentative skill and flourish for which he would become famous. His decisive attack was his 1858 paper "On the Theory of the Vertebrate Skull," which targeted the anatomical relationships that Owen said could only be understood in light of archetypes.

Huxley disagreed with some of Owen's empirical conclusions—does the spinal column share structural similarities with the skull?—but the main force of the essay was directed at the conception of nature that allowed such conclusions to be arrived at. He began by celebrating the advancements of recent times:

> The biological science of the last half-century is honorably distinguished from that of preceding epochs, by the constantly increasing prominence of the idea, that a community of plan is discernible amidst the manifold diversities of organic structure. That there is nothing really aberrant in nature; that

the most widely different organisms are connected by a hidden bond; that an apparently new and isolated structure will prove, when its characters are thoroughly sifted, to be only a modification of something which existed before,— are propositions which are gradually assuming the position of articles of faith in the mind of the investigators of animated nature, and are directly, or by implication, admitted among the axioms of natural history.[93]

This declaration of uniformity in the living world, sometimes hidden, was a common point agreed upon by both Huxley and theists. Beginning with this emphasis was a strategy that would become quite frequent for him, reiterating that everyone agreed on the importance of natural laws and uniform processes. Then came his paring away of Owen's work—his opponent's practice might *appear* to follow these rules, but in fact should be seen as quite different. He conceded that a diagram of general anatomical relationships could be useful, and there "is no harm in calling such a convenient diagram the 'Archetype' of the skull, but I prefer to avoid a word whose connotation is so fundamentally opposed to the spirit of modern science."[94] Thus, Owen was declared to be not simply wrong, but actually in violation of those grand rules with which Huxley opened the paper. "Archetype" implied a designer, and therefore exactly the sort of divine interference over which Huxley savaged *Vestiges*. Again using a strategy that he would repeat, Huxley explicitly labeled Owen's practice as antiquated and outmoded. He said he was

> unable to see the propriety and advantage of introducing into science any ideal conception, which is other than the simplest possible generalized expression of observed facts, and [views] with extreme aversion, any attempt to introduce the phraseology and mode of thought of an obsolete and scholastic realism into biology.[95]

Such theistic science was marked as being preempirical and dogmatic. And this was not simply a matter of terminology or metaphysical preference, Huxley warned. It led directly to incorrect scientific practice, such as seeing unwarranted similarities between skulls and vertebrae.

Adrian Desmond and other scholars have noted that Huxley's opposition to Owen's archetypes carried with it a side effect: hostility to the idea that species undergo progressive change and adaptation.[96] He came to associate the claim that creatures changed significantly over time with idealist theism, and worked hard to show that the fossil record demonstrated that animals largely persisted in form. His 1856 manifesto "On Natural History, as Knowledge, Discipline, and Power" closely tied that rejection of progressive change to a naturalistic understanding of laws of nature. As before, he begins by reminding his readers that everyone agrees that all the sciences rely on lawful regularities:

> The mathematician discovers in the universe a "Divine Geometry"; the physicist and the chemist everywhere find that the operations of nature may be expressed in terms of the human intellect; and, in like manner, among living beings, the naturalist discovers that their "vital" processes are not performed by the gift of powers and faculties entirely peculiar and irrespective of those which are met with in the physical world.

He then told the story of the thinkers who drew the wrong lessons from this. The complicated parts of living things were fit together in a way that suggested an artificer, which led to the conclusions of "Paley and the natural theologians, whose arguments may be summed up thus—that the structure of living beings is, in the main, such as would result from the benevolent operation, under the conditions of the physical world, of an intelligence similar in kind, however superior in degree, to our own." Huxley linked directly this divine inference with the idea of biological structures being adapted to a purpose. Now, again, he moved to declare that consideration of divinely inspired laws was fatal to scientific practice. "Utilitarian adaptation to purpose is not the greatest principle worked out in nature, and that its value, even as an instrument of research, has been enormously overrated."[97] These errors were blamed on Cuvier, and Huxley denied that the sort of anatomical laws he used were really laws at all.[98] He returned to the uniformity of true natural laws as an empirical, inductive, and deductive fact:

> Our method then is not the method of adaptation, of necessary physiological correlations; for of such necessities, in the case in question, we know nothing: but it is the method of agreement; that method by which, having observed facts invariably occur together, we conclude they invariably have done so, and invariably will do so; a method used as much in the common affairs of life as in philosophy.[99]

Natural laws were to have no higher meaning or purpose, and were synonymous with the uniformity of nature as observed in everyday life.

Huxley's struggle against teleology required an alternate articulation of the order of nature as found in living things. The laws of embryonic development were one route, in which he was particularly influenced by Karl von Baer's work.[100] His own innovation was "persistent types," those species such as crocodiles and ferns that the fossil record indicated were nearly unchanged since early times. He pointed to these as being more representative of species change, allowing only for small and subtle alterations over time that required no new biological classes. He defended this concept as being in harmony with "physical uniformity" and incompatible with a divine role in creation:

> It is difficult to comprehend the meaning of such facts as these, if we suppose that each species of animal and plant, or each great type of organization, was formed and placed upon the surface of the globe at long intervals by a distinct act of creative power; and it is well to recollect that such an assumption is as unsupported by tradition or revelation as it is opposed to the general analogy of Nature.[101]

The logic of persistent types as a weapon against divine action was straightforward. "The naturalist who takes a wide view of fossil forms, in connection with existing life, can hardly recognise in these results anything but strong evidence in favour of the belief that a general uniformity has prevailed among the operations of Nature, through all time of which we have any record."[102] Unchanging species meant unchanging laws. Huxley's repeated argument that a lack of species change was the best indicator of proper naturalistic, uniform laws suffered a shock from a book published the same year as he proposed persistent types—*On the Origin of Species.*

Darwin had sent an early copy of the tome to Huxley, and eagerly anticipated a response. Huxley's letter was not quite what he had hoped for:

> As for your doctrine, I am prepared to go to the stake, if requisite, in support of Chapter IX, and most parts of Chapters X, XI, XII, and Chapter XIII contains much that is most admirable, but on one or two points I enter a *caveat* until I can see further into all sides of the question.

> As to the first four chapters, I agree thoroughly and fully with all the prin-
> ciples laid down in them. I think you have demonstrated a true cause for the
> production of species, and have thrown the *onus probandi*, that species did
> not arise in the way you suppose, on your adversaries.
>
> But I feel that I have not yet by any means fully realised the bearings of
> those most remarkable and original Chapters—III, IV, and V, and I will write
> no more about them just now.[103]

His ambivalence came through clearly. He praised Darwin's ambitions to bring the world of life under the rule of natural law but was unwilling to commit to the utilitarian adaptation he had so recently eviscerated in print.

Any hesitancy he felt regarding the details of Darwin's theory was obscured in the ensuing public debate. Instead, Darwin became the emblem for his naturalistic worldview. His anonymous review in the *Times* laid out his plan to use the *Origin* to demonstrate what uniform, nontheistic science should look like. Quickly dismissing the possibility of a literally correct Genesis, he also declared that any direct act of creation was not part of science.[104] Divine action was not only placed in complete opposition to natural law, but the replacement of the first by the second was described as the chief marker of the development of science:

> But what is the history of astronomy, of all the branches of physics, of chem-
> istry, of medicine, but a narration of the steps by which the human mind has
> been compelled, often sorely against its will, to recognise the operation of
> secondary causes in events where ignorance beheld an immediate interven-
> tion of a higher power? And when we know that living things are formed of
> the same elements as the inorganic world, that they act and react upon it,
> bound by a thousand ties of natural piety, is it probable, nay is it possible, that
> they, and they alone, should have no order in their seeming disorder, no unity
> in their seeming multiplicity, should suffer no explanation by the discovery of
> some central and sublime law of mutual connection?[105]

Darwin was credited with finally bringing the living world under the jurisdiction of laws of nature, and extending uniformity to all things. Decades later, Huxley reminisced that his circle's interest in Darwin's theory was due to exactly this: "That which we were looking for, and could not find, was a hypothesis respecting the origin of known organic forms which assumed the operation of no causes but such as could be proved to be actually at work."[106]

Huxley's essay on the *Origin* in *Macmillan's Magazine* was essentially a defense of his own persistent types, followed by a very brief description of the operation of natural selection.[107] Why put the name of a book in the essay's title when it barely makes an appearance? What Huxley was concerned about was linking Darwin's work with what occupied most of the text—a defense of the reality and consequence of uniform natural laws. He named uniformitarian geology as the only route to true understanding of the Earth's past, explicitly linking it to physics' ability to understand motion:

> The physical philosopher who is accurately acquainted with the velocity of a cannon-ball, and the precise character of the line which it traverses for a yard of its course, is necessitated by what he knows of the laws of nature to conclude that it came from a certain spot, whence it was impelled by a certain force, and that it has followed a certain trajectory. In like manner, the student of physical geology, who fully believes in the uniformity of the general condition of the earth through geologic time, may feel compelled by what he knows of causation, and by the general analogy of nature, to suppose that our solar system was once a nebulous mass; that it gradually condensed, that it broke up into that wonderful group of harmoniously rolling balls we call planets and satellites, and that then each of these underwent its appointed metamorphosis, until at last our own share of the cosmic vapour passed into that condition in which we first meet with definite records of its state, and in which it has since, with comparatively little change, remained.[108]

He complained that this sort of reasoning was widely accepted as long as the subject was cannonballs and rocks, but apparently became outrageous when applied to living things. This defense of natural law in the biological realm could be read as a defense of Darwin, but a careful reading shows that it was in support of Huxley's own persistent types. The tension between his own ideas and Darwin's was real and manifest, though Huxley's literary skill presented both as allies under the aegis of uniformity. He continued to champion his own ideas at the same time he promoted Darwin, and three years later, even as Darwinian evolution was winning converts, Huxley's presidential address to the Geological Society showed that he still refused to accept progressive tendencies in the fossil record.[109]

This wariness in discarding such a useful weapon against theistic scientists was perhaps justified, given the speed with which Darwin's ideas were adopted and adapted by that group. Theistic evolution schemes have been well discussed elsewhere, and their details are not necessary here.[110] Briefly, belief in theistic evolution—that is, evolution guided, planned, or supported in some way by God—was widespread, and many theists commented on how Darwin's ideas had extended uniformity throughout the world of life. Temple stated: "Once more, the doctrine of Evolution restores to the science of Nature the unity which we should expect in the creation of God."[111] The Duke of Argyll argued that God's choice to create species via natural laws instead of direct fiat was no slight of his power: creation by process "is Creation still."[112] The essence of the theistic evolution position was that the creation of species was effected by laws of exactly the sort that made the Earth go around the Sun. Species appeared due to "Creation by Law," even if the precise details of that law were still in dispute.[113]

Unsurprisingly, the scientific naturalists rejected this interpretation completely. Huxley denied that theistic evolution was an explanation of any sort, and marked it as simply an extension of Owen's earlier ideas. Most important, since it involved divine action, theistic evolution could not even *be* a law in Huxley's sense. He dismissed theistic evolutionists' use of the term *law* as simply a linguistic strategy to catch unwary scientists.[114] Joseph Hooker wrote regarding Asa Gray's theory that "I see nothing antidarwinian" but that "he got deeper and deeper in to theological and metaphysical wanderings, and finally formulated his ideas in an illogical fashion."[115]

Near the end of his life, Huxley wrote a number of pieces in which he encapsulated his views on the importance of the uniformity of nature as a purely naturalistic framework. His 1887 *Introductory Science Primer* began with an extensive discourse on the "Order of Nature." While asserting the crucial nature of laws, he denied that they had any transcendental reality or compulsive force.

When we have made out by careful and repeated observation that something is always the cause of a certain effect, or that certain events always take place in the same order, we speak of the truth thus discovered as a **law of nature** [emphasis in the original]. . . . But it is desirable to remember that which is very often forgotten, that the laws of nature are not the causes of the order of nature, but only our way of stating as much as we have made out of that order. Stones do not fall to the ground in consequence of the law just stated, as people sometimes carelessly say; but the law is a way of asserting that which invariably happens when heavy bodies at the surface of the earth, stones among the rest, are free to move.[116]

In something of a departure from his earlier statements about laws being truly within nature, perhaps driven by his later study of Hume, Huxley admitted that they were simply observations of regularities that humans happened to have recorded. He often used the idea that laws were just regularities (and therefore had no ontological reality) as a weapon against theists presenting natural laws as God-given.[117]

His emphasis that laws were in an important sense a human creation opened the door to human fallibility, which had a significant consequence. Allowing this human element again provided a defense against those attacking uniformity. Someone claiming to find a disruption of the laws of nature or a violation of uniformity could never have a logical foundation:

> To speak of the violation, or the suspension, of a law of nature is an absurdity. All that the phrase can really mean is that, under certain circumstances the assertion contained in the law is not true; and the just conclusion is, not that the order of nature is interrupted, but that we have made a mistake in stating that order. A true natural law is an universal rule, and, as such, admits of no exceptions.[118]

A correct natural law was uniform and universal. A law that seemed to lack those characteristics was simply an error. Literally, a nonuniform world was impossible.

Battling for Uniformity

Huxley publically declared himself to be an adherent of evolution in 1868. As part of this, he presented a brief catechism. Those who believe in evolution believe that the world and everything on it used to be in a different form. What we see today is the result of gradual changes over vast time. Further, "One who adopts the nebular hypothesis in astronomy, or is a uniformitarian in geology, or a Darwinian in biology, is so far an adherent of the doctrine of evolution." Thus, someone who believed in the uniformity of nature also believed in evolution. And believing in evolution was believing in science.

In that speech Huxley also addressed an issue that seemed to question the uniformity that he said underlay science. The fossil record contained "great gaps" between distinct groups of plants and animals, which raised doubt about the uniform change expected by the laws of evolution. This was but one instance of a problem that bedeviled all men of science who held to uniformity: what to do when nature does not appear uniform? On the issue of the fossil gaps, Huxley replied:

> We, who believe in evolution, reply that these gaps were once non-existent; that the connecting forms existed in previous epochs of the world's history, but that they have died out . . . we account for this failure of the needful evidence by the known imperfection of the geological record. We say that the series of formations with which we are acquainted is but a small fraction of those which have existed, and that between those which we know there are great breaks and gaps.[119]

The apparent discontinuities in the record were not to be taken as accurate reflections of the discontinuities of natural law. Rather, it should be assumed that the laws of nature did not vary, only the fickle processes that recorded those laws' actions. The failure of uniformity was only apparent. Establishing the meaning of these gaps was key to applying naturalistic uniformity to the geological past, and it is interesting to note that the only section of the *Origin* to which Huxley swore immediate loyalty was that in which Darwin discussed the imperfections of geological record.

Some years later Huxley acknowledged that science, as a human enterprise, had human frailties. No explanation of a phenomenon could ever be complete due to human limitations.[120] But Huxley turned this limitation into a rebuke to critics of an orderly nature:

> But the more carefully nature has been studied, the more widely has order been found to prevail, while what seemed disorder has proved to be nothing but complexity; until, at present, no one is so foolish as to believe that anything happens by chance, or that there are any real accidents, in the sense of events which have no cause. And if we say that a thing happens by chance, ev-

erybody admits that all we really mean is, that we do not know its cause or the reason why that particular thing happens. Chance and accident are only *aliases* of ignorance.[121]

A critic who claimed to have found a disruption in uniformity, then, had only made a mistake based on his lack of knowledge. Conclusions of true complexity or disorder in nature were the result of human arrogance, of assuming that one could ever know all the causes involved in an event. Assuming that complexity will eventually be unraveled into simple laws was actually the *humble* stance. He pointed out that there was no logical prohibition against invoking unknown and mysterious forces to explain a phenomenon, but also that "philosophy has prospered exactly as it has disregarded such possibilities, and has endeavoured to resolve every event by ordinary reasoning." Use natural laws as guides; follow them and they will lead us somewhere. Anything else was virtually an appeal to final causation, which Huxley declared "a barren virgin."[122]

Investigators in physics and chemistry had found their simple laws underlying complexity, and Huxley was eager to show that the biological sciences could do the same. As discussed earlier, embryology and morphology provided some tools for him to find natural laws within living things. In his riveting discourse on a lobster, he argued:

> Thus the study of development proves that the doctrine of unity of plan is not merely a fancy, that it is not merely one way of looking at the matter, but that it is the expression of deep-seated natural facts. . . . These are wonderful truths, the more so because the zoologist finds them to be of universal application. The investigation of a polype, of a snail, of a fish, of a horse, or of a man, would have led us, though by a less easy path, perhaps, to exactly the same point. Unity of plan everywhere lies hidden under the mask of diversity of structure—the complex is everywhere evolved out of the simple.[123]

The natural laws hidden in the lobster were real, objective, and universal. Just as the physicist could point to falling apples and hurtling comets as instances of gravitation, the anatomist, properly guided by the uniformity of nature, could find his own laws buried under complexity.

Huxley had complicated ideas about how laws from different sciences should interact. On one hand, his views on uniformity committed him to the application of the laws of physics to biology:

Consider why is the skeleton of this horse capable of supporting the masses of flesh and the various organs forming the living body, unless it is because of the action of the same forces of cohesion which combines together the particles of matter composing this piece of chalk? What is there in the muscular contractile power of the animal but the force which is expressible, and which is in a certain sense convertible, into the force of gravity which it overcomes?[124]

He pointed to the experiments of Emil Du Bois-Reymond as demonstrating that nerves were simply electrical conduits. At times he was content to

reduce all scientific problems, except those which are purely mathematical, to questions of molecular physics—that is to say, to the attractions, repulsions, motions, and co-ordination of the ultimate particles of matter. Social phænomena are the result of the interaction of the components of society, or men, with one another and the surrounding universe . . . the phænomena of biology and of chemistry are, in their ultimate analysis, questions of molecular physics. Indeed, the fact is acknowledged by all chemists and biologists who look beyond their immediate occupations.[125]

He was certainly deeply committed to denying any barrier between the organic and inorganic world. In terms of method, too, Huxley wanted to make it clear that biology should share in all the prestige accorded to astronomy for its amazing predictive abilities.[126]

Despite these unities, Huxley warned that there were lurking dangers. The laws of physiology concerned the function of organs toward a common end, but if a careless investigator applied that kind of thinking to the organism as a whole (i.e., to morphology), dangerous conclusions could be reached.[127] The troublesome move was extending the study of function beyond the purely empirical:

No further reason for [this morphological law] can be given than for the law of gravitation. The whole object of morphology is to ascertain *what* structural peculiarities invariably coexist with one another: *why* these structural peculiarities coexist is a question with which it does not necessarily concern itself, and so far as the mere restorations of the palæontologist are concerned, it is a wholly irrelevant question. The empirical laws of morphology supply all that the palæontologist requires for this object.[128]

The trap was that a paleontologist could end up like Owen, seeing tele-ology where there was none. The laws of physiology could provide only "very useful hints and guesses" about the nature of extinct animals. Deeper claims about function could not, and should not, be justified based on the observations available. On precisely where this border was, Huxley was somewhat vague, but it was the critical point at which sci-ence stopped being properly naturalistic and starting being dangerously theistic. A naturalistic scientist had to be wary of teleologists sneaking divine action in through function, just as they could always claim that a given arrangement of atoms was preordained.[129]

Huxley's relentless offensive for Darwin and against theism in science drew the attention of William Thomson, later known as Lord Kelvin.[130] In the 1860s, Thomson had begun using the new laws of thermodynam-ics to estimate the likely age of the Earth. While his results were far from precise, he was quite certain that the Earth's surface had been very differ-ent (likely molten) about one hundred million years in the past. His work was based on cutting-edge physics and astronomy, and he became in-creasingly frustrated that certain "very influential" geologists continued to invoke vast, sometimes infinite, time to make their theories work.[131]

Thomson called for another wave of geological reform, as in the late eighteenth century when geologists freed themselves from "the dictation of authority and dogmatic hypotheses."[132] Uniformitarian geology based on current processes, he said, could not be justified in light of physics:

> It is quite certain that a great mistake has been made—that British popular geology at the present time is in direct opposition to the principles of natural philosophy. . . . There cannot be uniformity. The earth is filled with evidences that it has not been going on for ever in the present state, and that there is a progress of events towards a state infinitely different from the present.[133]

He argued that the only way to evade his dating was to claim that the Sun had somehow been shining (and would continue to shine) forever. He admitted that it was logically possible that God created the Sun to be exempt from thermodynamics:

> All things are possible to Creative Power. But we know also, that Creative Power has created in our minds a wish to investigate and a capacity for inves-tigating; and there is nothing too rash, there is nothing audacious, in ques-

tioning human assumptions regarding Creative Power. Have we reason to believe Creative Power did order the sun to go on, and shine, and give out heat
for ever? Are we to suppose that the sun is a perpetual miracle? I use the
word *miracle* in the sense of a perpetual violation of those laws of action between matter and matter which we are allowed to investigate here at the surface of the earth, in our laboratories and mechanical workshops.[134]

If experiment and theory were of any value, it was not philosophically
acceptable to extend geological laws and the continuity of life back more
than a few million years.

Huxley was not about to let Thomson demolish the uniformity of geology, and he dedicated his 1869 presidential address to the Geological
Society to attacking the Scottish physicist. He listed the main schools of
geological thought as catastrophism, uniformitarianism, and evolutionism. The first (which included the Mosaic cosmogony) was dismissed for
invoking "the operation of forces different in their nature, or immeasur-
ably different in power, from those which we at present see in action in
the universe."[135] The uniformitarians (meaning Lyell and Hutton) were
critiqued for not being willing to extend their premise to a time before
the formation of rocks. Only evolutionism was given the mantle of being truly uniform, for trying to deduce the history of the world "from the
known properties of the matter of the earth, in the conditions in which
the earth has been placed."[136] Huxley's defense of geology allowed that
there were those investigators who wandered from the true path of the
uniformity of nature, but vigorously asserted that geology was more and
more scientific as it hewed closer and closer to that path. Denying the
application of that principle to the history of the Earth, as Thomson
seemed to do, would prevent geology from being a science at all.

Thomson's counterattack insisted that "in some of the most recent
geological writings of the highest character I still find the same tendency
to overlook essential principles in thermodynamics."[137] He provided a
thorough list of geologists and paleontologists—including Huxley—referring to unlimited or vastly long periods of geological time. He warned
biologists that they had been misled by geologists into thinking that
they had effectively infinite time to support their speculations on species change. This was, of course, an attack on Darwin's theory. Thomson's short time scale was clearly insufficient for natural selection, and
was used explicitly to attack Darwin by Fleeming Jenkin and others.[138]

Thomson cautioned that the "ultra-uniformitarianism" used by supporters of evolution had no basis in proper science. Indeed, it seemed that
such speculators, in ignoring the first and second laws of thermodynamics, had completely departed from the rules of investigation: "Many geologists are contented to regard the general principles of natural philosophy, and their application to terrestrial physics, as matters quite foreign
to their ordinary pursuits."[139]

We are used to thinking of attacks on geological estimates of the
Earth's age as relying on attacks on uniformity: radioactive dating is unreliable, or uniformitarian processes were disrupted by directly creative
acts such as Noah's flood. It is easy to interpret Thomson's arguments
in this light, particularly with his statements such as "There cannot be
uniformity."[140] However, this is misleading. He was actually attacking a
particular kind of geological uniformitarianism that allowed for an extremely old Earth, not uniformity in general, and his position was essentially that geologists were not being uniform *enough*. Huxley, and others, he said, assumed that geological forces should be thought of as being
constant deep into the past, virtually forever. But this constancy of geology was made impossible by the constancy of physics: the second law of
thermodynamics demanded that deep in the past the surface of the Earth
would appear and behave quite differently from how it does today.

Far from attacking uniformity, Thomson was vigorously defending it.
He believed strongly that natural laws were permanent and universal.[141]
Indeed, he was actually attacked for this assumption.[142] He thought the
whole scientific enterprise relied on the idea that currently observed natural laws could be extended throughout time: "The essence of science,
as is well illustrated by astronomy and cosmical physics, consists in infer-
ring antecedent conditions, and anticipating future evolutions, from phenomena which have actually come under observation."[143] He completely
rejected the possibility of violations of natural laws, particularly supernatural intervention. Since natural laws were uniform, scientists must restrict themselves to explanations in terms of those laws: "If a probable
solution [to any scientific problem], consistent with the ordinary course
of nature can be found, we must not invoke an abnormal act of creative
power."[144] He explicitly denied that divine intervention could make the
Sun and Earth exempt from physics.[145]

Thomson was a Christian with a latitudinarian perspective, and
he was adamant that a uniform universe was still perfectly consonant

with a religious worldview. Writing with his collaborator P. G. Tait, he declared that

> we have the sober scientific certainty that heavens and earth shall "wax old as doth a garment" [Psalm 102:26]; and that this slow progress must gradually, by natural agencies which we see going on under fixed laws, bring about circumstances in which "the elements shall melt with fervent heat" [2 Peter 3:10].[146]

They made the case that a world running by "natural agencies" was an idea with strong scriptural support. The reason for this, of course, was that Thomson saw natural laws as coming from God. They were clearly the result of a creative, consistent intelligence, and it was this intelligence that allowed for uniformity in the first place.[147]

Huxley and Thomson's clash was in some sense puzzling. Both saw themselves as the defenders of the uniformity of nature, and each saw the other as misunderstanding this fundamental principle. While they agreed on uniformity as a guiding light, they disagreed completely about the origin and meaning of it. Their argument has typically been seen as being focused on Darwin, but natural selection was epiphenomenal to the core issue: who would be the spokesman for a uniform universe?

Uniformity as Antitheology

Over time Huxley attacked theistic laws far less, and increasingly targeted scriptural literalism in science as the chief enemy. "As the geologist of my young days wrote, he had one eye upon fact, and the other on Genesis; at present, he wisely keeps both eyes on fact, and ignores the pentateuchal mythology altogether."[148] This was likely a strategy to make his work more palatable to moderates, and to make uniform laws a wedge issue: men of science were pressured to either agree with him or ally themselves with the radical literalists. This also helped sharpen the distinction he often drew between religion and theology, fingering the dogmatic theologians as the real opposition. In his last two decades he structured his attacks more carefully to claim uniformity as a wholly naturalistic category fundamentally opposed to theological conceptions of nature.

Historical narratives became an important tool for him to articulate his vision of naturalistic uniformity. Consider, he said, the earliest "thinking men" trying to understand the world around them. These primitive humans found that

> there is order amidst the confusion, and that many events take place according to unchanging rules. To this region of familiar steadiness and customary regularity they gave the name of Nature. But, at the same time, their infantile and untutored reason, little more, as yet, than the playfellow of the imagination, led them to believe that this tangible commonplace, orderly world of Nature was surrounded and interpenetrated by another intangible and mysterious world, no more bound by fixed rules than, as they fancied, were [their own] thoughts and passions.

This was the world of the supernatural, and Huxley asserted that from the earliest times of which we have any knowledge, "Naturalism and Supernaturalism have consciously or unconsciously, competed and struggled with one another."[149]

He told the story of human progress as the elimination of the supernatural from man's thoughts, and pointed to his century as the era when naturalism finally gained the upper hand: "The stream of tendency toward Naturalism . . . has, of late years, flowed so strongly that even the Churches have begun, I dare not say to drift, but, at any rate, to swing at their moorings." Naturalistic science was an enemy that had surrounded supernatural theology, and was daily driving it back.[150] He increasingly presented naturalism and supernaturalism as a zero-sum game, making any allegiance to natural laws seem to be a division from religion.

The results of science were presented as having the inevitable effect of forcing one away from supernaturalism:

> No truths brought to light by biological investigation were better calculated to inspire distrust of the dogmas intruded upon science in the name of theology, than those which relate to the distribution of animals and plants on the surface of the earth. Very skilful accommodation was needful, if the limitation of sloths to South America, and of the ornithorhynchus to Australia, was to be reconciled with the literal interpretation of the history of the deluge; and with the establishment of the existence of distinct provinces of distribution, any serious belief in the peopling of the world by migration from Mount Ararat came to an end.[151]

This was the foundation of Huxley's highly strategic construction of the military metaphor.[152] Theology, not religious belief per se, was the enemy. Theology was the intruder on science. Slavish literal interpretation of the Bible was one of its weapons. And most dramatically, these absurd beliefs had *come to an end*, thus marking any further resistance as antiquated and ridiculous. Uniformity was proposed as synonymous with rationalism, progress, and naturalism.

Huxley's friend and ally John Tyndall also spoke vigorously of the power of uniformity to banish an intervening God, particularly in his famous Belfast Address in 1874. A valid natural law, he said, "asserts itself everywhere in nature."[153] The advance of science was steadily crushing the mutability of nature demanded by religious believers:

> Now, as science demands the radical extirpation of caprice and the absolute reliance upon law in nature, there grew with the growth of scientific notions a desire and determination to sweep from the field of theory this mob of gods and demons, and to place natural phenomena on a basis more congruent with themselves.[154]

Tyndall could hardly be clearer. The uniformity of natural laws left no room for divine action in science. A sterling example of the military metaphor of science and theology is seen in the martial language used:

> The impregnable position of science may be described in a few words. We claim, and we shall wrest, from theology the entire domain of cosmological theory. All schemes and systems which thus infringe upon the domain of science must, *in so far as they do this*, submit to its control, and relinquish all thought of controlling it. Acting otherwise proved disastrous in the past, and is simply fatuous to-day.[155]

Science, as a complete scheme of the universe, could have no interaction with theology other than accepting its surrender. Before the advent of science, Tyndall said the unlearned masses had no option other than filling the world with "witchcraft, and magic, and miracles, and special providences." The power of natural laws would simply squeeze the world until nothing else remained: "The law of gravitation crushes the simple worshipers of Ottery St. Mary, while singing their hymns, just as surely as if they were engaged in a midnight brawl."[156]

The subtext of these claims was not just that uniformity restricts theology from entering science but that uniformity can only be justified in a world without divine intervention. How can scientists plan and conduct an experiment if they must worry that Jehovah will change the constants of nature? Uniformity can only be justified, then, if everyone agrees a priori that there can be no divine interventions.

Miracles

Building on this reading of uniformity, the scientific naturalists thought they had one attack for which there was no counter. Miracles, they said, were the essence of Christianity. And a miracle, it seemed, must be a violation of a natural law, and therefore a violation of uniformity, and thus cannot be consonant with science. Taking a position on miracles, then, forced one into either the theistic or naturalistic camp. This was a maneuver emphasized repeatedly by Victorian scientific naturalists, many of whom were directly inspired by David Hume.[157]

It was not coincidental that miracles were chosen as a target for attack. The second half of the nineteenth century saw serious disruptions in Protestant positions on miracles and the miraculous, particularly in Britain. During and after the Reformation, the general Protestant view was that miracles had been limited to biblical times, and could no longer be found. This was a strategic move to differentiate them from superstitious Catholics calling for their saints' protection, and also to anchor the Bible's authority over ecclesiastical tradition—only it had true miracles. This position came under scrutiny for theological reasons from, among others, John Henry Newman and Horace Bushnell, who wanted to abandon the limitation of miracles to biblical times and accept modern ones. Strauss's *Life* and the infamous *Essays and Reviews* put more pressure on these questions by interrogating even biblical miracles, and the Reverend James Mozley contended that miracles should not be considered a primary mode of divine action.[158]

Huxley chose miracles as the subject for some of his famous battles with Gladstone, and developed his ideas further in his book on Hume. He argued that an observation of an apparently miraculous event (say, the unsupported floating of lead in midair) would be no evidence that the laws of nature had been suspended. A man of science would then simply investigate to find the hitherto-concealed laws that allowed such a thing to happen. Humanity's limited experience with the world would inevita-

bly lead to gaps, but those gaps could not be used to show divine intervention: "The day-fly has better grounds for calling a thunderstorm supernatural, than has man, with his experience of an infinitesimal fraction
of duration, to say that the most astonishing event that can be imagined
is beyond the scope of natural causes."[159] Usually even a small amount
of thought could provide naturalistic explanations for otherwise mysterious events, such as miraculous healings.[160] He warned against anyone
declaring an event to be impossible on the grounds of natural law, as that
would play into the hands of "ecclesiastical apologists" and their worship of absolute statements.[161] One could say an event was improbable,
but even the most extraordinary events did not threaten the laws of nature. He made this point while obliquely attacking one well-known miracle in particular:

> That is to say, there is an uniform experience against such an event, and
> therefore, if it occurs, it is a violation of the laws of nature. Or, to put the ar-
> gument in its naked absurdity, that which never has happened never can hap-
> pen, without a violation of the laws of nature. In truth, if a dead man did
> come to life, the fact would be evidence, not that any law of nature had been
> violated, but that those laws, even when they express the results of a very long
> and uniform experience, are necessarily based on incomplete knowledge, and
> are to be held only as grounds of more or less justifiable expectation. To sum
> up, the definition of a miracle as a suspension or a contravention of the order
> of Nature is self-contradictory, because all we know of the order of nature is
> derived from our observation of the course of events of which the so-called
> miracle is a part.[162]

Thus, Huxley said, the very concept of a miracle in the first place, regardless of evidence, could not stand on its own power.

Huxley's typical approach to addressing miracles was distinctly Humean—instead of questioning whether the miracle occurred, or could
have occurred, he questioned the witnesses and the reliability of their
account. He specifically chose to use the story of the Gadarene swine in
his debates with Gladstone because it was one of the best attested, with
many witnesses.[163] Elsewhere he recounted some medieval miracles associated with Eginhard that had mountains of supporting evidence. He
suggested that even these eyewitness accounts were not sufficient evidence. A medieval man, he said, had all his beliefs and his most cher-
ished hopes "bound up with the belief in the miraculous." One could not

expect such a person to perform any serious investigation of the miracles or to question them sincerely. This was not, he averred, an accusation of fraud. Rather, it was simply that people lent their support to claims they wished to be true.[164] Only an expert committed to naturalism could be a reliable witness.

The next move was a brilliant use by Huxley of the social context of his listeners. Of course, he said, everyone in his audience was a good Protestant who would never believe Catholic medieval miracles. And yet,

> if you do not believe in these miracles recounted by a witness whose character and competency are firmly established, whose sincerity cannot be doubted, and who appeals to his sovereign and other contemporaries as witnesses of the truth of what he says, in a document of which a MS. copy exists, probably dating within a century of the author's death, why do you profess to believe in stories of a like character, which are found in documents of the dates and of the authorship of which nothing is certainly determined, and no known copies of which come within two or three centuries of the events they record?

These "stories of a like character" were, of course, the Gospels. Huxley stressed that knowledge of Matthew, Mark, Luke, and John was essentially zero when compared to knowledge of Eginhard.[165] He thus trapped his audience into either siding with Catholic superstition or admitting the unreliability of the New Testament.

Huxley continued to remind Anglicans of their similarity to Roman Catholics in this regard. Both required biblical miracles to be genuine and to be representative of true divine (as opposed to diabolical or heretical) action. Anglicans then had to engage in a hypocritical activity of declaring all miracles after the presumed corruption of the Roman church to be "shams and impostures." Soon after the establishment of the Church of England, Huxley recounted, "torrents of theological special pleading about the subject flowed from clerical pens" trying to justify one set of miracles but not another.[166] These activities only made sense within the realm of orthodoxy; the question of whether the miracles actually happened, Huxley seized for science. Indeed, the unwillingness of men of science to accept miracle stories pointed to their moral superiority over theologians:

> Whether such events are possible or impossible, no man can say; but scientific ethics can and does declare that the profession of belief in them, on the

evidence of documents of unknown date and of unknown authorship, is im-
moral. Theological apologists who insist that morality will vanish if their dogmas are exploded, would do well to consider the fact that, in the matter of intellectual veracity, science is already a long way ahead of the Churches; and
that, in this particular, it is exerting an educational influence on mankind of
which the Churches have shown themselves utterly incapable.

Orthodox Christianity's foolish reliance on miracles would, he contended, lead to its own destruction. The fall would be "neither sudden
nor speedy," but it was inevitable.[167]

Huxley's strategy for attacking miracles was actually somewhat subtle. Its Humean approach gnawed around the edges of the miracles, un-
dermining confidence in the reports and our understanding of what
"miracle" might mean. It was conveniently moderate: instead of requiring a miracle to be accepted or rejected, it could simply be considered
for future investigation.[168] He rarely used (though sometimes his critics
thought he did) a frontal attack of the sort most often associated with
scientific naturalism: a simple declaration that miracles were impossible in light of the uniformity of nature. This tactic was used much more
often by John Tyndall, who was merciless in drawing the line between
naturalism and miracles, particularly in the infamous "prayer gauge"
debate.[169]

This debate came on the heels of increasing unease among liberal
Anglican clergy such as Charles Kingsley and A. P. Stanley regarding
prayers offered for national problems such as poor weather or livestock
disease. Tyndall supported those clergy on the grounds that divine action in the physical world implied a violation of the conservation of energy.[170] The debate grew in 1871 when the Prince of Wales was suffering
from typhoid, and apparently recovered after the queen asked Anglican
clergy to pray for his recovery. Tyndall contended that if God created
physical effects, then those effects were measurable, and he suggested
setting up an experiment with hospitals to see if prayer worked. He said
that private prayer was of no concern to a man of science, but when it became a public event, a physicist had the right to examine it with "those
methods of examination from which our present knowledge of the physical universe is derived."[171]

This 1870s controversy over whether one could scientifically determine the efficacy of prayer provided a high-profile public forum in which
Tyndall laid out the naturalistic case for the irrationality of miracles. He

said that once science had demonstrated the uniformity of nature, "the age of miracles is past."[172] There was no indication that natural law was ever suspended, and therefore, there was no possibility of miracles. The only way out, he said, was to retort: "How do you know that a uniform experience will continue uniform? You tell me that the sun has risen for six thousand years: that is no proof that it will rise tomorrow; within the next twelve hours it may be puffed out by the Almighty."[173] He said someone attacking uniformity in this way, however, could barely function in the normal world, and had no reason to believe that Jack and the beanstalk was not a true story, since perhaps the natural laws governing bean growth had been suspended at some time. The rhetorical move here was a clear one: someone who believes in the miracles of the Bible or that God will answer a prayer for his or her sick child was no different from someone who believed in fairy tales.

The Victorian naturalists felt they had hemmed the theists into an inescapable dilemma. For science to be done, the universe needed to be uniform. But uniformity forbade miracles, and without miracles, what was Christianity? Which, then, would the theists sacrifice: science or religion? As we have already seen, the theistic scientists refused to discard either, and had robust interpretations of uniformity in a religious framework. So it should be no surprise that they drew upon religious uniformity as a resource to explain miracles in an orderly world.

There was widespread agreement among theistic scientists that (as with Huxley) apparent violations of natural law were illusory. Many other Christians agreed—Frederick Temple declared, "There may be instances where this Order is apparently broken, but really maintained, because one physical law is absorbed in a higher."[174] That is, an event that appeared to be outside the laws of nature actually was lawful, but it simply obeyed a law of which humans were not yet aware. An analogy might be to consider someone who understood the law of gravity, but not that of buoyancy. A hot-air balloon would appear to that individual to be miraculous, but a better-informed observer would understand that no laws had been broken.

What, then, of the supernatural? Would not religious believers need violations of natural law to be assured of the existence of supernatural forces? One of the prices of this strategy was that, in an important sense, the category of the supernatural faded away (or was at least redefined). The Duke of Argyll acknowledged that if something happened in our world, then uniformity demands that it be the result of natural law:

The Reign of Law in Nature is, indeed, so far as we can observe it, universal. But the common idea of the Supernatural is that which is at variance with Natural Law, above it, or in violation of it. Nothing, however wonderful, which happens according to Natural Law, would be considered by any one as Supernatural. The law in obedience to which a wonderful thing happens may not be known; but this would not give it a supernatural character, so long as we assuredly believe that it did happen according to *some* law.[175]

If scientists had total knowledge of all natural laws, then nothing would ever appear supernatural. What seemed inexplicable was actually only temporarily obscured. Argyll pointed out that the technological advances of the Victorian period allowed completely normal humans to achieve feats that earlier generations would have called supernatural (such as sending a message instantly across the Atlantic Ocean).[176] Perhaps, then, Jesus's healings in the Gospels simply relied on laws of medicine not yet understood.[177] Argyll argued that the Bible and its writers did not separate natural and supernatural: "The vicious and unphilosophical distinction between 'natural' and 'supernatural' is absolutely unknown to them."[178]

So far, these theists were in almost complete agreement with Huxley and Tyndall. Did that not hem them into precisely the dilemma of choosing between uniformity and divine action? They replied strongly in the negative: God could still watch over his creation and enact his plans, but *through* natural laws, not with interruptions of the natural order. Argyll again asserted that there "is nothing in Religion incompatible with the belief that all exercises of God's power, whether ordinary or extraordinary, are effected through the instrumentality of means—that is to say, by the instrumentality of natural laws brought out, as it were, and used for a Divine purpose."[179] God created laws as the means by which he exercises power in the world, like a craftsman who builds his own tools. The deity could manipulate natural laws in a variety of ways without violating their essence, and could produce any of the fantastic events recorded in scripture. This idea had a long genealogy in Christianity going back to Augustine.[180] However, how could uniform, unchanging laws produce singular events that appear to be obvious disruptions of nature? Charles Babbage found a solution to this problem decades before in a parlor trick performed with his calculating machine. The machine, which of course ran on fixed rules, could produce a steady, regular sequence of numbers only to make suddenly a great jump—thus demonstrating that what appear to be exceptional events could be easily generated by fixed laws.[181]

But if God only worked through natural laws, in what sense could these events be miracles? Argyll argued that the marker of a miracle was not the presence of supernatural causes, but rather that it had its origin in divine intent. This view, he said, was perfectly harmonious with scripture and allows defense of all the essential events of Christianity.[182] Similarly, Frederick Temple argued that even if science were to someday give an explanation of all the miracles in the Bible, it would not at all change their role in revelation. The miracle could be in their timing, or intent, or effect, rather than in their breach of uniformity.[183] This fit well with a traditional Protestant distinction between miracles, which required an objective witness to provide proof of supernaturalism, and special providence, which appeared to be normal events—except when viewed through the eyes of faith.[184] So this move would essentially eliminate the category of formal miracles and subsume all divine actions under special providence. Miracles in a uniform universe might no longer be particularly miraculous, but the critical issue could be resolved. A providential God did not have to be incommensurable with uniformity, and therefore with science:

> Science will continue its progress, and as the thoughts of men become clearer it will be perpetually more plainly seen that nothing in Revelation really interferes with that progress. It will be seen that devout believers can observe, can cross-question nature, can look for uniformity and find it, with as keen an eye, with as active an imagination, with as sure a reasoning, as those who deny entirely all possibility of miracles and reject all Revelation on that account. The belief that God can work miracles and has worked them, has never yet obstructed the path of a single student of Science.[185]

The important religious function of miracles was retained, along with the power and potential of science. Some twenty-first-century theologians are proposing similar strategies.[186]

Conclusion

The Victorian scientific naturalists made a straightforward argument: only naturalism allowed for the uniformity of natural laws necessary for the practice of science. Even for Huxley, though, this was not always the case. Adrian Desmond has pointed out that Huxley's work in evolution was more similar to that of his archenemy Owen than he would have liked to admit. Once Huxley had fully converted to evolution, he focused on the development of the horse as "The Demonstrative Evidence of Evolution." His diagrams demonstrating equine evolution were effectively identical to the diagrams made by Owen to illustrate the horse's archetype. This continuity of practice was not a matter of hidden theism on Huxley's part, or precocious naturalism on Owen's part. Rather, each of them was adhering to the need for uniform laws in thinking about species change. This adherence led to similar practices, even though each individual would have argued strongly that the practice only made sense in his own theistic or naturalistic worldview. In a rare rhetoric-less moment, Huxley grudgingly admitted this:

> All who are competent to express an opinion on the subject are, at present, agreed that the manifold varieties of animal and vegetable form have not either come into existence by chance, nor result from capricious exertions of creative power; but that they have taken place in a definite order, the statement of which order is what men of science term a natural law. Whether such a law is to be regarded as an expression of the mode of operation of natural forces, or whether it is simply a statement of the manner in which a supernatural power has thought fit to act, is a secondary question, so long as the existence of the law and the possibility of its discovery by the human intellect are granted.[187]

The uniformity of natural laws was key to the practice of Victorian science. While both naturalists and theists claimed it as their own, both groups were able to deploy it in actual research without difficulty.

In addition to the basic premise of uniformity, there were two other important aspects that both groups successfully grappled with. First: the problem of what to do with apparent interruptions of uniformity. If men of science could not find a way to reconcile these with natural laws, scientific practice would be very difficult. The theists were able to work around these because they had a divine guarantee that uniformity existed and that apparent disruptions were illusory. Maxwell's evangelical emphasis of mystery was a particularly effective version of this. The naturalists had a harder time justifying apparent disruptions as illusory, but they still succeeded. Huxley's strategy was to assert that natural laws were simply human constructions, allowing for a pair of solutions—human frailty, and the claim that natural laws were only assemblies of observations, and therefore that no observation could ever be outside a law.

The second aspect that both groups embraced was scientific progress. Naturalists and theists agreed that it was critical for science to advance into the unknown and expand its reach. Theists argued for evidence that God had blessed scientific inquiry and demanded further exploration. Huxley and his allies saw the progress of science as pushing back the possibility of divine intervention, boosting their own cultural authority (as we will see in later chapters) and justifying the initial gamble on uniformity itself.

This chapter has argued that both theism and naturalism allowed for the use of the uniformity of nature in effective and productive ways. This overlap was needed for the emerging scientific naturalists to make the case for their new outlook on science. A theistic guarantee for uniformity was the standard, and when Huxley's circle pushed for a novel approach, they were the ones who needed to justify that it could provide the same foundations as the old approach. They argued that naturalistic practice adhered just as closely to uniformity as theistic, even as they stressed that theism was incompatible with uniformity. By following existing practices (though they claimed to be completely new), the naturalists could move into existing scientific institutions, projects, and journals without serious disruptions. The shared value of uniformity allowed for a transition between the two groups, but was surely not sufficient. The naturalists' move to capitalize on this will be discussed in chapter 7.

The Limits of Science

Everyone agreed that everyone else was doing science wrong. The Duke of Argyll accused Huxley of being "on the warpath," and of claiming that "science can decide by the microscope and the dissecting needle, whether the Sadducee was right in denying either angel or spirit, and the Pharisee was a fool in confessing both." Argyll said such "unbounded knowledge" was completely absurd.[1] Huxley, on the other hand, complained that the orthodox, in their terror that science would prove their cosmogony wrong, wanted to limit all philosophical inquiry to that which could be touched and seen here and now.[2]

This widespread concern with the limits of science—what can science know? what kind of entities can science investigate? where is the border between speculation and scientific claims?—was commonly used as a rhetorical bludgeon by both naturalists and theists. Each group argued that their opponents misunderstood what science could do. Exactly what was wrong about their claims varied—sometimes Huxley was attacked for claiming too much for science's domain, sometimes for claiming too little. But what remained steady was the accusation from both sides that it was their opponents' views on religion that led them to incorrect and misleading ideas about the limits of science. However, in practice, both groups of scientists used very similar ideas about what qualified as a scientific statement or subject.

Victorian science saw many dramatic shifts in what counted as "science," and figures such as Huxley and Maxwell were under constant pressure to justify their work as valid and reliable.[3] Both of them, in rather different ways, struggled to clearly articulate what they saw as the proper limits of science and how their claims fell within them. For Huxley, this took the form of his agnosticism; for Maxwell, his development of scientific models. This chapter will closely examine each figure's position regarding the limits of science, and show how they were put into play in one critical case: scientific discussion of the beginning of the universe. In many ways the ideas dealt with here are the converse of those presented in the previous chapter—the validity of hypotheses was often tied to considerations of uniformity. There are two important aspects of the limits question that will not be addressed here: whether science can speak to moral issues (discussed in chapter 4) and whether the human mind and consciousness fall within the limits of science (discussed in chapter 6). The analysis here will largely focus on methodological questions: What can science investigate meaningfully? What tools can men of science use? How can we be sure we are not fooling ourselves?

Huxley's Limits

Huxley's extensive writing on the limits of science is well known, much of it having been elevated to a formal philosophical position: agnosticism. His sensitive and subtle discussions of agnosticism were often in tension with his much broader statements about science, such as "In strictness all accurate knowledge is **Science**; and all exact reasoning is scientific reasoning."[4] Many of his critics complained that his razor-sharp skepticism was not deployed with equal power against scientific claims, and that agnosticism was simply a way of covering himself from attack. Historians have sometimes made this case as well. Adrian Desmond argues that agnosticism was largely a strategic innovation that let Huxley portray scientists as nonaligned and deflect inquiry about his own beliefs.[5] While the strategic role of agnosticism is clear, I suggest that we also take it seriously as a philosophical stance that can provide insight into Huxley's thinking about the limits of science.

The definitive work on agnosticism is Bernard Lightman's *Origins of Agnosticism*. Lightman locates the roots of agnosticism at the intersection of Hume, Kant, and Dean Henry Mansel's 1858 Brampton lectures.[6] Mansel's attempt to ward off historical and literary analysis of the Bible by emphasizing man's inability to apprehend the divine was flipped by Huxley to argue that humans could make no positive statements about God, and thus that theology could never have the persuasive force of science. Following this reasoning, he took the position that if no certain knowledge of God could be attained, then there could also never be a positive denial of God's existence. Taking Hume and Kant's warnings about theological certainties into account, he defended "the limitation of all knowledge or reality to the world of phenomena revealed to us by experience."[7] Huxley claimed to have coined the term in 1869 (though it was not used in print until 1876) during his time at the Metaphysical Society in order to firmly distinguish his position from the atheists, materialists, and the innumerable forms of theism.[8] He described the etymology "as suggestively antithetic to the 'gnostic' of Church history, who professed to know so much about the very things of which I was ignorant; and I took the earliest opportunity of parading it at our Society, to show that I, too, had a tail, like the other foxes."[9] Bernard Lightman has suggested that Huxley's long period without using the term publicly was due to his inability to control its use and meaning, particularly by Spencer.[10]

Huxley happily acknowledged his debt to both John Locke and Hume. In his book on the latter philosopher he spent a significant amount of time framing his own (rather than Hume's) position:

> Fundamentally, then, philosophy is the answer to the question, What can I know? and it is by applying itself to this problem, that philosophy is properly distinguished as a special department of scientific research. What is commonly called science, whether mathematical, physical, or biological, consists of the answers which mankind have been able to give to the inquiry, What do I know? They furnish us with the results of the mental operations which constitute thinking; while philosophy, in the stricter sense of the term, inquires into the foundation of the first principles which those operations assume or imply.[11]

Note his distinction between science and philosophy, and also his isolation of the problem of first principles from science per se. He had long since accepted that first principles were not amenable to the kind of analysis that made science productive, and accordingly marginalized those questions. He approvingly quoted Locke's warning against universal knowledge, lest we "raise questions and perplex ourselves and others with disputes about things to which our understandings are not suited, and of which we cannot frame in our minds any clear and distinct perception, or whereof (as it has, perhaps, too often happened) we have not any notion at all."[12] Huxley was concerned to show that Hume, Locke, Kant, and he were of one mind on the proper boundaries of philosophy:

If, in thus conceiving the object and the limitations of philosophy, Hume shows himself the spiritual child and continuator of the work of Locke, he appears no less plainly as the parent of Kant and as the protagonist of that more modern way of thinking, which has been called "agnosticism," from its profession of an incapacity to discover the indispensable conditions of either positive or negative knowledge, in many propositions, respecting which, not only the vulgar, but philosophers of the more sanguine sort, revel in the luxury of unqualified assurance.[13]

He worked to show that agnosticism had a genuine philosophical pedigree, and that it had genuine enemies. Those who possessed "unqualified assurance" were explicitly targeted. In particular, theologians who claimed absolute knowledge were placed as the targets for the arrows forged by Hume, Locke, and Kant.

Huxley also included as an ancestor of agnosticism Sir William Hamilton, whose essay "On the Philosophy of the Unconditioned" he claimed to have read as a child. While he was sure his youthful mind did not understand the depths of it, "It stamped upon my mind the strong conviction that, on even the most solemn and important of questions, men are apt to take cunning phrases for answers; and that the limitation of our faculties, in a great number of cases, renders real answers to such questions, not merely actually impossible, but theoretically inconceivable."[14]

Acknowledging these limitations, he said, was "the essence of science, whether ancient or modern. It simply means that a man shall not say he knows or believes that which he has no scientific grounds for professing to know or believe." However, stating that science was about only accepting ideas that had scientific grounds was more circular than helpful.[15] He tried to express positively the limits claimed by agnosticism:

> In matters of the intellect, follow your reason as far as it will take you, without regard to any other consideration. And negatively: In matters of the intellect do not pretend that conclusions are certain which are not demonstrated or demonstrable. That I take to be the agnostic faith, which if a man keep whole and undefiled, he shall not be ashamed to look the universe in the face, whatever the future may have in store for him.[16]

Reason was key. The considerations that he imagined might restrict reason were clearly implied to be theological. He warned against thinking conclusions to be "certain" in the sense of immutability or permanence.

The continual progress of science could overturn any accepted idea, and the agnostic sense of man's limitations implied an obligation to always have one's mind open for new evidence. And in a rare moment of humility, he admitted that perhaps he did not always live up to his own expectations: "The apostolic injunction to 'suffer fools gladly' should be the rule of life of a true agnostic. I am deeply conscious how far I myself fall short of this ideal, but it is my personal conception of what agnostics ought to be."[17]

His efforts to articulate these limits of reason often remained vague, though he offered frequent examples of assertions or ideas that fell outside them. Miracles, as discussed in chapter 2, were the classic example. He argued that the very concept of a miracle was so confused, and human knowledge of the world so limited, that a miracle could never rise to the level of scientific knowledge.

> On trial of any so-called miracle the verdict of science is "Not proven." But true Agnosticism will not forget that existence, motion, and law-abiding operation in nature are more stupendous miracles than any recounted by the mythologies, and that there may be things, not only in the heavens and earth, but beyond the intelligible universe, which "are not dreamt of in our philosophy."

Miracles violated the key principle of agnosticism—"that we know nothing of what may be beyond phenomena"—because they tried to link experience to something beyond human reason; namely, divine action.[18]

While he was happy to use agnosticism as a tool to deny theists their fundamentals, he was equally committed, in principle, to applying it to his allies: "To my mind, atheism is, on purely philosophical grounds, untenable. That there is no evidence of the existence of such a being as the God of the theologians is true enough; but strictly scientific reasoning can take us no further. Where we know nothing we can neither affirm nor deny with propriety."[19] Indeed, he claimed to be more offended by atheists than by theists: "Consequently Agnosticism puts aside not only the greater part of popular theology, but also the greater part of anti-theology. On the whole, the 'bosh' of heterodoxy is more offensive to me than that of orthodoxy, because heterodoxy professes to be guided by reason and science, and orthodoxy does not."[20] Hooker praised Huxley's ability to show the pointlessness of atheistic arguments: "This you can do better than anyone, briefly and effectively. The haziness of ordinary people's minds in regard to both Theism and Atheism, and the idea that either can be supported or negatived by reasoning—e.g. from little fishes—is wonderful."[21] Writing to Charles Kingsley, Huxley refused the prospect of meaningful discussion of extreme philosophical positions where he did not see "the possibility of obtaining any evidence as to the truth and falsehood" of the claim. "My fundamental axiom of specula- tive philosophy is that *materialism and spiritualism are opposite poles of the same absurdity*—the absurdity of imagining that we know any- thing about either spirit or matter."[22] He did not stipulate what it was that made these subjects impossible to assess scientifically, but he did suggest that the historical trajectory of these ideas provided evidence of their intractability:

> Materialism and Idealism; Theism and Atheism; the doctrine of the soul and its mortality or immortality—appear in the history of philosophy like the shades of Scandinavian heroes, eternally slaying one another and eternally coming to life again in a metaphysical "Nifelheim." It is getting on for twenty- five centuries, at least, since mankind began seriously to give their minds to these topics.[23]

Thus, he said, even if the a priori reasons for avoiding these topics were unclear, the failure of humans to gain even the slightest purchase on their resolution revealed their extrascientific nature.

Huxley's clearest exposition of the limits of inquiry was probably in his famous 1868 "On the Physical Basis of Life," in which he sought to refute the archbishop of York's claims about the "New Philosophy" of the scientific naturalists.[24] The archbishop's concern was the limits of science. He was attempting to make the case that the naturalists limited science to only what the senses could observe, discarding anything else as impure metaphysics. Huxley repositioned the argument onto Humean territory, attributing to the Scottish philosopher the essentials of his own position. Hume, he said, laid out the standards of worthwhile inquiry:

> If a man asks me what the politics of the inhabitants of the moon are, and I reply that I do not know; that neither I, nor any one else, has any means of knowing; and that, under these circumstances, I decline to trouble myself about the subject at all, I do not think he has any right to call me a sceptic. On the contrary, in replying thus, I conceive that I am simply honest and truth- ful, and show a proper regard for the economy of time. So Hume's strong and

subtle intellect takes up a great many problems about which we are naturally curious, and shows us that they are essentially questions of lunar politics, in their essence incapable of being answered, and therefore not worth the attention of men who have work to do in the world.[25]

He continued to evade a clear articulation of limits by invoking an economy of effort—he had better things to do than to concern himself with questions about which he could gain no information. Finally, however, he offered a quote from Hume as the distillation of the issues at hand:

> "If we take in hand any volume of Divinity, or school metaphysics, for instance, let us ask, *Does it contain any abstract reasoning concerning quantity or number?* No. *Does it contain any experimental reasoning concerning matter of fact and existence?* No. Commit it then to the flames; for it can contain nothing but sophistry and illusion."[26]

Science, then, required a particular blend of abstract and experimental reasoning. Without those firm foundations, humans could gain no knowledge. Huxley stated that this meant a man of science needed to subscribe to only two beliefs: "the first, that the order of Nature is ascertainable by our faculties to an extent which is practically unlimited; the second, that our volition counts for something as a condition of the course of events." Each of these, he said, could be "verified experimentally" and therefore provide "the strongest foundation upon which any belief can rest."[27]

While maintaining that this point of view was compatible with either material or spiritual outlooks, the progress of science much preferred "the materialistic terminology." Material phenomena were more easily and uniformly accessed, and had proved much easier to connect to other phenomena. The alternative, "spiritualistic" terminology, was dismissed as utterly barren, and leading to nothing but obscurity and confusion of ideas. Huxley argued that a critical feature of scientific validity for an idea was whether it allowed for progress. Was it helpful for further investigation? Or did it only cloud the issues? With this standard, he said, it was clear that material inquiries were the only option: "Thus there can be little doubt, that the further science advances, the more extensively and consistently will all the phænomena of Nature be represented by materialistic formulae and symbols."[28] These symbols should not, he warned, be mistaken for reality. The boundaries of human understanding, as put forth by Kant and Hume, meant that one could never truly know what lay beyond them:

> But the man of science, who, forgetting the limits of philosophical inquiry, slides from these formulae and symbols into what is commonly understood by materialism, seems to me to place himself on a level with the mathematician, who should mistake the x's and y's with which he works his problems, for real entities—and with this further disadvantage, as compared with the mathematician, that the blunders of the latter are of no practical consequence, while the errors of systematic materialism may paralyse the energies and destroy the beauty of a life.[29]

We can now offer a provisional summary of Huxley's conception of the limits of science. Science should restrict itself to statements and ideas that were accessible to experimentation, quantification, provided for further investigation, and could be represented in materialist terms without complete allegiance to materialism.

Hypothesis

It is not difficult to see why many contemporaries casually confused Huxley's position with that of the positivists or the British empirical tradition.[30] The scientific naturalists relentlessly trumpeted their rejection of a priori, idealist, or overly speculative reasoning in science. Huxley and his allies argued that their work, unlike that of the natural theologians, was grounded in hard empirical facts about the physical world. They constantly asserted the need to limit science to claims that grew directly out of observation and experiment, thus allowing no room for theological interference or distraction. Huxley's frequent insistence on science's close relation to everyday experience was part of this. Describing science as "*trained and organised common sense*" was not only an argument for democratizing science (though it certainly was that); it was also an argument that basic human experience was the essential foundation for science.[31]

> The method of **observation** and **experiment** by which such great results are obtained in science, is identically the same as that which is employed by every one, every day of his life, but refined and rendered precise. If a child acquires

a new toy, he observes its characters and experiments upon its properties; and we are all of us constantly making observations and experiments upon one thing or another.[32]

Scientific observation was distinct only because it was "free from unconscious inference."[33]

Huxley and other scientific naturalists were often attacked for their insistence on solely empirical evidence. W. S. Lilly's 1886 attack in the *Fortnightly Review* was a particularly clear example:

> But, as it appears to me, the doctrines of Professor Clifford, of Professor Huxley, of Mr. Herbert Spencer, in their ultimate resolution, are substantially at one with this. Whatever differences divide these illustrious men from one another, they all agree in putting aside, as unverifiable, everything which the senses cannot verify; everything beyond the bounds of physical science; everything which cannot be brought into a laboratory and dealt with chemically.[34]

Huxley was not, of course, naïve about what could be achieved through pure empirical observation. His fiery response to Lilly in "Science and Morals" provided a thorough illustration of his views on the limits of experience and observation. He wondered about Lilly's accusation: "Can such a statement as this be seriously made in respect of any human being?"[35] He professed his belief in causality, philology, history, and mathematics, though their truths seemed to be beyond the reach of the laboratory. The Victorian debate over empiricism was vigorous, and Huxley was quite aware of the limits of observation.

A crux of this debate was the proper role of hypotheses. Many self-identified Baconians tried to argue that their use was outside science, a claim that Huxley rejected completely: "I do protest that, of the vast number of cants in this world, there are none, to my mind, so contemptible as the pseudoscientific cant which is talked about the 'Baconian philosophy.'"[36] His deep involvement with the methodological debates surrounding Darwin's ideas necessitated that he develop a keen sense of the role that hypothesis could and could not play in scientific reasoning. In the late 1880s he showed his mature position on these issues. The writing of his *Introductory Science Primer* and his essay on "The Progress of Science" for Victoria's jubilee provided him with important opportunities to systematize many of the ideas he had developed piecemeal over his career. The early sections of the primer discussed the limits and difficulties of observation:

> But those who have never tried to observe accurately will be surprised to find how difficult a business it is. There is not one person in a hundred who can describe the commonest occurrence with even an approach to accuracy. That is to say, either he will omit something which did occur, and which is of importance; or he will imply or suggest the occurrence of something which he did not actually observe, but which he unconsciously infers must have happened.[37]

So while observation was necessary for science, it was limited and therefore not sufficient. He admitted that observation was often impossible past a certain point (temporally, spatially, conceptually). The question was: what then? Some said this was the natural limit to science. Huxley defended the right to move beyond:

> When our means of observation of any natural fact fail to carry us beyond a certain point, it *is* perfectly legitimate, and often extremely useful, to make a supposition as to what we should see, if we could carry direct observation a step further. A supposition of this kind is what is called a **hypothesis**, and the value of any hypothesis depends upon the extent to which reasoning upon the assumption that it is true, enables us to explain or account for the phenomena with which it is concerned.[38]

Note that Huxley described a hypothesis as a way to *extend* observation, not replace it. It was supposed to be a tool that functioned analogously to empirical experience. Hypotheses were to be evaluated on the grounds of their ability to provide explanation for physical phenomena that actually were directly observed. He argued that "everybody" was "compelled to invent" hypotheses whenever they are confronted with phenomena with obscured causes. If there was no direct evidence for a cause, there was no shame in invoking a hypothetical factor. This was "just as legitimate and necessary in science as in common life." But, again, he understood the dangers of speculation:

> Only the scientific reasoner must be careful to remember that which is sometimes forgotten in daily life, that a hypothesis must be regarded as a means and not as an end; that we may cherish it so long as it helps us to explain the

order of nature; but that we are bound to throw it away without hesitation as soon as it is shown to be inconsistent with any part of that order.[39]

The trap was becoming attached to a hypothesis as an idea with its own value beyond its utility in solving a particular problem. Men of science needed to be wary of a hypothesis gaining a life of its own and leading them astray.

Huxley even made the case that the progress of science required "the invention of verifiable hypotheses" to go beyond direct observations. He again methodologically linked hypothesis to observation, making the case that the willingness to hypothesize was a marker of the ability to find facts:

> It is a favourite popular delusion that the scientific inquirer is under a sort of moral obligation to abstain from going beyond that generalisation of observed facts which is absurdly called "Baconian" induction. But any one who is practically acquainted with scientific work is aware that those who refuse to go beyond fact, rarely get as far as fact.[40]

Hypotheses always needed to be verifiable (in the sense that their consequences had to be observable), but they did not need to be strictly *true*. That is, the purpose of a hypothesis was to help push an investigator past stumbling blocks, not to uncover ontological realities. Huxley accepted that progress could come from ideas that were later conclusively disproved:

> Any one who has studied the history of science knows that almost every great step therein has been made by the "anticipation of Nature," that is, by the invention of hypotheses, which, though verifiable, often had very little foundation to start with; and, not unfrequently, in spite of a long career of usefulness, turned out to be wholly erroneous in the long run.[41]

The critical issue, he said, was the gradual willingness to put aside *unverifiable* hypothesis as outside the limits of science. Even verifiable hypotheses needed to be considered "not as ideal truths, the real entities of an intelligible world behind phenomena, but as a symbolical language, by the aid of which Nature can be interpreted in terms apprehensible by our intellects." In celebrating fifty years of the queen's reign, he was un-

ambiguous in crediting the power of modern science to a proper understanding of the use of hypothesis.[42]

Unsurprisingly, Huxley argued that Darwin's ideas were the perfect
example of scientific hypothesis. Natural selection was described as having all the benefits discussed so far, as well as illustrating a critical issue:
no hypothesis should be expected to explain everything.

> In the first place, all human inquiry must stop somewhere; all our knowledge
> and all our investigation cannot take us beyond the limits set by the finite and
> restricted character of our faculties, or destroy the endless unknown, which
> accompanies, like its shadow, the endless procession of phenomena. So far as
> I can venture to offer an opinion on such a matter, the purpose of our being
> in existence, the highest object that human beings can set before themselves,
> is not the pursuit of any such chimera as the annihilation of the unknown; but
> it is simply the unwearied endeavour to remove its boundaries a little further
> from our little sphere of action.[43]

This was, of course, also a defensive move to protect Darwin from those
who asked for the evolutionary explanations for various facts. It was inevitable, Huxley said, that a hypothesis would leave some area unexplained. As a human construct, no hypothesis could be perfect. But, he
warned, we must be careful not to mistake an unexplained feature for a
fatal objection:

> There is a wide gulf between the thing you cannot explain and the thing that
> upsets you altogether. There is hardly any hypothesis in this world which has
> not some fact in connection with it which has not been explained, but that is a
> very different affair to a fact that entirely opposes your hypothesis.[44]

A hypothesis could be incomplete and still survive the process of verification. And when a man of science declared that he "accepted" a hypothesis, he meant that he accepted it "provisionally."

> Men of science do not pledge themselves to creeds; they are bound by articles
> of no sort; there is not a single belief that it is not a bounden duty with them
> to hold with a light hand and to part with cheerfully, the moment it is really
> proved to be contrary to any fact, great or small. . . . So I say that we accept
> this view as we accept any other, so long as it will help us, and we feel bound

to retain it only so long as it will serve our great purpose—the improvement of Man's estate and the widening of his knowledge.[45]

Claiming such detachment from beloved ideas was certainly a rhetorical move, but given Huxley's genuine and deep-seated reservations about natural selection (chapter 2), we can take many of his statements more seriously. Huxley did not accept Darwinian evolution for a long time, and yet defended it vigorously, in a way described fairly accurately by his rhetoric about hypotheses. He promoted natural selection as an epochal idea not because he was fully convinced of its truth—he certainly was not. Rather, he saw it as a powerful hypothesis with the potential to change the very way people thought about the living world. Huxley's position on hypotheses gives us a way to understand his Darwinian defenses somewhere between the extreme poles of full-fledged convert (bringing truth to the unbelievers) or conniving propagandist (promoting ideas he did not believe). Conversations about hypothesis, Huxley might say, were where the real work of science was done.

Huxley's friend John Tyndall provided some of the scientific naturalists' most eloquent and spirited defenses of the use of hypothesis. His physics practice was deeply intertwined with theoretical entities such as molecules and the ether, and he vigorously justified their value to science. Like Huxley, he located the beginning of science in everyday inquiry, even in the nursing of an infant. But once we have taken "our facts from Nature we transfer them to the domain of thought: look at them, compare them, observe their mutual relations and connexions, and bringing them ever clearer before the mental eye, finally alight upon the cause which unites them."[46] These mental operations required hypotheses, which were not simply speculation. The ether, for instance, "must not be considered as a vague or fanciful conception on the part of scientific men. Of its reality most of them are as convinced as they are of the existence of the sun and moon."[47] It was perfectly acceptable for physicists to discuss its properties and behavior.

Tyndall's famous essay "Scientific Use of the Imagination" was a manifesto for the value of moving "beyond the boundary of mere observation, into a region where things are intellectually discerned." He plainly acknowledged that not all men of science agreed with this: "There are Tories even in science who regard Imagination as a faculty to be feared and avoided rather than employed."[48] Imagination, Tyndall told us, was the architect of theory—it was the vehicle for carrying our physical experience into a new region. He clarified that scientific imagination was very different from unfettered speculation. It needed to be built on experience, and united with reason, to lead us "into a world not less real than that of the senses, and of which the world of sense itself is the suggestion and, to a great extent, the outcome."[49] This was not absolute knowledge, nor was it certain. But it was essentially a simple extension of reasoning by analogy, so we should believe in the ether for the same reasons we believe other people have minds.

He held up Darwin as the exemplar of someone who properly and skillfully used the scientific imagination to reach conclusions forever hidden from direct experience. Tyndall admitted that the theory of natural selection and its manifold implications invoked many unseen things, and that Darwin had "drawn heavily upon the scientific tolerance of his age." Despite this, one should not accuse him of transgressing the limits of science unless "we are perfectly sure that he is overstepping the bounds of reason, that he is unwittingly sinning against observed fact or demonstrated law—for a mind like that of Darwin can never sin wittingly against either fact or law." While Tyndall offered this nod to Darwin's outstanding moral character, he also called for caution in restricting scientific ideas generally: "If there be the least doubt in the matter, it ought to be given in favour of freedom of such a mind."[50] Intellectual freedom, he said, demands a generous interpretation of the limits of science, rather than a precautionary restriction. Those outside science could be reassured by the "the hard discipline which checks licentiousness in speculation" within the community.[51] Overenthusiastic speculation was a real danger, but the benefit of the doubt should be given to all honest men of science.

Tyndall's Belfast Address and its ensuing controversies dealt with many of these same issues, though it struck a significantly more aggressive tone. He combined an appeal to uniformity with an embrace of a Kantian limitation of human senses to justify the use of theory and hypothesis: "Believing, as I do, in the continuity of nature, I cannot stop abruptly where our microscopes cease to be of use. Here the vision of the mind authoritatively supplements the vision of the eye." This allowed meaningful discussion of invisible entities such as molecules, as well as of events in the distant past. The linking of all events by cause and effect provided a bridge to the unseen. On these grounds he defended his right to talk about the "antecedents of the solar system." Anything between the primordial nebula and the present was a valid target for science.[52] He

warned that once scientific investigation entered this realm, it was naïve to expect straightforward experiential confirmation of a theory such
as natural selection. The strength of Darwin's theory came "not in an
experimental demonstration (for the subject is hardly accessible to this
mode of proof), but in its general harmony with scientific thought."[53]

Much of the criticism Tyndall received for the address was based on
the claim that he had gone beyond the limits of science. In particular, he
was attacked for "crossing the boundary of the experimental evidence."
He replied that this was

> the habitual action of the scientific mind—at least of that portion of it which
> applies itself to physical investigation. . . . in physics the experiential inces-
> santly leads to the ultra-experiential; that out of experience there always
> grows something finer than mere experience, and that in their different pow-
> ers of ideal extension consists, for the most part, the difference between the
> great and the mediocre investigator.

Moving beyond naked empiricism was not only allowed, it was a mark of
greatness—what would Darwin have been had he done no more than describe barnacles? Tyndall insisted that crossing "the boundary of experience, therefore, does not, in the abstract, constitute a sufficient ground
for censure."[54]

Tyndall was comfortable stating that there "seems no limit to the insight regarding physical processes"—with the important emphasis on the
physical.[55] By this he meant to indicate that a physicist must restrict himself to "matter and force" alone.[56] The danger was in "being irresistibly
led beyond the bounds of inorganic nature."[57] Tyndall's sweeping claims
along these lines sometimes became grandiose: "It is perfectly vain to
attempt to stop enquiry in this direction. Depend upon it, if a chemist
by bringing the proper materials together, in a retort or crucible, could
make a baby, he would do it. There is no law, moral or physical, forbid-
ding him from doing it."[58]

Against those critics who read this as a claim of absolute knowledge,
he demurred that science could say nothing about the true nature of atoms or the interaction of mind and matter.[59] Indeed, while he swore "a
determination to push physical considerations to their utmost legitimate
limit," this very action led to "an acknowledgement that physical considerations do not lead to the final explanation of all that we feel and

know."[60] Even in the Belfast Address, Tyndall clearly stated that humans have extrascientific needs: "Physical science cannot cover all the demands of [their] nature."[61] For example, literature provided insight and meaning that science could not. Civilization, he said, needs "not only a Darwin, but a Carlyle. Not in each of these, but in all, is human nature whole."[62] He particularly enjoyed Ralph Waldo Emerson for his ability to discuss scientific concepts in a literary way. Tyndall suggested that activities such as literature could have beneficial influences on science, even though they were outside the limits of science: "Some of [science's] greatest discoveries have been made under the stimulus of a nonscientific ideal."[63] Even religion had something to contribute to the full human experience, specifically "in the region of *poetry* and *emotion*, inward completeness and dignity to man," but not, however, in the realm of objective knowledge.[64]

Critics

Theist critiques of scientific naturalism often targeted the problem of limits. A. J. Balfour's famous assault on naturalism in *The Foundations of Belief* laid out many of these attacks particularly clearly. Balfour was a major intellectual and political figure, and his book was representative of conservative responses to challenges to the Anglican establishment.[65] Coming in 1895, after the scientific naturalists had spent nearly a generation formulating their arguments, it constituted a thorough and powerful retort against the limits of knowledge as conceived by Huxley and his group. Balfour defined naturalism as being based on the doctrine that "we may know 'phenomena' and the laws by which they are connected, but nothing more."

> "More" there may or may not be; but if it exists we can never apprehend it: and whatever the World may be "in its reality" (supposing such an expression to be otherwise than meaningless), the World for us, the World with which alone we are concerned, or of which alone we can have any cognisance, is that World which is revealed to us through perception, and which is the subject-matter of the Natural Sciences. Here, and here only, are we on firm ground. Here, and here only, can we discover anything which deserves to be described as Knowledge.[66]

Balfour actually did not disagree that human senses and reason were limited and certainly imperfect for understanding the world—he acknowledged that both naturalists and theists accepted this.[67] His complaint was that naturalism, on *these grounds*, demanded "terms of surrender to every other system of belief."[68] The limits of naturalistic science, he argued, thus demanded too much from too little.

Further, those limits did not allow naturalism to provide many aspects of what Balfour considered to be an adequate worldview, such as morality and aesthetics. It had no *"emotional* adequacy."[69] Imagine, he said, a "catechism of the future, purged of every element drawn from any other source than the naturalistic creed."[70] Its inadequacy was obvious. And worse, the limits of science imposed by naturalism allowed no justification for the principles that made science possible: reliable experience of the world, the rationality of phenomena, and the uniformity of nature. He argued persuasively that science could not provide its own first principles, and therefore was not a stand-alone epistemology of nature.[71] Naturalistic limits would prevent the use of theory and still gave no justification for thinking that there was a correspondence between the real world and our perception of it.[72] Balfour argued that this impossibility of justifying science with only scientific methods would eventually destroy naturalism, but leave science untouched. "Science preceded the theory of science, and is independent of it. Science preceded naturalism, and will survive it."[73] The principles that allowed scientific practice would always be outside the limits of science, and must be found in a larger worldview.

As Lightman has shown, Huxley's response to Balfour (written in the last days of his life), was not particularly strong. Balfour's sophisticated understanding of the practice of science and his disavowal of easy targets such as biblical literalism or Paleyan natural theology made him a difficult opponent.[74] Huxley fell back on the strategy of trying to control the terms of the debate: he objected to the conflation of naturalism, agnosticism, and materialism, and denied that he or anyone else held the positions being attacked.[75] As he had successfully done in the past, he accused his attacker of using straw men. He did not try to show how naturalism and agnosticism did allow for the use of theory and accepted the impossibility of establishing first principles, perhaps because such a nuanced position was a rather poor weapon.

As we have seen in the details of Huxley's and Tyndall's positions, many of Balfour's criticisms were poorly aimed. The scientific naturalists did not reject the use of theory and creative hypothesis. But we need to notice that his criticism along those lines was an explicitly religious one: that the naturalists were so committed to denying supernaturalism that they chose limits to science that destroyed critical methodological values of science. Theistic scientists had to shoulder this burden as well, however. Justifying useful and meaningful limits of *theistic* science was an equally complicated endeavor.

Maxwell's Limits

Maxwell was not hesitant to declare something beyond the limits of science. When reviewing *Paradoxical Philosophy*, written by his friends Balfour Stewart and P. G. Tait, in *Nature*, he mused:

> *Nature* is a journal of science, and one of the severest tests of a scientific mind is to discern the limits of the legitimate application of scientific methods. We shall therefore endeavour to keep within the bounds of science in speaking of the subject-matter of this book, remembering that there are many things in heaven and earth which, by the selection required for the application of our scientific methods, have been excluded from our philosophy.[76]

In particular, Maxwell was concerned about their suggestion that the ether constituted the "material organism of beings exercising functions of life and mind." His friends' ideas about the afterlife were interesting, but not part of science: this "is a question far transcending the limits of physical speculation."[77] Maxwell pointed out that even with modern science, humans knew nothing more about death than our earliest ancestors. We remained, he said, just where the psalmist left us: "His breath goeth forth, he returneth to his earth; in that very day his thoughts perish."[78]

What were the criteria Maxwell used to declare the afterlife ether unscientific? Among others, he argued that science must begin its investigations with the material world. Early in his career, he sent a letter to his friend Richard Buckley Litchfield in which he wrote:

> With respect to the "material sciences," they appear to me to be the appointed road to all *scientific* truth, whether metaphysical, mental or social. The knowledge which exists on these subjects derives a great part of its value

from ideas suggested by analogies from the material sciences, and the remaining part, though valuable and important to mankind is not *scientific* but aphoristic.[79]

Scientific knowledge was therefore linked to, but different from, other kinds of social and philosophical knowledge. Maxwell's description of these other kinds of knowledge as "aphoristic" was telling. It was not that these things were not true or important, but they were built on different foundations. Science required extreme precision:

> In *physical* speculation there must be nothing vague or indistinct. The truths with which we deal are far above the region of mist and storm which conceals them from the popular mind and yet they are solidly built upon the foundations of the world and were established of old, according to number, and measure, and weight.[80]

That final phrase is drawn from the deuterocanonical Book of Wisdom, referring to the manner in which God created the world. Maxwell was showing the scriptural basis for the importance of numerical precision: exact measurement was sanctioned in the actions of the deity.[81]

Like the majority of Victorian scientists, Maxwell saw observation and experience as the foundation of the practice of science. He expected his students to begin their education by simply becoming familiar with the phenomena of the natural world.[82] His Aberdeen inaugural lecture placed observation and measurement central to science: "Nothing that we can say or think here can escape from the ordeal of the measuring rod and the balance. All quantities must be exact quantities, and all laws must be expressed with reference to exact quantities, so that we have a most effectual means of discovering error, and an absolute security against vagueness and ambiguity."[83] Many theistic men of science had confidence that experience could give reliable information due to the assumption of a divinely preestablished harmony between the world and the senses.[84]

Maxwell was not, of course, a complete empiricist. As one of the great theorists of the century, he used mathematics and speculative analysis in amazingly successful ways. However, he thought deeply and carefully about how and when science could move beyond observation while still remaining reliable. He most often articulated this concern in terms of connecting ideas to observations. For example, in an essay on the contributions of Hermann Helmholtz, he wrote:

> The great work for the men of science of the present age is to extend our knowledge of the motion of matter from those instances in which we can see and measure the motion to those in which our senses are unable to trace it. For this purpose we must avail ourselves of such principles of dynamics as are applicable to cases in which the precise nature of the motion cannot be directly observed, and we must also discover methods of observation by which effects which indicate the nature of the unseen motion may be measured.[85]

This passage touched on several critical issues for Maxwell. Science should begin with what could be seen and measured, and then *extend* that knowledge beyond what could be observed. Investigators should look for principles that were likely to hold good in those unseen regions, and use them to find new effects tied to the hypothetical, invisible phenomena. This process could be fruitful, but also quite dangerous, lest one unjustly reify the unseen entities or improperly extend their knowledge. The key was to properly create and use the ideas that linked the visible and invisible—hypotheses.

Just as the scientific naturalists did, Maxwell argued that hypothesis was essential to modern science. In his early work on color vision, he denied the possibility of proceeding simply with observations; instead, "Our reasonings on them must be guided by some theory."[86] Maxwell made his reputation with sophisticated theoretical models in both electromagnetism and molecular physics. Over the course of his career he developed a wide variety of models with very different functions and intentions, and was quite reflective on their significance and role in science. There has been a large historiographic conversation about how "real" Maxwell considered his models to be, and I will not try to resolve that debate here.[87] Neither will I attempt a comprehensive survey of his work in electromagnetism. Rather, the purpose of this section is to assess where he saw the limits of science, and his models will be used as tools to help find them. Intricate theoretical models using hypothetical entities were, as discussed earlier, somewhat suspect to many Victorian natural philosophers, and Maxwell constantly felt the need to discuss their scientific legitimacy. The way he framed these justifications provides a great deal of insight into his thinking on the limits of science. I will restrict my comments here to his electromagnetic models, addressing his molecular models in chapter 6.

Maxwell's first foray into electromagnetic theory, his "On Faraday's Lines of Force," began with a warning that "the present state of electrical science seems peculiarly unfavourable to speculation." He suggested that the first move in an effective investigation should be "one of simplification and reduction of the results of previous investigation to a form in which the mind can grasp them." This simplified form could be a mathematical formula or a physical hypothesis, each of which carried its own danger: the former could distract from the observed phenomena; the latter could yield a partial explanation that hid important facts.[88] He immediately established a concern readily recognizable to his Victorian colleagues, that theory would carry one away from the grounding of real experience. To prevent this, he gave the following advice:

> We must therefore discover some method of investigation which allows the mind at every step to lay hold of a clear physical conception, without being committed to any theory founded on the physical science from which that conception is borrowed, so that it is neither drawn aside from the subject in pursuit of analytical subtleties, nor carried beyond the truth by a favourite hypothesis.

He wanted the reassurance of "physical ideas without adopting a physical theory," which he saw as a primary benefit of his method of analogies.[89] The problem here was clearly human failings, either distraction by mathematics or personal attachment to an idea. The best way to keep within proper limits was constant contact with a "physical conception," implying that physical intuition (and presumably the senses) was more reliable than our mental faculties.

These physical conceptions were hypotheses that took many different forms, with different characteristics. And not every acceptable hypothesis was worth pursuing. At one point he discussed an assumption regarding Ampère's law, declaring that although it was "most warrantable and philosophical in the present state of science, it will be more conducive to freedom of investigation if we endeavour to do without it."[90] Thus, a hypothesis might be allowed, but not fruitful. Maxwell considered Faraday's electrotonic state (certainly an unobservable entity) in a similar way:

> The conjecture of a philosopher so familiar with nature may sometimes be more pregnant with truth than the best established experimental law discovered by empirical inquirers, and though not bound to admit it as a physical truth, we may accept it as a new idea by which our mathematical conceptions may be rendered clearer.[91]

It was perhaps not true, or real, but was again valued for its utility in refining and improving other ideas. Concepts such as the electrotonic state were clearly on the boundary of acceptable hypotheses. He anticipated the objection that since that state had no clear physical referent, it was unphilosophical. "I would answer, that it is a good thing to have two ways of looking at a subject, and to admit that there *are* two ways of looking at it."[92] Any hypothesis of this sort was only one possible way of thinking about the problem, and humans could not assess the correct one with just theoretical means.

He warned that the tentative results of these papers did not contain "even the shadow of a true physical theory; in fact, its chief merit as a temporary instrument of research is that it does not, even in appearance, *account for* anything."[93] That is, the lack of professed physical explanation meant there was little danger of the conclusions being unscientific due to unwarranted conceptions. He hoped that the "temporary theory" that he had produced would be of use to experimenters without impeding their work, and hopefully give rise to "a mature theory, in which physical facts will be physically explained, [that] will be formed by those who by interrogating Nature herself can obtain the only true solution of the questions which the mathematical theory suggests."[94] He clearly intended his theoretical investigation to be of service to experiential investigation, and repeatedly voiced skepticism about the ability of pure reason to give reliable scientific conclusions.

In the years after these early explorations, Maxwell pursued several avenues of attack on the phenomena of electricity and magnetism. He eventually took hold of energy physics as the most useful theory for understanding them.[95] The value of dynamics was that it provided a method for analyzing a system without needing to specify all of its inner workings, thus avoiding the danger of misleading or inappropriate hypotheses. When Maxwell reviewed his friends Thomson and Tait's *Natural Philosophy*, he explained that in "the study of any complex object, we must fix our attention on those elements of it which we are able to ob-

serve and to cause to vary, and ignore those which we can neither observe nor cause to vary." Dynamics allowed physicists to focus on just
those observable elements. In a metaphor that echoed Maxwell's childhood questions, he described this process as a belfry:

> In an ordinary belfry, each bell has one rope which comes down through a
> hole in the floor to the bellringers' room. But suppose that each rope, instead
> of acting on one bell, contributes to the motion of many pieces of machin-
> ery, and that the motion of each piece is determined not by the motion of one
> rope alone, but by that of several, and suppose, further, that all this machin-
> ery is silent and utterly unknown to the men at the ropes, who can only see as
> far as the holes in the floor above them.

Like men of science investigating the invisible world of electromagnetism, the bell ringers could not directly observe the mechanisms governing their world. That did not mean they were helpless. Their "scientific
duty" was to investigate as thoroughly as possible through their manipulation of the ropes:

> They can give each rope any position and any velocity, and they can estimate
> its momentum by stopping all the ropes at once, and feeling what sort of tug
> each rope gives. If they take the trouble to ascertain how much work they
> have to do in order to drag the ropes down to a given set of positions, and to
> express this in terms of these positions, they have found the potential energy
> of the system in terms of the known co-ordinates. If they then find the tug on
> any one rope arising from a velocity equal to unity communicated to itself
> or to any other rope, they can express the kinetic energy in terms of the co-
> ordinates and velocities.

This thorough investigation would allow knowledge of the movements of
all the ropes, but would never provide total knowledge about the inner
workings of the belfry:

> These data are sufficient to determine the motion of every one of the ropes
> when it and all the others are acted on by any given forces. This is all that the
> men at the ropes can ever know. If the machinery above has more degrees of
> freedom than there are ropes, the co-ordinates which express these degrees
> of freedom must be ignored. There is no help for it.[96]

The bell ringers could never know what was "really" happening above them. They could postulate any number of intricate systems that would replicate what little they could observe, but they could never have any assurance that those systems were true. Instead, dynamics provided a way to gain a great deal of knowledge of the system's behavior without ever needing to take the risk of speculative ideas.

This strategy allowed Maxwell to confidently place his electromagnetic investigations well within the limits of science. It ensured that his theory could always keep hold of a physical conception (energy) even while drifting away from directly observable phenomena. The frailties of the human imagination and reason could be outflanked.

Maxwell put these methods on display in his *Treatise on Electricity and Magnetism*, famous for both its insight and its difficulty. The book's extended argument showed him carefully, and usually explicitly, making the case that each step was scientifically allowable. His constant concern was to demonstrate that the theory had maintained close connection with the observed world and not wandered into a nest of unverifiable hypotheses. Dynamics was his touchstone, allowing progress to be made even through particularly dangerous theoretical traps.[97]

The *Treatise* began with Maxwell's praise of German electrical investigators, but he warned that using their mathematics would necessarily result in imbibing their physical hypotheses as well. He said, in contrast, that his theory, "though in some parts it may appear less definite, corresponds, as I think, more faithfully with our actual knowledge, both in what it affirms and in what it leaves undecided."[98] He presented his work as superior not only in spite of it leaving some areas unexplored, but *because* of its gaps, which were evidence that it had not transgressed proper scientific limitations in pursuit of a fanciful completeness.

The first section of the text was all experiential: how to electrify a body, how to use basic measuring tools. Maxwell warned that while thinking of electricity as a physical thing, one "must not too hastily assume that it is, or is not, a substance, or that it is, or is not, a form of energy, or that it belongs to any known category of physical quantities."[99] Concepts such as electrical tension were acknowledged as strange, although their introduction was still "a warrantable step in scientific procedure."[100] Maxwell promised to develop his mathematics on the smallest possible number of hypotheses while trying to apply them to the widest realm of phenomena.

As it became useful to introduce non–directly observable entities, he would bring up possible hypotheses and consider their status carefully—sometimes introducing multiple possibilities. The belfry metaphor can be seen behind many moves:

> By establishing the necessity of assuming these internal forces in the theory of an electric medium, we have advanced a step in that theory which will not be lost though we should fail in accounting for these internal forces, or in explaining the mechanism by which they can be maintained. . . . We have seen that the internal stresses in solid bodies can be ascertained with precision, though the theories which account for these stresses by means of molecular forces may still be doubtful. In the same way we may estimate these internal electrical forces before we are able to account for them.[101]

Solid knowledge was often placed immediately next to total mysteries without destroying its value. As long as the theoretical results could be brought back to observables (or at least conceivable tests), they were still allowable.

In the more speculative sections Maxwell was happy to play with possibilities and then discard them, showing how they transgressed his limits. In the second volume he raised the possibility of a theory of magnetic fluids (in analogy to theories of electrical fluids). He ended up dismissing it on several grounds. He acknowledged that such a theory could, in a purely mathematical sense, explain any phenomena so long as ad hoc rules were freely allowed.

> This theory of magnetism, like the corresponding theory of electricity, is evidently too large for the facts, and requires to be restricted by artificial conditions. For it not only gives no reason why one body may not differ from another on account of having more of both fluids, but it enables us to say what could be the properties of a body containing an excess of one magnetic fluid. It is true that a reason is given why such a body cannot exist, but this reason is only introduced as an after-thought to explain this particular fact. It does not grow out of the theory.

The hypothesis explained too much, a clear sign of runaway speculation. Restrictions necessary to fit the observations needed to come naturally from within the theory; otherwise, an investigator was simply forcing his pet theory onto the facts. A good hypothesis disciplined the man of science as much as mathematics did. Maxwell suggested the idea of polarization as a substitute for magnetic fluids, since it would "not be capable of expressing too much, and which shall leave room for the introduction of new ideas as these are developed from new facts."[102]

The most controversial part in the *Treatise*—the electromagnetic theory of light—was saved until the very end. That theory required an electromagnetic ether that could support waves, and Maxwell anticipated the objection that he was simply inventing hypothetical entities:

> To fill all space with a new medium whenever any new phenomenon is to be explained is by no means philosophical, but if the study of two different branches of science has independently suggested the idea of a medium, and if the properties which must be attributed to the medium in order to account for electromagnetic phenomena are of the same kind as those which we attribute to the luminiferous medium in order to account for the phenomena of light, the evidence for the physical existence of the medium will be considerably strengthened.

Generating new entities at will was precisely the sort of unscientific move that Maxwell was working so hard to avoid. He needed to convince his readers that his imagination was not out of control, speculating beyond acceptable limits. He said we should be reassured by evidence from multiple scientific disciplines pointing the same way, providing good reason to be confident in the hypothesis: "The combination of the optical with the electrical evidence will produce a conviction of the reality of the medium similar to that which we obtain, in the case of other kinds of matter, from the combined evidence of the senses."[103] This was one of the strongest statements Maxwell was willing to make about the ether, and note that it came only after hundreds of pages of technical arguments. He was gently proposing that a sufficiently supported hypothesis could be considered to have a similar status to empirical observation. This was sure to irritate some men of science, but Maxwell had worked hard in the *Treatise* to articulate the limits of investigation that justified such a view.

As to the structure of the ether, he was even more cautious. It seemed that the Faraday effect suggested some kind of structure with angular velocity, and molecular vortices would fulfill that need. However, Maxwell realized this was treading near the edge of acceptable speculation. The hypothesis of molecular vortices did produce some useful mathematical results, but he warned that the theory proposed in the preceding pages is evidently of a provisional kind, resting as it does on unproved hypotheses relating to the nature of molecular vortices, and the mode in which they are affected by the displacement of the medium. We must therefore regard any coincidence with observed facts as of much less scientific value in the theory of the magnetic rotation of the plane of polarization than in the electromagnetic theory of light, which, though it involves hypotheses about the electric properties of media, does not speculate as to the constitution of their molecules.[104]

He carefully distinguished between hypotheses with different levels of validity. The electromagnetic theory of light was marked as acceptable because it did not require specific statements about the composition of the ether. The hypotheses regarding the molecular vortices themselves, however, were set aside as too speculative, even though both were connected to the same experimental evidence. This suggests that Maxwell was thinking of a sort of graduated ladder of speculation—the more steps of hypothesis necessary to link an idea to observation, the less scientific the idea. As such ideas drifted from the anchor of experience, and became more reliant on reason and imagination, they could become misleading. He cautioned the reader not to confuse well-verified concepts and illustrative models:

> The attempt which I then made to imagine a working model of this mechanism must be taken for no more than it really is, a demonstration that mechanism may be imagined capable of producing a connexion mechanically equivalent to the actual connexion of the parts of the electromagnetic field. The problem of determining the mechanism required to establish a given species of connexion between the motions of the parts of a system always admits of an infinite number of solutions. Of these, some may be more clumsy or more complex than others, but all must satisfy the conditions of mechanism in general.[105]

The belfry could always be filled with any number of possible systems, and it was the duty of the natural philosopher to vigilantly patrol the limits of speculation to prevent imagination from being mistaken for reality.

However, this did not mean that the ether was an unreasonable speculation. He noted that among Continental physicists there seemed to be "some prejudice, or à priori objection, against the hypothesis of a medium in which the phenomena of radiation of light and heat, and the electric actions at a distance take place. . . . Hence the undulatory theory of light has met with much opposition, directed not against its failure to explain the phenomena, but against its assumption of the existence of a medium in which light is propagated."[106] He understood that it was once the case that ethers were invented and multiplied whenever convenient, which was certainly beyond the limits of acceptable speculation. But his ether was held up as quite different. It only relied on hypotheses that were closely related to direct observation or were based on robust theories (e.g., dynamics). Parts of ether theory that were more speculative (e.g., molecular vortices) were sliced off and isolated, leaving the strong core to stand on its own. Direct measurement of all predictions, at least in principle, was required. The final sentence in the *Treatise* was "If we admit this medium as an hypothesis, I think it ought to occupy a prominent place in our investigations, and that we ought to endeavour to construct a mental representation of all the details of its action, and this has been my constant aim in this treatise."[107]

From this brief exploration of Maxwell's electromagnetic theories, we can see what issues formed the boundary by which a hypothesis was considered scientific. It needed to stay close to experience, observation, and experiment—it needed to at least be analogical to observable processes. It should not be apt to mislead or encourage fanciful extensions. It should be restricted in scope so as to restrain the human tendency to pursue favorite avenues. It did not need to be complete. Having poorly defined sections, lingering uncertainties, and remaining mysteries was *not* a mark of being unscientific. Indeed, such gaps were sometimes good, as they indicated that the hypothesis had not been pushed too far. However, the relevant phenomena still needed to be explorable despite these unknowns: a belfry without any ropes would be pointless.

A major feature of Maxwell's scientific boundaries was that a hypothesis must be provisional.[108] This was an issue he brought up in many contexts, including molecular theory.[109] A man of science must be ready to construct "model after model of hypothetical apparatus" until a useful one was found.[110] Even inelegant hypotheses could be welcomed, since they would not be expected to be permanent: he celebrated an early electrical hypothesis of William Thomson's "which, though rough and clumsy compared with the realities of nature, may have served its turn as a provisional hypothesis."[111] When reviewing Joseph Plateau's book on soap bubbles, he wrote that in its speculative section, "there is of course room for improvement. . . . In such matters everything human, at least in our century, must be very imperfect, but for the same reason any real progress, however small, is of the greater value."[112]

Limiting Resources

That passing reference to human imperfection was a clue to the substantial philosophical and religious resources that Maxwell used to craft his sense of the limits of science. He was vocal that science could not function solely with observation and theory. Those needed to be welded together with other tools. In his 1870 address to the Mathematical and Physical Sections of the British Association, he spoke of the difficulty of understanding this relationship:

> I have been carried by the penetrating insight and forcible expression of Dr Tyndall into that sanctuary of minuteness and of power where molecules obey the laws of their existence, clash together in fierce collision, or grapple in yet more fierce embrace, building up in secret the forms of visible things. . . . But who will lead me into that still more hidden and dimmer region where Thought weds Fact, where the mental operation of the mathematician and the physical action of the molecules are seen in their true relation? Does not the way to it pass through the very den of the metaphysician, strewed with the remains of former explorers, and abhorred by every man of science?[113]

Metaphysics was necessary to make sense of precisely the boundary that he was so concerned to articulate in his electromagnetic modeling, where the mental and the physical met. Maxwell did not deliver a discourse on metaphysics at that meeting, and he was elsewhere rather critical of such enterprises. ("I have read some metaphysics of various kinds and find it more or less ignorant discussion of mathematical and physical principles, jumbled with a little physiology of the senses. The value of the metaphysics is equal to the mathematical and physical knowledge of the author divided by his confidence in reasoning from the names of things.")[114] So his point was not to advocate for an actual system of metaphysics, but instead to draw attention to the need to grapple with those sorts of questions. Men of science, he said, already dealt with those issues on a regular basis: "In our daily work we are led up to questions the same in kind with those of metaphysics; and we approach them, not trusting to the native penetrating power of our own minds, but trained by a long-continued adjustment of our modes of thought to the facts of external nature."[115]

This distrust of the "power of our own minds" was a key issue for Maxwell's understanding of the limits of science. In that same lecture he discussed how to make sense of deviations between theory and observation, suggesting that they were due to our limited understanding of the natural world. Further, humans regularly make errors for a variety of reasons, and there were many more ways to think wrongly than to think correctly.[116] Maxwell mused that George Boole's work in this vein led one to "another of those points of view from which Science seems to look out into a region beyond her own domain."[117] Kevin Lambert has shown how Boole provided Maxwell with important resources for understanding how preexisting laws of thought could constrain science and limit the use of models and analogies.[118]

Much has been written on the sources of Maxwell's use of analogies and articulation of limits of human understanding. Richard Olson's classic *Scottish Philosophy and British Physics* argues for the importance of Maxwell's training in the Scottish Common Sense tradition, in particular its emphasis on the importance of understanding the human mind before understanding the mind's approach to nature.[119] This tradition debated the foundations and limits of scientific knowledge—what can be known, what cannot, and what kind of reasoning was reliable. Maxwell studied with William Hamilton at Edinburgh, and Olson finds echoes of Hamilton's Kantian Common Sense in his writings. His position that epistemological, psychological, and metaphysical questions were worth the time of a natural philosopher was quite close to Hamilton's.[120] And as discussed in chapter 2, Maxwell's considerations of analogies and relational knowledge show a likely debt to that philosopher. The Common Sense concern of losing sight of phenomena during speculation was indeed quite similar to what we have seen in Maxwell's struggle with his models.

However, David B. Wilson and Peter Harman have made the case for the additional importance of considering Maxwell's Cambridge training.[121] Harman acknowledges that Maxwell took important epistemological considerations from Hamilton, but emphasizes that the Cambridge mixed mathematics tradition also provided resources for his work with analogies and geometrical representation. Harman points out that William Whewell, whom Maxwell studied closely, also discussed relational knowledge and analogical reasoning.[122] Maxwell often used Whewell's categories explicitly, such as his references to the "fundamental ideas" of force and mass underpinning natural philosophy.[123]

Maxwell was clearly a highly synthetic thinker, and the threads of Common Sense, Whewellian Cambridge, and Boolean philosophy can all be seen in his articulation of the limits of science. However, there is at least one further resource on which he likely drew: his evangelical theology. From looking at Maxwell's models, we can see that an overriding issue was the reliability of human perception, reason, and knowledge. And as we have already seen in chapter 2, his evangelicalism had a great deal to say about those issues. A fundamental premise of Victorian evangelical thought was the fallen nature of man, which carried with it deep limitations of the ability of humans to truly understand the universe around them.[124] John Brooke's "archetypal evangelical scientist" was averse to purely speculative hypothesis due to humility before God.[125] The general features of evangelical thinking on science were quite similar to the issues Maxwell flagged as critical for understanding the limits of science.

Further, many of the specific features of evangelical theology emphasized by Maxwell shared these concerns. His vision of a divine creator who had wrapped the natural world in mystery while revealing selected portions, which was so important for his pursuit of unified laws, can be seen here as well. His evangelical God, wholly other, chose to make some aspects of the world understandable to humans despite their fallen and fallible nature. Maxwell's sense of the divine gave him a deep appreciation that much of the world would always remain unknown while maintaining the prospect of sound knowledge. Thus, his theology of nature placed human knowledge just on the edge of possibility. It was deeply limited by human pride but saved by God's grace in making some things known and available. And these were exactly the issues that motivated his deep introspection regarding his scientific models: What parts of phenomena are accessible to us? What can be known with certainty? What tendencies of the human mind can lead us astray? How can we discipline our speculation? How can a philosopher be both humble and confident in his work?

Maxwell's views on the limitations of scientific investigation and knowledge were complex and subtle. Those views have origins in many places and schools of thought, and a monocausal account is surely insufficient. I do not want to try to claim that his ideas about analogy and modeling were simply outgrowths of his religious thinking, any more than they were simply recapitulations of Hamilton. Maxwell had many resources relevant to the problem of limits, and he likely drew on them all in different ways. I am here suggesting that his evangelical theology was a crucial ingredient in this philosophical stew.

The result of this mélange was a position on the limits of science remarkably close to that of Huxley and the scientific naturalists. Both Maxwell and Huxley stressed the importance of observation and empiricism while acknowledging the inherent limitations of human perception. Each was, in an important sense, a channel for Kantian-influenced epistemology.[126] Both worried about the danger of rampant speculation and unverified hypotheses while also accepting the need for theory. Both saw theory as the tool that allowed science to provide deeper understanding of the unseen world, whether microscopic, invisible, or far in the past. Scientific knowledge was reliable, but necessarily incomplete, and was always vulnerable to the weakness of human reason and emotion. Natural philosophers needed physical and conceptual tools to discipline their minds and senses. Huxley and the scientific naturalists presented these views as methods for preventing theological and dogmatic intrusion into proper science. However, these limits were, at the least, deeply compatible with Maxwell's evangelical outlook. Their similar limitations of hypothesis and scientific statement were arrived at via quite different approaches and justifications.

Beginnings

The religious and naturalistic aspects of limits can be seen clearly in the debates around one issue of great concern for Victorian scientists: the beginning of the universe. We can see the philosophical positions we have discussed so far put into play as both theistic and naturalistic men of science tried to articulate what could be known about the origin of all things. Despite aggressive maneuvering by both groups to outflank the other, they arrived at very similar statements about what could be known.

Maxwell presented a number of articles and lectures that dealt with what science could know about the beginning of the universe. These presented two separate arguments: the first based on the characteristics of molecules, the second based on the second law of thermodynamics. His best-known presentation of the first argument was his 1873 lecture on molecules, but it first appeared in his 1870 address to the BA. There, he noted the universality and precision of the uniformity of molecules:

> when we find that here, and in the starry heavens, there are innumerable multitudes of little bodies of exactly the same mass, so many, and no more, to the grain, and vibrating in exactly the same time, so many times, and no more, in a second . . .

As discussed in chapter 2, Maxwell saw religious implications in this uniformity. However, the argument continued by stressing the impossibility of changing these uniform characteristics:

> When we reflect that no power in nature can now alter in the least either the mass or the period of any one of them, we seem to have advanced along the path of natural knowledge to one of those points at which we must accept the guidance of that faith by which we understand that "that which is seen was not made of things which do appear."[127]

He made the case that this realization brought one to a limiting edge of science. The final quote was from Hebrews 11:3, a passage that stresses the creation of the world by God. More specifically, it describes the creation of the visible world by that which is invisible. Invisible here does not refer to simply things beyond our vision—Maxwell had no problem investigating unseen things, and even defined molecular science as a study of "things invisible and imperceptible by our senses, and which cannot be subjected to direct experiment."[128] Rather, he used this biblical passage to evoke the sense of a realm completely distinct from the human ability to gather knowledge. It was the divine mystery. The limit of science Maxwell drew here was marked by two factors: features that cannot be created by natural laws ("no power in nature"), and processes inaccessible to investigation ("not made of things which do appear").[129]

The 1873 Bradford lecture was based on these same molecular properties. Maxwell again suggested that their uniformity implied manufacture, attributing the idea to Herschel. He stressed that the path to this conclusion was "strictly scientific" but had led to "the point at which Science must stop." The issue was not, he said, that science could not study the inaccessible interior of the molecule—the *Treatise* had shown how to investigate a closed belfry. Instead, the lack of any natural process that could affect molecular properties gave no traction with which a natural philosopher could begin work. "Science is incompetent to reason upon the creation of matter itself out of nothing. We have reached the utmost limit of our thinking faculties when we have admitted that because matter cannot be eternal and self-existent it must have been created."[130] Human reason, and therefore science, could not understand the moment of creation itself.

This passage appeared again in his *Encyclopædia Britannica* article on atoms. The slight changes he made to the phrasing helped clarify why he placed molecular origins outside science:

> The formation of the molecule is therefore an event not belonging to that order of nature under which we live. It is an operation of a kind which is not, so far as we are aware, going on on earth or in the sun or the stars, either now or since these bodies began to be formed. It must be referred to the epoch, not of the formation of the earth or of the solar system, but of the establishment of the existing order of nature, and till not only these worlds and systems, but the very order of nature itself is dissolved, we have no reason to expect the occurrence of any operation of a similar kind.

The "order of nature" was certainly a reference to the uniformity of nature that was necessary for science to function. Maxwell clearly allowed for natural processes in the formation of the Earth and the stars, acknowledging that their characteristics were of the sort that could be understood with natural laws. It was only when a feature seemed to have no link with natural law that it could be placed beyond the limits of science. Natural philosophers could only address those parts of nature accessible to senses and reason, not the essential nature of a molecule: "It is only when we contemplate not matter in itself, but the form in which it actually exists, that our mind finds something on which it can lay hold."[131]

The second argument about science's approach to the creation relied on the use of the concept of entropy from thermodynamics. In the early 1850s William Thomson began articulating his vision of a progressive universe in which an original storehouse of energy was gradually dissipated.[132] The laws of thermodynamics suggested to him that the creation of energy was reserved for divine power only—God had created the universe in a primeval state of potential energy that humans were now using for their own purposes. Thomson advocated the broad use of the equations of thermodynamics to discern the state of the universe in either the past or the future. However, there was no way for science to consider what was present before the initial distribution of energy:

> [All living things are] organized forms of matter to which science can point no antecedent except the Will of a Creator, a truth amply confirmed by the evidence of geological history. But if duly impressed with this limitation to the certainty of all speculations regarding the future and pre-historical periods of the past, we may legitimately push them into endless futurity, and we can be stopped by no barrier of past time, without ascertaining at some finite epoch a state of matter derivable from no antecedent by natural laws.[133]

Thomson's line that the laws of physics should be applied as widely as possible until there was no natural antecedent was very close to what we have already seen of Maxwell's position. Indeed, his 1870 address pursued Thomson's ideas. He stated that the irreversible increase in entropy implied a "conception of a state of things which cannot be conceived as the physical result of a previous state of things, and we find that this critical condition actually existed at an epoch not in the utmost depths of a past eternity, but separated from the present time by a finite interval." That is, there was evidence of a period before which the current order of nature did not exist. He emphasized that the "physical researches of recent times" had therefore shown the importance of a beginning.[134]

In a letter to Mark Pattison, the rector of Oxford's Lincoln College, Maxwell expanded on what entropy could reveal about the creation of the universe. He described this as essentially a basic problem in thermodynamics: one can track back the changing heat of a cooling body, until "we arrive at an epoch at which the temperature varied in a discontinuous manner, and if we seek for the state of the body at any time still farther back we arrive at an impossible result."[135] Hence, that body was not part of the uniform natural world before that point—that is, a beginning. Maxwell also tried to explain exactly what kind of status these claims had: "This is a purely mathematical result founded however on experimental data."[136] That is, no one could ever measure such a thing, but it was a sound theoretical result deduced from solid observations. He talked briefly about whether such a discontinuity could ever appear again, as some materialists suggested, only to discard it. Materialist speculation about the beginning of the universe was a failure because of their lack of metaphysics: "The practical relation of metaphysics to physics is most intimate . . . the effect of the absence of metaphysics may be traced in most physical treatises of the present century."[137] Maxwell closed the letter with a brief defense of his discussion of these issues: "I happen to be interested in speculations standing on experimental & mathematical data and reaching beyond the sphere of the senses without passing into that of words and nothing more."[138] That final defense was strikingly consonant with what we have already seen of Maxwell's views on the limits of science: speculation was acceptable as long as it maintained contact with observation and was restrained by mathematics, even if it dealt with nonobservable entities. The danger—reasoning based on "only words," or pure thought—was the same as that of electromagnetic theory.

Maxwell's double argument led to his position that the *idea* of a beginning was suggested by science, but the nature of the creation itself was outside the limits of science. About the other side of the thermodynamic discontinuity, science could tell us nothing. But why does Maxwell mark off some mysterious goals as beyond investigation (the beginning of the universe) while he encourages investigation of others (the electromagnetic belfry)? The issue was whether the target was part of the world of uniform natural laws. If it was mysterious, but clearly connected to the natural system as the ether was, it should be investigated. It could be interacted with, poked, prodded, and analyzed. If it was mysterious, but disconnected from the natural system as the creation was, it could not be investigated. No experiment or mathematics could be expected to provide reliable information if it reached into a place where our present natural order did not exist. There was no connection to the tools men of science needed to make sense of the world.

A critic might complain that Maxwell and similar theistic scientists were disingenuously warding off scientific inquiry to reserve the creation as a theological category. And yet, the scientific naturalists tended to discuss the beginning of the universe in terms very similar to Maxwell. Clifford warned that "we do not know anything at all of the beginning of the universe" because one could only extrapolate from experience, which clearly had nothing to say about creation.[139] Tyndall was the naturalist who most frequently touched on these issues. When attacked by critics who claimed that he sought to explain the emergence of all phenomena with recourse to evolution, he responded that evolution "does not solve—it does not profess to solve—the ultimate mystery of this universe. It leaves, in fact, that mystery untouched."[140] He asserted that sci-

ence had much to say about the past of the Earth, Sun, and stars, but that
theory was limited in its gifts of retrodiction. Specifically, he said, physicists could extrapolate backward from the current observed condition
of the universe to primordial nebulae. But beyond this was mere speculation, with no foundation in the current state of the world: "You have
been thus led to the outer rim of speculative science, for beyond the neb-
ulæ scientific thought has never hitherto ventured."[141]

Science could only begin its analysis once matter, force, and energy
were present. Without those elemental categories, Tyndall could not find
anything meaningful for science to say. In his essay "Matter and Force"
he wrote that the question of where the original matter came from "remains unanswered, and science makes no attempt to answer it. As far as
I can see, there is no quality to the human intellect which is fit to be applied to the solution to the problem. It entirely transcends us." Just as
with Maxwell, the problem was placed beyond the limits of human understanding. Tyndall embraced matter and force as comprehensible, and
everything beyond as unreachable:

> The phenomena of matter and force lie within our intellectual range, and as
> far as they reach we will at all hazards push our enquiries. But behind, and
> above, and around all, the real mystery of the universe lies unsolved, and, as
> far as we are concerned, is incapable of solution.

He warned against humans pretending to know that which was beyond
their innate capabilities. And to make it clear who he thought was violating those rules, he added: "Be careful, above all things, of professing to
see in the phenomena of the material world the evidences of the Divine
pleasure or displeasure."[142] It was the theists, he said, who tried to go beyond the limits of science.

Huxley rarely discussed the beginning of the universe per se. When
briefly engaging with such issues in his "Progress of Science," he dismissed the idea that it could be a meaningful conversation: "For myself,
I must confess that I find the air of this region of speculation too rarefied
for my constitution, and I am disposed to take refuge in 'ignoramus et
ignorabimus.'"[143] However, as the discussion of his views on uniformity
showed, he marked discontinuities such as beginnings as unscientific precisely because they were discontinuous. Without uniformity there could
be no science, so the very concept of a beginning was outside the limits

of science and could not be considered. Any event that was not the product of natural law was "out of the domain of science altogether."[144]

Both groups agreed that the moment of the creation was not something to be discussed scientifically. This was not a religious prejudice intended to safeguard Genesis—an unlikely goal for Tyndall—but rather a natural conclusion to draw from the uniformity assumptions on which they all agreed. The creation, one way or another, was clearly a moment when uniformity no longer applied, and therefore forbade science from speaking. Both groups, in their own way, were following Herschel's suggestion: "But to ascend to the origin of things, and speculate on the creation, is not the business of the natural philosopher."[145]

Conclusion

Maxwell and Huxley agreed broadly on what was needed to make an investigation or statement scientific: rooting in observed phenomena, the careful use of hypothesis, and an understanding of the inherent limitations of human reason and knowledge. Anything that was not part of the web of uniform natural laws could not be part of science. Knowledge could be found outside science that was useful and significant, but it was another category completely. Ideas that transcended the limits of science could be considered and discussed without mistaking them for scientific statements.

By the late Victorian period these limits had broad recognition through the scientific community, though there was still significant divergence. It is interesting to note that there was much more variance about the limits question than there was regarding the uniformity of nature, even within the naturalist/theist groups. Huxley and Tyndall were moderate among the scientific naturalists, whereas Spencer had a very broad vision. Maxwell was more expansive than Thomson, but less adventurous than Tait.

And even when there was agreement about what the limits were, there was often disagreement about whether a given idea violated them. Maxwell placed the development of molecules outside natural law, and therefore not accessible to scientific hypothesis. Huxley responded (some years after Maxwell's death) that speculation that molecules "may be indefinitely altered, or that new units may be generated under conditions yet to be discovered, is perfectly legitimate. Theoretically, at any rate, the transmutability of the elements is a verifiable scientific hypothesis." The bounding issue of the question, whether natural processes could affect molecules, was agreed on by both. Maxwell pointed to the lack of any such known processes as evidence that the molecules emerged in a preuniformity era; Huxley retorted that humans had simply not discovered those processes yet. As he often did, Huxley appealed to the narrative of scientific progress to imagine a time when there would be newly understood laws that would move molecular transformation back inside the boundaries of science. "It seems safe to prophesy that the hypothesis of the evolution of the elements from a primitive matter will, in future, play no less a part in the history of science than the atomic hypothesis, which, to begin with, had no greater, if so great, an empirical foundation."[146]

More so than most of the issues discussed in this book, the question of limits gave rise to large gaps between conceptual idea and practical application. Perhaps this was because, by definition, discussions of limits took place at the edge of what was known about the natural world. The subsequent vagueness of terms, definitions, and rules inevitably made the application of these ideas, even when agreed upon, far from simple.

CHAPTER FOUR

The Goals of Science Education
The Working Men's College

One of the Supreme Court decisions that has shaped the presence of creationism in the American classroom is *Edwards v. Aguillard*, which applied the so-called Lemon test. That test required, among other factors, that government activities—in this case the public school curriculum—have a "legitimate secular purpose." The *Edwards* decision determined that teaching creationism in the classroom was done with religious intent and therefore failed the test. While procreationist groups have since adapted their strategies, this test remains a core tool for science educators protecting their curricula.[1]

This idea that a religious intent is incompatible with science education is a major part of the educational side of modern scientific naturalism. The claim is that the *goals* of science teaching are incompatible with theist religion. The particulars of religion and law in the United States make this a useful strategy in court, but it has grown into an axiom on its terms: science teaching must have no religious intent.

But it has not always been this way. The Victorians grappled with the same issues—the relationship of religious intent and science education—but came to quite different conclusions. An illustrative counterexample can be found in the Victorian era in the Working Men's College. This institution, the brainchild of the controversial theologian F. D. Maurice, was the site of vigorous, first-class science education that was profoundly religious. Even stranger, it was also the site of science education that was profoundly agnostic and anticlerical. The Working Men's College, designed to provide a route to self-improvement for shop workers and craftsmen, operated with a largely volunteer faculty that included world-

class intellectual figures. Two of these teachers were Maxwell and Huxley. Each taught for the college for a significant period of time, despite
having completely different views on religion and society. And this was
not simply a matter of having a job—neither needed the work, and both
had well-developed ideas about the role and benefits of science education that were closely bound to their religious views. How, then, could
they each see that institution as an opportunity to further their divergent
goals? What was it about the Working Men's College that made it a place
where science, religion, and secularism could all thrive together? What
attitudes toward science allowed for its teaching in such a potentially ex-
plosive mixture?

This chapter is primarily concerned with this dynamic overlap. It will
look closely at the role of science education from three perspectives:
Maurice and the Working Men's College itself, Maxwell, and Huxley.
For each party, the significance of science education was closely bound
to the larger concerns about Victorian religion, society, and politics. The
question is how, despite very different foundations, these three came
together as part of a single movement. We will see that the values and
goals of science education for both theists and naturalists found common ground in the classrooms of the working classes.

F. D. Maurice and Christian Socialism

The key figure in the formation of the Working Men's College was Frederick Denison Maurice, the son of a Unitarian minister. Maurice studied at Cambridge from 1823 to 1826 and credited these years with shap-
ing his entire life's work. He was particularly active with the discussion
group known as the Apostles, and his time there saw the formation of
many of his distinctive ideas, including education reform and frustration
with sectarianism.[2]

Around 1828–29, having left Cambridge, Maurice underwent a conversion from Unitarianism to orthodox Trinitarianism. He subsequently
took orders in the Church of England and became ordained in 1834. He
emerged from his conversion with an intense personal religiosity and a
sense of man as an "utterly dependent being" who needed a relationship
with God to realize his own personality. In particular, he felt the need to
highlight the role of Christ (one of the sources of his dissatisfaction with

Unitarianism). He emphasized "Christ in you," a heterodox position that brought him alarmingly close to the Quakers.[3]

Despite his unusual theological positions, Maurice argued that the Church of England was "the only firm, consistent witness" of the "Universal Church," and that the Thirty-Nine Articles supported him. This placed him as something of a "conforming rebel"—promoting thorough heterodoxy but in the framework of the established Church. This precarious position fit well with his "dread of parties or factions" that allowed him to move among numerous religious groups. He supported everyone from the evangelicals to the Tractarians on different issues. Maurice emphasized that various religious groups should not be kept apart when "together they could do a mighty work." For him, Christianity was a crucial social glue: Christ preached that the middle and higher classes had an obligation to help the lower, and the lower classes needed an individualizing faith in order to break down the apparent separation from their fellow man.[4]

In 1840 he was elected to a professorship of English literature and modern history at King's College London, and six years later offered a chair of theology.[5] King's emphasis on Christianity as an indispensible part of modern education sat well with Maurice's outlook. His classes were so popular that he asked for and received as an assistant the young Charles Kingsley.[6] Although Maurice was not a particularly gifted preacher, his charisma worked on a personal level. The conservative and radical aspects of his theology did not sit easily with each other. This was even more complicated by his sense of having a religious calling to bring delicate issues to a head.

His infamous *Theological Essays* were the most public manifestation of this urge. They were intended to turn British Christianity in a new direction and emphasized God as love, an epic struggle between good and evil, and a rejection of the doctrine of eternal punishment. This led quickly to his dismissal from King's in 1853 and ferocious attacks on his ideas from the orthodox religious press. There was tremendous public interest in the case, and a great deal of support for Maurice. He remained an important figure in British religious life and in 1866 was appointed to the Knightsbridge Professorship of Casuistry, Moral Theology, and Moral Philosophy at Cambridge.

More important than his formal theology for this chapter's argument was his response to the social turmoil of 1840s Britain. The revolutions

of 1848 on the Continent and the resurgence of Chartism at home convinced him that the world had ignored "the manifestation of Christ as
the actual King of Men."[7] That is, the political and social instability had
a moral root that could best be solved by healing class divisions with the
one institution that cut across all classes—the national Church. The solution to Chartism was Christian brotherhood. Maurice thought that it
took a Divine Spirit to awaken all of man's faculties, and only a Christian
movement would unite all classes with a sense of common humanity.

Maurice was persuaded by John Malcolm Ludlow that the key to such
a movement was to Christianize one of the sources of class tension—socialism.[8] Their initial efforts were built around cooperative societies and
workshops that were an almost complete failure. Maurice and his friends
attributed this failure to the workers not understanding the principles of
cooperative production.[9] Education was thus the remedy, and they published a series of inexpensive newspapers and magazines to spread their
message, including the *Christian Socialist* and *Politics for the People*.
Ludlow was adamant:

> A new idea has gone abroad into the world. That Socialism, the latest-born of
> the forces now at work in modern society, and Christianity, the eldest born of
> those forces, are in their nature not hostile, but akin to each other, or rather
> that the one is but the development, the outgrowth, the manifestation of the
> other, so that even the strangest and most monstrous forms of Socialism are
> at bottom but Christian heresies.[10]

Or more simply: "They are Socialists, not *although*, but *because* they are
Christians."[11]

Maurice's acolyte Charles Kingsley wrote extensively for this new
Christian Socialist press under the pseudonym Parson Lot.[12] He argued
directly against radical claims that Christianity was inherently a tool of
oppression and the establishment. Rather, "The Bible throughout is the
history of the People's cause."[13] Christian Socialism was explicitly placed
as an alternative to Chartism: "There will be no true freedom without
virtue, no true science without religion, no true industry without the fear
of God, and love to your fellow-citizens."[14]

Both difficulties and opportunities for the movement were caused by
an enduring question in Britain: what kind of Christianity? Maurice rejected sectarianism and sought to create an ecumenical group that would
find allies both inside and outside the Church of England.[15] Even beyond

welcoming Dissenters, he made efforts to invite those outside the faith completely: "The writers of this paper, we have said, need not all be professed Christians; still less need they be professed Churchmen."[16] The *Christian Socialist*'s regular attacks on the hypocrisy of bishops and the "tyranny of the Establishment" no doubt helped secure the movement's credibility among those on the fringes of British society.[17]

While the group aimed its rhetoric largely at the working classes, it was mostly definitely a top-down organization. It was driven by members of the middle and upper classes and stood against revolution (and often any political reform at all).[18] Maurice and his followers spent a great deal of energy arguing that Christianity demanded that those with wealth and special skills use them for the benefit of the less fortunate.[19] Class structures were not targeted for destruction and no revolution was anticipated.

The purpose of Christian Socialism was not solely to improve the material conditions of the working classes, "but to raise them morally, to help them to feel themselves members of a redeemed race, with a vocation to fill in Christ's Church."[20] The great problem of the industrial era, Maurice said, was the tendency to reduce men to simple tools for profit.[21] This tendency must be fought, and competition and capitalism were rejected as incompatible with God's law. In Victorian Britain competition was widely seen as a natural law, but that did not mean it should be yielded to. Perhaps the model of science could resolve the issue:

> Every fresh discovery in Science helps man to live more in accordance with nature, and to obviate some at least of the evils, which were supposed inseparable from the action of her laws. Are the evils of capitalism an unavoidable result of the law of supply and demand, or once we understand the law can we avoid those evils the way knowledge of the weather lets sailors avoid storms?[22]

This was not solely a metaphor. Education and knowledge were Maurice's silver bullets. For him, the best way to achieve socialist goals was by raising the minds of the workers. He began offering night classes for shop workers in 1848, much to the displeasure of the leaders of King's College.[23]

Maurice thought his superiors at the college should have been pleased. He framed his educational efforts as being distinctively Christian, though it would be impossible, he said, to separate education into religious and secular paths. The crucial factor was that there would be instruction for everyone, regardless of class or wealth: "We shall all agree, probably, that our Universities must be universal in fact as well as in name; must cease to be monopolized for the benefit of one or two privileged classes . . . our aim is rather the material and spiritual enfranchisement of the labourer."[24] Many Christian churches, particularly the Methodists, were being persuaded to become involved with educational movements, arguing that education was an affair of the spirit.[25]

Maurice was hardly the first to consider providing higher education for the laboring classes. His ideas were preceded by the creation of the "Mechanics' Institutes." These were schools founded to offer schooling to skilled artisans, and grew rapidly in the early decades of the nineteenth century. Unsurprisingly, they appeared first in heavily industrialized areas such as Birmingham, London, and Scotland, but by 1826 every large town and many small ones had their own institute.[26] The pursuit of knowledge within the working classes had always been complicated by social and economic barriers, but the rapid expansion of the cheap press made literacy and education seem within reach.[27] The Mechanics' Institutes and the Society for the Diffusion of Useful Knowledge helped with these goals by providing inexpensive books, lending libraries, social connections, and sources of psychological encouragement.[28] The primary strategy of the institutes was to teach practical science with the goal of improving invention and efficiency, and they did maintain respectable scientific libraries.[29] Unfortunately, within a decade it was clear that the plan simply was not working. The science being taught had little economic benefit.[30] Enrollment plummeted, and most of the students were of rather higher class than intended.[31] The Mechanics' Institutes were widely seen as a failure, and Maurice had a plan to improve on their performance.

The Working Men's College

Maurice's educational mission gradually eclipsed his general socialist concerns. As early as 1849 he was helping organize evening classes for young men run by local clergy and other gentlemen. The instruction had an "antiutilitarian" approach intended to awaken common humanity rather than improve marketable skills. At first he was excited not so much by the students as by the teachers. Their participation was "a clear

disclaimer by scholars, physicians, and clergymen, of the monstrous notion that their gifts or offices separate them from the rest of the community; a solemn assertion that it is their business and their privilege to
claim fellowship with men as men."[32] As with Christian Socialism, the
act of helping the workingmen was supposed to also help the upper and
middle classes themselves.

These classes became formalized into the Working Men's College in
1854 (and classes for women began the next year). It was first set up in
Red Lion Square, Holborn, with Maurice as principal. The original class
offerings covered Shakespeare, English literature, the Gospel of John,
public health, English geography and history, grammar, arithmetic and
algebra, geometry, mechanics, natural philosophy, astronomy, and drawing (taught by none less than John Ruskin).[33] Notably absent were classes
on trades and engineering. Maurice had no vision of teaching utilitarian
skills.[34] He thought an education may incidentally improve one's material well-being, but only as a result of improving one's character. Pursuing wealth only reinforced the industrialist notion of man as automaton:

> We do not profess to make you a more effective or useful machine; we do try,
> and we hope to go on trying, to teach you to value more and more your true
> manhood. For, after all, do you really believe that the wages you received, or
> any advantage which belongs to you separately, and which distinguished you
> from your fellows, truly express the best and deepest part of you? . . . The
> more a man is raised out of his selfishness, we may generally affirm the more
> successful he will be in his daily business.[35]

Students were to be given whatever liberal learning they might want.
Maurice criticized the Mechanics' Institutes for trying to decide what
knowledge was safe to give to workers. He said they should instead be
given access to the whole intellectual and artistic world that one could
have received at Oxford or Cambridge.

The teachers were all unpaid, which caused obvious difficulties, but it
"at once put the relation between teachers and taught on a high ground:
emphasizing the principle of mutual help and mutual obligation on which
the College was founded." The reward of teaching was supposed to be
spiritual and internal, not monetary and material. Maurice reminded the
instructors that in an important sense the college was for them, not the
students. It was "established first for the benefit of us the Teachers, sec-
ondly for the benefit of those whom we taught. . . . Young professional

men felt that they wanted . . . co-operation with human beings who were not of their class." Class animosity needed to be halted from both the bottom and the top—a goal of the earlier Mechanics' Institutes as well.[36]

As part of this project, Maurice insisted on the name "college" as indicating a common membership brought together for purposes other than profit. This came from the Christian Socialist value of guiding one to "rejoice in binding class to class and man to man by ties of sympathy, instead of placing them in conflict."[37] The experience of education was to bond student to teacher, and student to student, at a time of great divisiveness.[38] The college was held together from the very beginning by Maurice's personal charisma and energy.

The first year's students, averaging about thirty a night, were drawn from a variety of professions. Clerks were a plurality; those in the book trade, warehousemen, and various craft trades made up the bulk of the remainder. The instructors worked hard to distinguish the college from the common sort of lectures that were focused more on entertainment than education. The educational mission was taken quite seriously, but with the important secondary mission of providing a place "for social and mental intercourse" among the members of different classes.[39] Despite many early struggles, the college grew and thrived, with branches appearing all over Britain. By 1859, there were Working Men's Colleges in London, Manchester, Cambridge, Halifax, Oxford, Salford, Sheffield, and Wolverhampton.[40]

The education offered was seen as "political," but not in the sense of the Chartists. By "politics," Maurice meant "the great witness for the fellowship of man with man"—essentially, Christian love.[41] J. N. Langley, a Latin teacher at the Wolverhampton branch, wrote:

> The London Working Men's College was the first institution of the kind that formally recognized the principle that politics should form an essential element in all adult education. Our men come to us with some kind of political creed ready formed: surely any education which is to be of any real service in the daily battle of life must furnish them with some test by which they may examine that creed,—may see on what principles or on what facts it is based and to what consequences it leads.[42]

The instruction was not, then, intended to provide a political ideology or recruit the students into a political party. Rather, it was intended to drive them to evaluate the political beliefs they already had. The implication here was one of persuading radicalized workers to discard dangerous beliefs they may have picked up. Anxiety about what working people may have learned in their unstructured educations was widespread.[43] Maurice said the question of politics could not be avoided: "If the nettle is apt to sting, we must grasp it hard, that it may not sting us."[44] A proper liberal education was the best antidote to Chartist tracts.

Along these lines, the college was explicitly against political reform. Langley wrote that it could do no possible good, a position that he asserted was widely shared by the working classes: "They have told me, over and over again, that the £6 franchise would let in a great many good men of their class, but a great many more bad ones."[45] Perhaps in many years, after the college had succeeded in its mission, the working classes could participate fully in political life. But now it would simply be absurd to give such responsibilities to a shoemaker:

> To invest such a man with the power of saying his "yea" or "nay" to great political questions, would be as absurd as to ask an Englishman of ordinary capacity to elect a Professor of Sanscrit or an Equity judge. Our College can have no higher ambition than that of giving working men a right and reasonable claim to a share in the responsibility of guiding the destinies of the country; and any means that contribute to this end—cricket and the "manly art," or geology and the "Principia"—they may legitimately employ.[46]

Part of teaching the students how to be good citizens meant encouraging them to defend queen and country. In 1859, the college formed a rifle corps, which eventually became the first three companies of the Nineteenth Middlesex. Unsurprisingly, some observers were uncomfortable watching the working classes drill with weapons, but Maurice was confident that their education had instilled in them unimpeachable values of loyalty.[47]

Through all of these economic and political concerns, the Christian roots and motivation of the college remained quite clear. Its leaders referred to it as "fundamentally a *religious* Institution. Everything which tends to improve the condition of men in this world—to raise their views in regard to that condition, and their hopes in regard to another—ought to be considered strictly a religious thing."[48] Its mission could not be separated from Christianity, but Maurice did not want that to restrict the material taught, the instructors, or the students. In an obituary for Maurice, it was noted that the college adopts the principle of perfect and unrestrained liberty of secular teaching; many of its students, some of its teachers, would (we fear) acknowledge themselves pure secularists; but both its constitution and its practice enforce in the strongest way the supremacy of the moral element of life, and give the place of honour to religious teaching; and, while Maurice was its President, there was no fear that its influence would ever become merely secular.

The college was built for "persons of all sects and no sect at all," and there was not the slightest intent to require any religious test. But all the same, its leaders could not conceive of how it could function without divine guidance. The tension between the moral framework of Christianity and the frequently unorthodox beliefs of the intended students would remain for the life of the institution. Bringing together such different parties certainly caused difficulties, but just as often was productive. The Working Men's College became a surprising site of rich interaction among Christians, atheists, and agnostics.

The Role of Science Education

A critical fulcrum on which the relationship between the secular and sacred wings of the Working Men's College would turn was the teaching of science. Education about the natural world was a basic part of the curriculum, and one that was seen as fundamental to the college's broader goals. The traditional foundations of the English liberal education—the classics—were dismissed as no longer being a route to universal knowledge.[49] Mathematics and science were to be the replacements. The instructors at the college sought to instill knowledge of cutting-edge work, using texts such as John Herschel's *A Treatise on Astronomy* and Augustus De Morgan's *Elements of Arithmetic*.[50] Such studies had some practical applications, of course, and also helped instruct a student in deductive reasoning and how to arrive at a sound conclusion.

But those benefits were trivial compared to what Maurice saw as the true significance of learning about the order of the natural world. In the very first issue of the *Working Men's College Magazine*, he explained that education would reveal nature to be "not a dead machine cast forth from the hands of an artificer, but a mighty harmony of living powers, to be renewed and quickened every hour by the Mind and Spirit of a Creator." That is, the study of nature would bring students closer to the actions of God. This natural theological approach to teaching science was extremely widespread in Victorian Britain. Maurice asserted that it was frankly "impossible" not to include investigations into the natural world in a Christian education.[51] This was even more important for the intended audience of artisans and craftsmen because they would learn that their own creative actions were reflections of one original Creator. Students would learn not to fear the innovations of science, but rather to "value every discovery, and hope for more."[52]

Unlike some Anglican clerics who had been concerned that the science teaching of the Mechanics' Institutes had led workers away from the Church, Maurice was sure that it would strengthen true religion.[53] A further benefit was that an education in science made the Bible more plausible, and therefore Christianity more robust. Charles Kingsley, writing under his pseudonym, attacked those who said the scriptures were unreliable due to numerous miracles. Such a skeptic might say: "They represent God as breaking the laws of nature—our reason revolts against them." He modestly reminded his readers that

> I too may know somewhat of the laws of nature; and my reason too may be somewhat cultivated; and yet it may not revolt against these miracles, or others beside them. I may have good cause for denying, as I do deny, that these miracles break the laws of nature; I may have good cause for believing, as I do believe, that the Bible miracles are the very strictest and deepest fulfillment of the laws of nature.

His understanding of the laws of nature gave him an understanding of how God could use those laws to his own ends, or suspend them as needed. In particular, natural laws were part of God's higher moral law, which was committed to freedom and liberation:

> Do you think I will give up the story of God's interfering in the People's cause, because some thousand years ago, He may have been forced to break the laws of matter to do it? Not I!—I will rather put this down as a fresh proof how glorious the People's cause is—that the very laws of matter must bow before it.[54]

The argument that God was a slave to the laws that he himself created was rejected completely. Those making such claims only revealed the shallowness of their knowledge of both science and religion. An education in science was thus a crucial part of rescuing the working classes from the linked dangers of ignorance and political radicalism.

Maxwell

Over the years of its existence, the mission of the Working Men's College attracted a number of Victorian luminaries, including the already-mentioned John Ruskin and the pioneering physiologist William Carpenter. One of the early recruits was the young Cambridge student James Clerk Maxwell. Maxwell became a member of the Apostles discussion group at Cambridge in 1852, which was then firmly under the spell of Maurice and his ideas.[55] Maxwell's personal religiosity increased suddenly with his evangelical conversion experience in the summer of 1853, which was also when he was following Maurice's theological difficulties at King's. He read the *Theological Essays*, and it seems likely that his personal religious beliefs were shaped by the experience.[56] He was highly attracted to Maurice's "combination of intense Christian earnestness with universal sympathy" and later met the theologian through the Apostles. Maxwell, like many others of Maurice's followers, was not fully convinced of all the aspects of the presented theology, but was deeply impressed by his mission and passion.[57]

Maxwell was particularly moved by Maurice's arguments for the obligation of the higher classes to the lower. He was himself a Scottish aristocrat with well-developed paternalist attitudes, and this message seemed to resonate. He was pleased that Maurice was committed to an understanding of Christianity that cut across class lines, and Maxwell enjoyed close contact with workers and their "active and practical habits of mind."[58] He quickly found himself deeply involved in Maurice's educational movement. No doubt he was heavily influenced by some of his Cambridge colleagues who were instrumental in the early days of the Working Men's College.

Of particular importance was his close friend Richard Buckley Litchfield. Litchfield was born to an evangelical family and went to Cambridge to study mathematics. He did well in mathematics, but poorly in classics, disappointing his father. His passions drew him to the sciences, particularly geology, and he ended up marrying Charles Darwin's daughter Henrietta. Sometime while he was still a student, between 1849

and 1854, he fell in with Maurice's movement and working-class education became the dominant interest of his life.[59] He became the editor of
the *Working Men's College Magazine* and was a regular teacher of arithmetic, algebra, and singing. His goal was to make the working classes
"rational, intelligent, and sober-minded," not simply more productive.[60]
And one of his major tools for this was instruction in natural science. He
called the neglect of science by Oxford "a monstrous anomaly" and was
delighted that the Working Men's College had no obligation to follow
suit. Science presented the central idea of supreme, unchangeable law
and was the sole source of positive, reliable, true knowledge.[61] This was
of particular importance for the working classes:

> They cannot afford to neglect that which is more than anything else the power
> that now rules the world. . . . Man's command over nature is in direct propor-
> tion to the amount of this knowledge, which bears not only on the satisfaction
> of his material wants, but upon the moral, social, and political advancement
> of the race. . . . Science has been of vast service to the cause of progress by be-
> ing always the great witness for the right of intellectual freedom; the great foe
> of all tyranny, and most of all of the tyranny of authority over opinion.

The investigations of science would even break down that great bugbear
of Victorian religion, sectarianism: "It may be safely said that but for
Science, the doctrine of Toleration would never have been accepted by
the world. Toleration has sprung from the reaction of a strong mass of
enlightened opinion standing apart from the profession of any creed."[62]
Religious persecution along the lines of the Test Acts was inconceivable
in a world where everyone understood science. Education in science was
therefore a near-invincible method of social improvement.

Litchfield drew Maxwell into the new Cambridge branch of the Working Men's College. This branch was founded soon after the London original (being modeled explicitly on it), and was overseen by Harvey Goodwin, the Hulsean Lecturer at Cambridge University. Maxwell, along with
his friend C. J. Monro, was on the first Council of Teachers.[63] The school
drew 186 students in its first year despite holding classes in "a sort of loft"
over a gymnasium.[64] There was some controversy over whether the word
college should be used, and in the end Maurice expressed his personal pleasure that it had been included. Goodwin argued the special significance
of having this institution in the heart of one of the ancient universities:

In looking at the noble buildings of the university, founded by men who had the civilization and education of the people at heart, they (the working men) meant to say, "We claim a right to have a share in the blessings which those buildings were designed to effect: we claim to be members of the British nation, members of the same society as you, and therefore we like to have the same name which you have gloried in."[65]

Maurice's vision of the college as a social glue was fully in the minds of the instructors: "If as much progress in learning has not been made as could be desired, there was one thing very evident—that the most kindly feelings had grown up between the members of the college and those gentlemen belonging to the University, as well as others, who had come forward to give them instruction."[66] The mathematics class had the largest enrollment, but the subject was carefully taught to show its benefit for self-improvement rather than practical utility. There was also a free Bible class taught by Goodwin himself.[67] A useful example of the sort of students attracted to the Cambridge college was Josiah Chater, a bookkeeper and draper who attended classes there for many years. A deeply religious man, he studied Latin, German, history, mechanical drawing, and natural philosophy. He took full advantage of the library, eventually joined the council, and helped run the college until its closing in 1871.[68]

Maxwell devoted a significant amount of time and energy to his teaching and examining. When he was a student, he regularly reported the state of his work at the college to his family and friends, writing to his father that he intended "to addict myself rather to the working men who are getting up classes, than to pups., who are in the main a vexation."[69] He also described his excitement at Maurice's personal visit:

Maurice was here from Friday to Monday, inspecting the working men's education. He was at Goodwin's on Friday night, where we met him and the teachers of the Cambridge affair. He talked of the history of the foundation of the old colleges, and how they were mostly intended to counteract the monastic system, and allow of work and study without retirement from the world.[70]

Maxwell was clearly inspired by the theologian, particularly with respect to the need for active work in the community. He acknowledged that he was "loath to say nay to" Maurice, which may have led to his involvement in agitation to establish early closing hours for shops such that their employees might attend classes.[71]

Maxwell's first professorship was far to the north, in Aberdeen. There was as yet no Working Men's College there, but he was fully dedicated to Maurice's mission and searched for other ways to carry it out. He began working with the Aberdeen Mechanics' Institution, founded in 1824 to instruct tradesmen in useful science.[72] It had a generous library of five hundred books focused almost completely on mathematics and science, including Leslie's *Geometry*, Babbage's *Economy of Machinery and Manufactures*, Herschel's *Preliminary Discourse* and *Astronomy*, and Lyell's *Principles of Geology*.[73]

Virtually upon his arrival in 1856, Maxwell was elected to the Council of the Institution. He taught natural philosophy classes on Mondays, which covered statics, mechanics, hydrostatics, hydraulics, acoustics, optics, heat, the steam engine, and the electric telegraph. Each term his students answered 227 exercises on these subjects.[74] His teaching featured prominently in his letters to friends and family.[75] In letters to R. B. Litchfield he asked after the fortunes of the London branch of the Working Men's College.[76] When his later position at King's College London ended in 1865, he stayed in the city to finish his teaching at the Working Men's College there.[77] Even decades after he started teaching as a student, he apparently still regularly worked with the school, apologizing when he was unable to lecture.[78]

Despite his many projects, Maxwell devoted a great deal of effort to working-class education over the course of his career. Why? Part of this was certainly noblesse oblige from his role as a Scottish laird. He had a genuine sense of a need to make the most of his elevated social role.[79] But another powerful force was one that Maurice cultivated in his followers: teaching was part of God's plan. Maxwell was not a particularly gifted teacher. He often lectured far above his students, and comprehension was further exacerbated by his complicated relationship with the English language. Even his often-celebratory biographers Campbell and Garnett acknowledged his difficulties: "Had Maxwell the qualities of a teacher? That he was not on the whole successful in oral communication is an impression too widespread to be contradicted without positive proof. Yet his letters bear sufficient evidence that in many respects he had a true vocation as an educator." He worked hard at teaching, especially to overcome his tendency to be obscure and his difficulties at the blackboard.[80]

He was independently wealthy, and did not need professorships to earn a living. And yet he saw teaching as his "appointed business."[81] He wrote to Litchfield that "this college work is what I and my father looked forward to for long, and I find we were both quite right—that it was the thing for me to do."[82] Upon his evangelical conversion, Maxwell committed himself to live as an instrument of God's will, and he seems to have latched onto teaching as an expression of that.[83] His feelings of everyday work being part of a divine plan were quite strong. Writing to a friend, he described it thus:

> Happy is the man who can recognise in the work of To-day a connected portion of the work of life, and an embodiment of the work of Eternity. The foundations of his confidence are unchangeable, for he has been made a partaker of Infinity. He strenuously works out his daily enterprises, because the present is given him for a possession. Thus ought Man to be an impersonation of the divine process of nature.[84]

So teaching was not merely a job, it was a calling, of which his time with the Working Men's College was a particularly vivid example.

And Maxwell had high expectations for the results of his science teaching. Practical benefit and more efficient engineering were useful side effects of science, but they were not the important parts of science *education*. In his inaugural lecture at Aberdeen he declared that science education helped grow "that well ordered steady frame of mind and manners which belongs to educated men and by which they are distinguished from the undisciplined."[85] Further, it teaches confidence and how to overcome difficulties. But most important, a science education provided students with new ways of thinking. Scientific and mathematical discourse modeled "healthy and vigorous thinking" that provided "absolute security against vagueness and ambiguity."[86]

He said this clarity of thought was crucial for the modern age and was the only reliable route to spotting error. When he was inaugurated as a professor at Cambridge, he encouraged students to strive to emulate those men of science "who by aspiring to noble ends, whether intellectual or practical, have risen above the region of storms into a clearer atmosphere, where there is no misrepresentation of opinion, nor ambiguity of expression, but where one mind comes into closest contact with another at the point where both approach nearest to the truth."[87] The actual truths of science were secondary to the demonstration of strict reasoning that allowed a student to uncover those truths for himself, rather than simply accepting received wisdom. The methods of science could be acquired by no other means, and had a profound effect "on the mind of the worker."[88] Indeed, the act of pursuing and contemplating such truths would lead to "the formation of a manly character" and a "noble frame of mind."[89] Maxwell did not expect his students to dedicate their lives to science, but rather to take the mental and moral benefits that came from learning science and apply those to all professions. Their goal should be "infusing a scientific spirit into the affairs and phraseology of business life."[90] If they could impress the clarity of scientific thought into their work as a wheelwright or accounting clerk, who knew what progress might be made? There were profound lessons to be taken from the study of nature, which were particularly helpful for freeing a student from the limits of his own mind: "The education of man is so well provided for in the world around him, and so hopeless in any of the worlds which he makes for himself."[91]

For Maxwell, the most important truth that physical science had to offer his students was the statement that the world was governed by exact principles. These basic principles could then be used to understand a myriad of other problems. This ability to extend one truth to illuminate many others showed that the world was a connected whole, not a scattered collage.

> And is not that one thing to learn something of the first principles of the works on nature, the rudiments of the creation so that we may no longer be oppressed with the magnitude of wonders, but rather confident that as we have recognized the operation of general principles in some instances, so in due time we shall discover more and more that the whole system of nature is disposed according to a wonderful *order*.[92]

Physical science education thus had the "peculiar value" of demonstrating how to use one fundamental idea to prove vast conclusions. Even better, in science, one can immediately compare ideas to facts and see if a wrong turn has been taken.[93] This procedure (beginning with basic principles, then checking conclusions against observations) was to be engendered through science, but adapted for a myriad of situations:

> The student of physical science will find that the method or mode of procedure by which knowledge has been accumulated, and even the process by

which he himself masters it from day to day, will furnish him with facts relating to the conditions of human knowledge which he may take with him as
guides to the study of other and more complicated subjects.[94]

These more complicated subjects were, chiefly, the moral sciences and
theological questions. Maxwell tried to invest his students with experience of what truth looked like, and how to recognize it in new contexts.
And once his students had seen truth in science, they would also look for
"it in the nature of Man and not . . . rest satisfied with mere opinion in
any department of knowledge."[95]

A proper science education would arm them with a "scientific faculty" that would serve the critical function of sorting true science from
statements that only looked scientific.[96] Maxwell's concern was about
popular treatises in which

> whatever shreds of the science are allowed to appear, are exhibited in an ex-
> ceedingly diffuse and attenuated form. . . . In this way, by simple reading, the
> student may become possessed of the phrases of the science without having
> been put to the trouble of thinking a single thought about it.[97]

He was amazed at how powerful this effect could be: "We are daily receiving fresh proofs that the popularisation of scientific doctrines is pro-
ducing as great an alteration in the mental state of society as the material applications of science are effecting in its outward life." This was
certainly very impressive, but he worried that this influence had created
a situation in which people would believe anything as long as it sounded
scientific. The responsibility of the science educator was to equip his students to deal with this:

> If society is thus prepared to receive all kinds of scientific doctrines, it is our
> part to provide for the diffusion and cultivation, not only of true scientific
> principles, but of a spirit of sound criticism, founded on an examination of
> the evidences on which statements apparently scientific depend.

Lectures and experiments should teach men to think abstractly and with
their own senses, to ground them against misleading ideas.[98]

Maxwell was clear about what doctrines and popularizations he was
addressing. In the context of Victorian worries of political instability, he
saw people thinking that men of science "are supposed to be in league

with the material spirit of the age, and to form a kind of advanced Radical party among men of learning." He was very concerned about those who claimed science as a weapon in favor of revolutionary politics or against Christianity, and explicitly framed his own science teaching as a remedy for those dangers.[99] For example, studying physics teaches about cause and effect, and how to use those concepts for prediction:

> It is by a familiarity with such reasonings and predictions that you begin to understand the position of man as the appointed lord over the works of Creation and to comprehend the fundamental principles on which his dominion depends which are these—To know, to submit to, and to fulfil, the laws which the Author of the Universe has appointed. Attend to these laws and keep them, you succeed, break them, you fail and can do nothing.[100]

Rather than helping the radicals, natural science was a chief weapon against them, and that was a powerful reason to study it.

Maxwell also expected the truths of religion more generally to benefit from learning science. He stated that the purpose of the university was not only to impart philosophical knowledge, but to produce men qualified to "serve God both in Church and State."[101] The astounding order of the universe was a powerful tool for this: "As Physical Science advances we see more and more that the laws of nature are not mere arbitrary and unconnected decisions of Omnipotence, but that they are essential parts of one universal system in which infinite Power serves only to reveal unsearchable Wisdom and eternal Truth."[102] As with natural laws (chapter 2), it was unity that showed divine work. Theology students

> will also find the benefit of a careful and reverent study of the order of Creation. They will learn that though the world we live in, being made by God, displays His power and goodness even to the careless observer, yet that it conceals far more than it displays, and yields its deepest meaning only to patient thought.[103]

Thus, deeper knowledge of God's plan for the world was accessible only to those who made a careful study of nature—that is, students of science. The study of creation was virtually a commandment: "I have also thought it unnecessary to tell you that the study of the world in which we live is our obvious duty as a condition of our fulfilling the original command *'to subdue the earth and have dominion over the creatures.'*"[104] Understanding the laws of nature was a way of understanding God's mind—one could not truly live in accordance with a divine plan if one did not understand the instruments of that plan.[105]

Science was therefore a critical element of Maxwell's self-identity as a teacher fulfilling God's will. His goals as a teacher were fundamental to the values that drew him to Maurice and to the Working Men's College. But he had no illusion that his classes in natural philosophy could compose an entire education. He still followed the traditional British model of the liberal education, and he argued strongly for the importance of other subjects such as moral philosophy and the classics.[106] Maxwell expected science education to take place in the context of a university, in which all subjects were aimed to the "cultivation of some one or more of those faculties which are given us to win our livelihood to help our fellow men to do our duty, and to worship God, in short to equip us completely for the business of life."[107] Even when setting up the Cavendish lab, he warned against those who thought that science alone could be a complete liberal education: "It must be one of our most constant aims to maintain a living connexion between our work and the other liberal studies of Cambridge, whether literary, philological, historical, or philosophical."[108]

Maxwell's views on science education focused on the intellectual, moral, and humanistic benefits. These goals, and even his justification for being a teacher at all, were closely bound to his religious beliefs and thinking. And although the Working Men's College was an unusual institution, the ideals that attracted Maxwell to it were quite common. In the era before pragmatic economic goals became the justification for science education, this religious drive was a standard paradigm for teaching the study of nature.

Huxley

Huxley's interest in education stemmed from a deep-seated dissatisfaction with the existing school system in Britain, which he saw as being particularly unfair to the working classes. At best, it was inaccessible to them; at worst, it actively reinforced their subservience to an archaic authoritarian social structure. He felt victimized personally by this system, having been closed out of the Oxbridge system and its resulting career boost. The son of a schoolmaster, he saw himself as a member of the

working class, stating that "I am a plebian and I stand by my order."[109]
He argued that he was a craftsman as much as any blacksmith—dissec-
tion, experimentation, and drafting scientific illustrations meant that he
earned a living with his hands. In 1877 he presented himself to the working classes as an ambassador of a special relationship:

> One of the grounds of that sympathy between the handicraftsmen of this
> country and the men of science, by which it has so often been my good for-
> tune to profit, may, perhaps, lie here. You feel and we feel that, among the
> so-called learned folks, we alone are brought into contact with tangible facts
> in the way that you are. You know well enough that it is one thing to write a
> history of chairs in general, or to address a poem to a throne, or to speculate
> about the occult powers of the chair of St. Peter; and quite another thing to
> make with your own hands a veritable chair, that will stand fair and square,
> and afford a safe and satisfactory resting-place to a frame of sensitiveness and
> solidity.[110]

Facts were the meeting ground of the working classes and the men of
science. Adrian Desmond has argued that Huxley fashioned a low-class
Dissenting image of science: no priesthood had "access to her deepest
secrets"; they were accessible to anyone.[111]

He rejected any arguments based on an inherent distinction between
upper and working classes: "Some inborn plebeian blindness, in fact, pre-
vents me from understanding what advantage the former have over the
latter. I have never even been able to understand why pigeon-shooting
at Hurlingham should be refined and polite, while a rat-killing match in
Whitechapel is low." One newspaper quoted him as saying "There is not
a pin to choose between 'your average artisan and your average country
squire.'"[112]

This also gave rise to Huxley's hostility against those men of science
who had used their gifts to rise in the social ranks—Newton's knight-
hood became a shameful sign of participation in an unjust system.[113]
Overturning the British class system and replacing it with a social meritocracy became an enduring theme for Huxley's career: "There are two
things I really care about—one is the progress of scientific thought, and
the other is the bettering of the condition of the masses of the people
by bettering them in the way of lifting them out of the misery which has
hitherto been the lot of the majority of them."[114]

For Huxley, the class system's chief weapon was the traditional British education. Its greatest sin was its instillation of blind, unthinking worship of authority. Those who defended the old ways were fighting for nothing less than a caste system of permanent and enforced poverty.[115] The curriculum, he said, was carefully chosen to enforce religious and class boundaries and to prevent the working classes from discovering the true roots of their misery:

> Not only does our present primary education carefully abstain from hint-
> ing to the workman that some of his greatest evils are traceable to mere
> physical agencies, which could be removed by energy, patience, and frugal-
> ity; but it does worse—it renders him, so far as it can, deaf to those who could
> help him, and tries to substitute an Oriental submission to what is falsely de-
> clared to be the will of God, for his natural tendency to strive after a better
> condition.[116]

Huxley's distrust of the established Church of England is well known, and he blamed it for his own difficulties in procuring an advanced education. He was highly suspicious of clergy in education, seeing them as serving their own authority at the expense of their students. Science, and the devotion of oneself to the physical world, was the perfect solution— "Nature having no Test-Acts."[117]

Even beyond the way British education reified class and religion, Huxley argued that its methodology was flawed as well. In short, teachers spent too much time with books. The traditional liberal education focused on the classics and literature, which Huxley protested vehemently. He engaged in a highly public, long-running debate along these lines with Matthew Arnold, who claimed that to be educated was by definition to focus on literature, languages, and classics. Paul White has argued that their views were more similar than it might seem, and that both were making the case for certain kinds of cultural authority.[118] As was characteristic, Huxley framed his values in a historical perspective. He made the case that the Victorian school system was essentially identical to that of ancient times:

> I suppose that, fifteen hundred years ago, the child of any well-to-do Roman
> citizen was taught just these same things; reading and writing in his own, and,
> perhaps, the Greek tongue; the elements of mathematics; and the religion,
> morality, history, and geography current in his time. Furthermore, I do not
> think I err in affirming, that, if such a Christian Roman boy, who had finished

his education, could be transplanted into one of our public schools, and pass through its course of instruction, he would not meet with a single unfamiliar line of thought; amidst all the new facts he would have to learn, not one would suggest a different mode of regarding the universe from that current in his own time.[119]

Huxley said a classical education was fine, as long as you had no interest in learning anything new. For that, you needed science. Science would teach "that the ultimate court of appeal is observation and experiment, and not authority."[120] Teaching students subservience to ancient epics would produce no novelty in thought or deed. He pointed to this over-reliance on well-worn texts as the source of the intellectual decline of the English schools, particularly Oxford.[121] Those schools were characterized as essentially medieval in their outlook:

> The cardinal fact in the University question, appears to me to be this: that the student to whose wants the mediaeval University was adjusted, looked to the past and sought book-learning, while the modern looks to the future and seeks the knowledge of things. The mediaeval view was that all knowledge worth having was explicitly or implicitly contained in various ancient writings; in the Scriptures, in the writings of the greater Greeks, and those of the Christian Fathers.

This was in contrast to his vision of a modern university, which embraced science and the values of intellectual progress. At such a school, the emphasis would not be on students learning what was already known, but rather on training students to be able to learn more:

> The modern knows that the only source of real knowledge lies in the application of scientific methods of inquiry to the ascertainment of the facts of existence; that the ascertainable is infinitely greater than the ascertained, and that the chief business of the teacher is not so much to make scholars as to train pioneers.[122]

As White points out, these values were in harmony with many liberal Anglican reformers of the day. While Huxley explicitly allied his science teaching with the working classes and political reform, the same educational values were justified from a number of different ideological positions.

Literature per se was not the problem for Huxley. Instead, it was the mind-set with which literature was typically taught:

> Let me repeat that I say this, not as a depreciator of literature but in the interests of literature. The reason why our young people are so often scandalously and lamentably deficient in literary knowledge, and still more in the feeling and the desire for literary excellence, lies in the fact that they have been withheld from a true literary training by the pretence of it, which too often passes under the name of classical instruction.

He was attacking education based around Latin and Greek here, not literature generally. The worshipful study of dead languages was no substitute for improving the expression of ideas. It was certainly important for men of science to appreciate style and clarity of communication, to learn to convey their ideas with vividness and power. "But this is exactly what our present so-called literary education so often fails to confer."[123] The established educational system was portrayed as so sclerotic and arthritic that it failed even to provide the most powerful lessons of its favored materials.

All of this led Huxley on a lifelong quest to develop new schools, institutes, and forms of science education. He helped found a number of organizations, and participated in even more—among them, the Working Men's College. Even before he engaged with the college, Huxley was mobilizing to educate London's workers. Shortly after he began at the School of Mines at Jermyn Street in 1854, he started a series of "Lectures to Working Men."[124] He described his project to his friend Frederick Dyster:

> I enclose a prospectus of some People's Lectures (*Popular* lectures I hold to be an abomination unto the Lord) I am about to give here. I want the working class to understand that Science and her ways are great facts for them—that physical virtue is the base of all other, and that they are to be clean and temperate and all the rest—not because fellows in black with white ties tell them so, but because these are plain and patent laws of nature which they must obey "under penalties." I am sick of the dilettante middle class, and mean to try what I can do with these hard-handed fellows who live among facts. You will be with me, I know.[125]

His appeal to the working classes was immediate. Adrian Desmond has described how the young Huxley spoke the language of the radicals and convinced them of the importance of science for their values. He saw their cause as his cause.[126]

Dyster was close with the Christian Socialists, and he put Huxley in touch with Maurice and Kingsley.[127] Huxley was immediately interested in the Working Men's College and was invited to give a lecture there in 1857 (he ended up speaking to an audience of five, including Maurice). He saw that the college could be a great assistance to his own goals, but he could not accept its Christian Socialist foundations. In a letter to Kingsley (with whom he became a regular correspondent): "I don't profess to understand the logic of yourself, Maurice, and the rest of your school, but I have always said I would swear by your truthfulness and sincerity, and that good must come of your efforts."[128] Over the years he repeatedly confessed that he could not comprehend Maurice's theological positions, but the alliance remained.[129]

Huxley formalized his flirtation with Maurice's school in 1868, when he became principal of his own branch of the Working Men's College, located in South London, with Lubbock and Tyndall on the council. Huxley combined his own iconoclasm with Maurice's godly mission.[130] One of his most famous lectures, "A Liberal Education; and Where to Find It," was given to the students at his college. There, he laid out his vision of what the Working Men's College was and what it had to offer. As he had many times before, he assaulted the notion of inherent class difference, and rejected claims that the working classes should be educated for reasons of political stability or more efficient manufacturing. Instead, he confessed sympathy for the idea that "the masses should be educated because they are men and women with unlimited capacities of being, doing, and suffering, and that it is as true now, as it ever was, that the people perish for lack of knowledge."[131] Huxley refused to mold his college into the form of any existing schools, be they English or German.[132] What would set his students apart was that they would be educated in what he saw to be the one problem that every human must confront: the implacable laws of nature. He explained to them that

> education is the instruction of the intellect in the laws of Nature, under which name I include not merely things and their forces, but men and their ways; and the fashioning of the affections and of the will into an earnest and loving desire to move in harmony with those laws. For me, education means neither more nor less than this. Anything which professes to call itself education must be tried by this standard, and if it fails to stand the test, I will not call it

education, whatever may be the force of authority, or of numbers, upon the other side.[133]

There was certainly room for literature and other subjects under Huxley's tutelage, but the thing that would make his curriculum special was the central place of the sciences. But what, precisely, would his students gain from studying the natural world?

The practical benefits of science education often promulgated by the supporters of working-class education—that is, manufacturing advantage—were of little interest to Huxley.[134] When he did refer to "practical benefit," he typically meant something quite different. In an 1854 lecture defending the educational value of the sciences, he argued that such benefits were on the level of the individual: science "teaches them how to avoid disease and to cherish health, in themselves and those who are dear to them."[135] Learning the laws of nature provided the foundation of health and happiness.[136] Focusing too much on the industrial benefits of science education was a mistake with serious social consequences:

> [I] venture to doubt whether the glory, which rests upon being able to undersell all the rest of the world, is a very safe kind of glory—whether we may not purchase it too dear; especially if we allow education, which ought to be directed to the making of men, to be diverted into a process of manufacturing human tools, wonderfully adroit in the exercise of some technical industry, but good for nothing else.[137]

So the value of learning science was not to be found in the balance of trade, but in the minds and actions of those being taught.

Practicing science required, Huxley said, great moral strength. All the virtues associated with moral rectitude—patience, perseverance, self-denial—were demanded and taught by science. Something as simple as reliable observation was a moral act:

> Let those who doubt the efficacy of science as moral discipline make the experiment of trying to come to a comprehension of the meanest worm or weed, of its structure, its habits, its relation to the great scheme of nature . . . [and see] if they do not rise up from the attempt, in utter astonishment at the habitual laxity and inaccuracy of their mental processes, and in some dismay at the pertinacious manner in which their subjective conceptions and hasty preconceived notions interfere with their forming a truthful conception of ob-

jective fact. There is not one person in fifty whose habits of mind are sufficiently accurate to enable him to give a truthful description of the exterior of a rose.[138]

The methodology of observation was rigorous and difficult, and broke down all biases and erroneous reasoning. Huxley wanted all students to have the depth of character and strength of mind to find the details for themselves, regardless of what they might have been told by authorities.

For Huxley, the beneficial effects of science education came from encountering *facts*. In the process of scientific training, "The mind of the scholar should be brought into direct relation with fact, that he should not merely be told a thing, but made to see by the use of his own intellect and ability that the thing is so and not otherwise."[139] It was the conditioning of the mind to deal with, apprehend, and appreciate facts that was valuable. He credited learning anatomy as the best way to gain these skills. Lectures, demonstrations, and examinations were all necessary to get students grappling with facts. Indeed, it was the job of the science teacher to get his students to internalize facts until they were second nature:

> Therefore, the great business of the scientific teacher is, to imprint the fundamental, irrefragable facts of his science, not only by words upon the mind, but by sensible impressions upon the eye, and ear, and touch of the student, in so complete a manner, that every term used, or law enunciated, should afterwards call up vivid images of the particular structural, or other, facts which furnished the demonstration of the law, or the illustration of the term.[140]

These were the habits of mind that could never be taught only through texts and dead paper. Experiments and personal experience were essential.

Huxley was concerned that an overemphasis on the practical applications of science would distract from this essential aspect, and he worked to change the curriculum at the School of Mines from applied science to more general science education for precisely this reason.[141] Focusing on applications would also fail to capture the imagination of those with the true spirit of scientific investigation:

> In fact, the history of physical science teaches (and we cannot too carefully take the lesson to heart) that the practical advantages, attainable through its

agency, never have been, and never will be, sufficiently attractive to men inspired by the inborn genius of the interpreter of Nature, to give them courage
to undergo the toils and make the sacrifices which that calling requires from
its votaries. That which stirs their pulses is the love of knowledge and the joy
of the discovery of the causes of things sung by the old poet—the supreme de-
light of extending the realm of law and order ever farther towards the unat-
tainable goals of the infinitely great and the infinitely small, between which
our little race of life is run.[142]

It so happens, he said, that sometimes this pursuit of truth and new
knowledge stumbled across something of practical value. That was critical for the development and improvement of industry, but was not the
proper field of interest for a philosopher. Indeed, Huxley said that by the
time industrialists began exploiting a discovery, the man of science had
long since moved into a new investigation.

Straightforward training in a trade belonged in the workshop, while
the science classroom was the place for training of the mind. Biology and
physiology were presented as particularly useful for this:

There is no side of the human mind which physiological study leaves unculti-
vated. . . . Physiology is . . . in the most intimate relation with humanity; and
by teaching us that law and order, and a definite scheme of development, regulate even the strangest and wildest manifestations of individual life, she pre-
pares the student to look for a goal even amidst the erratic wanderings of
mankind, and to believe that history offers something more than an enter-
taining chaos—a journal of a toilsome, tragi-comic march nowither.[143]

This scientific thinking was not supposed to be qualitatively different
from ordinary life. Huxley famously declared that science was "nothing but *trained and organised common sense*, differing from the latter
only as a veteran may differ from a raw recruit."[144] Thus, science education provided nothing new; rather, it only refined that which was already present in every human. Huxley leveraged this argument for his
working-class students: if science was simply common sense, then it was
available to everyone and anyone. No aristocratic gatekeeper could bar
them from an education, even if Oxford and Cambridge were off-limits.
All the restriction of higher learning accomplished was harming science
by preventing the cultivation of the innumerable fine minds imprisoned

in poverty. Huxley squarely aimed his science education at the mental improvements that would allow workers to advance themselves.

These mental improvements were also supplemented by *moral* improvements. The traditional British liberal education aimed at the formation of character, and Huxley was unwilling to concede that ground to his enemies. He argued that science could be a moral discipline as easily as literature was.[145] Moral lessons could follow from direct encounters with facts and the laws of nature:

> Science seems to me to teach in the highest and strongest manner the great truth which is embodied in the Christian conception of entire surrender to the will of God. Sit down before facts as a little child, be prepared to give up every preconceived notion, follow humbly wherever and to whatever abysses nature leads, or you shall learn nothing.

Nature had its own moral order implicit in it, if one was only willing to study it. "Nature is juster than we. . . . The absolute justice of the system of things is as clear to me as any scientific fact. The gravitation of sin to sorrow is as certain as that of the earth to the sun, and more so—for experimental proof of the fact is within reach of us all—nay, is before us all in our own lives, if we had but the eyes to see it."[146]

Huxley, of course, demanded that these mental and moral outcomes should be available to all citizens regardless of class or creed. This would have the further social effect of allowing gifted and exceptional citizens to rise to their potential. He described the purpose of technical education as a "ladder, reaching from the gutter to the university, along which every child in the three kingdoms should have the chance of climbing as far as he was fit to go."[147] This ladder of science was Huxley's great scheme to bypass the barriers erected by the upper classes and the established Church that had so restricted his own life. He also tried to sell the moral benefits to the factory owners—surely a morally educated workforce would be less likely to revolt?[148]

While Huxley's anticlerical rhetoric was impassioned and genuine, he did not seek an educational system that completely banished religious ideas. Exactly what aspects of religion Huxley was willing to include in education and which he wanted to discard will be discussed in detail in chapter 5, but it is important here to note that he acknowledged that the British culture of his day simply could not accept a strictly secular system. The breaking point, however, was Christian *doctrine*. No "theolog-ical dogmas" whatsoever could be allowed in the classroom. The impo-sition of belief via authority was absolutely unacceptable. He thought this absence of doctrine could still leave room for "a deep sense of reli-gion."[149] What he was completely unwilling to permit was claims that sci-ence should not be taught for theological reasons. He spoke eloquently and pointedly against those who made that case:

> And if any of the ecclesiastical persons to whom I have referred, object that they find it derogatory to the honour of the God whom they worship, to awaken the minds of the young to the infinite wonder and majesty of the works which they proclaim His, and to teach them those laws which must needs be His laws, and therefore of all things needful for man to know—I can only recommend them to be let blood and put on low diet. There must be something very wrong going on in the instrument of logic if it turns out such conclusions from such premises.[150]

He reminded those of the "clerical profession" that if they professed theories that bore on the physical universe, they would certainly bene-fit from studying the facts of that universe. Perhaps, even, they would un-derstand why some people reject their doctrines:

> Why do not the clergy as a body acquire, as a part of their preliminary edu-cation, some such tincture of physical science as will put them in a position to understand the difficulties in the way of accepting their theories, which are forced upon the mind of every thoughtful and intelligent man, who has taken the trouble to instruct himself in the elements of natural knowledge?[151]

Huxley hardly hid his opinion that science education would pry the mis-led away from the authority of the established Church, but he certainly saw room for the survival of religion.[152] He saw his science as a danger only to the flimsy, self-serving doctrines of antiquated theology.

Huxley's enthusiasm for teaching the facts of the natural world did not mean that he saw science as the sole element of a proper education: "It is by no means necessary that [a student] should devote his whole school existence to physical science: in fact, no one would lament so one-sided a proceeding more than I." Education must have two ends, one of science and one of the arts and literature. Huxley clearly did not weigh these two equally—science was his priority. But in 1869, while defending science education, he made the case that the two mutually aided each other:

> In these times the educational tree seems to me to have its roots in the air, its leaves and flowers in the ground; and, I confess, I should very much like to turn it upside down, so that its roots might be solidly embedded among the facts of Nature, and draw thence a sound nutriment for the foliage and fruit of literature and of art.[153]

Thus, the existing content of the British educational system did not have to be discarded. It simply needed to be reframed, and put into the wider context revealed by the study of nature.

Conclusion

A priori, it would be difficult to imagine an organization and social mission that would draw both the evangelical, aristocratic Maxwell and the agnostic, iconoclastic Huxley. And yet, they both enthusiastically participated in Maurice's Working Men's College. How could it be that such a pair, with such completely different motivations and backgrounds, ended up working as part of the same movement?

Certainly Maurice would not have been surprised by this. He firmly believed in the impossibility of separating religious and secular truth, and welcomed whatever help came from each realm.[154] The mixing of Christian Socialism with radical politics inherent in the Working Men's College caused tensions right from the start within the organization. It also brought criticism from outside, such as when the Church of England attacked Maurice's followers for "fraternization" with atheists.[155]

The examples of Maxwell and Huxley can help us understand how these very different groups functioned together. Their social commitments were, at best, tangential to each other. Both were interested in training the working classes to think about truth, but Maxwell's had a capital *T*—scientific truth was mainly practice for thinking about the truths of God, and the truths of man as laid down by God. Huxley's truth, on the other hand, had no divine pedigree and its revelation was reliant on messy, human common sense. But clearly, on some issues, they overlapped quite strongly as valence values.[156] What elements of the Working Men's College functioned as valence values between them?

First, both took seriously the idea of self-improvement inherent in the Working Men's College scheme. Maxwell was genuinely persuaded by Maurice's arguments for an obligation toward the lower classes. He felt God's hand pushing him in this direction, and spent most of his professional life working to give his students the tools he thought would ameliorate the tensions of Victorian Britain. For him, learning science directly led to self-improvement. In particular, it turned his students inward, to the extrascientific nature of man, and to the role of God in the natural world and in human lives. Huxley was no less committed to the notion of self-improvement, but with a very different end vision. He wanted his students to improve themselves as a first step to rejecting the ossified class structures of the day. While he did not advocate revolution, he saw science education as a critical step to creating a freer, more open society. Huxley's self-improvement turned one outward, to improve one's place in society, and to shatter the pretensions of established authority. The college was a valuable opportunity for both of them to enact the strategies that they saw as so important. That they each saw it as aiming toward different end points was no reason not to support it.

Second, both prized science education as a moral process. That is, they stressed the value of learning science for the effect it had on the mind and character of the student, and only secondarily emphasized the practical, economic benefits. This framework was essential to the curriculum of the Working Men's College and set it firmly apart from many other educational initiatives of the day. Maxwell and Huxley agreed with Maurice that a student leaving a science class should in some sense be a better person, but naturally they all disagreed on the trajectory that improvement should take. Huxley argued for the need of literally uncovering the natural laws that enforced moral behavior and developing an evidence-based outlook, but Maxwell saw his students turning their science-sharpened faculties to questions of moral philosophy and theology. Their agendas were different, but overlapped enough to make cooperation worthwhile. This was possible due to what Frank Turner called the secularization of the social vision of natural theology.[157] The social benefits traditionally associated with natural theology gradually became co-opted over the second half of the nineteenth century by those espousing a purely naturalistic cosmology. I have here stressed how Huxley did this through his presentation of science education's benefits, but he also pursued this strategy through literary and political modes.[158]

Complete agreement on all points was not necessary for the college to become a lively, active contribution to Victorian education. The key was having a critical mass of shared values. The forces that could drive someone to support a cause, in this case science education, were varied and could come from any number of sources. Often, those forces remained self-contained, but sometimes they managed to connect to other compatible, but still distinct, value structures. As strange as it may seem to twenty-first-century observers still watching fundamentalists and secularists battle over evolution in the classroom, the Working Men's College shows how science education brought Christians and agnostics together. Rather than being a wedge between the religious and the secular, science education was a glue that bound them together in a common cause.

It is important to be clear that theism and agnosticism were not accidental features of the groups involved here. For both Maxwell and Huxley, their views on religion were absolutely essential to the way they thought about science education. They were fully steeped in these views while acting as science educators. Religion was powerfully involved, yet these groups did not have to shed their distinct identities to cooperate — Huxley became no less agnostic due to his association with Maurice, and Maxwell did not have his faith threatened by associating with radical craftsmen. The Working Men's College was a profoundly religious institution, but Huxley did not have to take off his agnostic hat to be a principal there. The science taught there did not suffer from having Bible study in the next room. Conversely, Maurice's movement was no less Christian due to having geology and biology on the curriculum. The association of science and religion at the Working Men's College harmed neither. Part of what allowed it to function in this way was its nonsectarian environment. Natural theology had shown for centuries how science teaching could be ecumenical, and at the college its model seems to have had room for agnostic approaches as well. Certainly, the lack of direct Church of England influence made the college a welcoming space for both Huxley's iconoclasm and Maxwell's evangelicalism.

The college still exists today, but its dynamism and momentum lasted barely two decades. It had been inspired by Maurice's charisma and energy, which was also its fatal weakness. Held together by his personality, it did not build a functioning bureaucracy for many years. There was no formal constitution, and most of the administrative work was performed by the executive committee despite its technically having no powers. The governing body was a council with sixty members, though rarely would enough of them attend meetings to have a functional quorum.[159] Various attempts were made over the years to reinvigorate the school, including offering free classes. Maurice's death in 1872 shook the college, but did lead to the formation of a board of trustees that kept it functioning despite lessened enthusiasm.[160] In the years after the decline of the Working Men's College and the demise of the Mechanics' Institutes, working-class education moved away from these paternalist schemes and toward more self-organized structures such as the Mutual Improvement Societies and the Workers' Educational Association.[161]

The Working Men's College was a remarkable product of very specific developments in British politics, religion, and social history. Its unusual characteristics made visible a crucial transition in Victorian science. The story of science and religion in the age of Darwin is usually told as a sudden break from dogmatic theology toward enlightened secular science. But we can see in the college an important corrective to this tale. Even at the height of the Darwinian controversies, Huxley, the greatest champion of unfettered science, could teach his ideas comfortably among fervent Christians. Advocates of the new professional science certainly aimed for complete detachment from the old ways, but for some decades their attitudes toward education fit neatly with the religious frameworks of Maxwell and his allies. A student at the Working Men's College would have received an outstanding science education from two outstanding scientists, and he would surely have known that that education was meant to improve his character and mind. But whether his professors wanted him to turn to Christ or assail bishops would have been far less clear. Science education was a tool for many purposes—personal, political, religious, doctrinal—all of which were richly present in the college's classrooms. The high-stakes battle was over what the students, and British society, would build with that tool.

CHAPTER FIVE

Intellectual Freedom

The narrative presented by the scientific naturalists was one of liberation. Only with the escape from dogmatic theology was science able to pursue truth and accuracy. Men of science were rebels battling for freedom of thought against oppressors working in the name of gods and prophets.[1] We can see this in one critical aspect of Huxley's iconoclasm. Even beyond the religion-theology distinction established in chapter 1, he aimed his attacks at what he variously called "Parsonism," "clericalism," "Ecclesiasticism," and most commonly "orthodoxy." This idea roughly corresponded to organized religion, but had further implications that are essential to understanding Huxley's educational work and views of science. A provisional definition of *orthodoxy* for Huxley might be "set theological doctrines enforced by specially endowed authorities." He railed against orthodoxy chiefly because he saw it as endangering something critical to scientific practice: intellectual freedom, the right to come to any conclusion without compulsion or restriction.

But this value was also widely held by religious figures, including Maxwell and his fellow theistic scientists. They agreed completely with Huxley that intellectual freedom and the right of individuals to pursue ideas were fundamental to science. However, they linked these values to true religion while Huxley defined them as opposite to false theology. Intellectual freedom touched on numerous important issues in Victorian science, and brought together naturalists and theists on matters ranging from the Bible to peer review.

Antiorthodoxy

Orthodoxy was at the core of most of Huxley's attacks, and he often celebrated his particularly effective presentations of "Science *versus* Parsonism."[2] His most militant language was usually reserved for these targets, and continually attested to the need for war:

> I am very glad you see the importance of doing battle with the clericals. I am astounded at the narrowness of view of many of our colleagues on this point. They shut their eyes to the obstacles which clericalism raises in every direction against scientific ways of thinking, which are even more important than scientific discoveries. I desire that the next generation may be less fettered by the gross and stupid superstitions of orthodoxy than mine has been. And I shall be well satisfied if I can succeed to however small an extent in bringing about that result.[3]

When Huxley spoke of orthodoxy, none of his occasional reverence for religion appeared. Destruction and combat were his usual metaphors. He used a rather biological image in a letter to Lankester: "Do you see any chance of educating the white corpuscles of the human race to destroy the theological bacteria which are bred in parsons?"[4] His X-Club friends cheered whenever he managed to "frighten the parsons."[5]

Huxley proclaimed his "untiring opposition to that ecclesiastical spirit, that clericalism, which in England, as everywhere else, and to whatever denomination it may belong, is the deadly enemy of science."[6] In his essay "Agnosticism and Christianity" he clarified that it was not that theology was inherently dangerous. Rather, it was theology in the hands of clergy: "For Theology, the science, is one thing; and Ecclesiasticism, the championship of a foregone conclusion as to the truth of a particular form of Theology, is another."[7] Theological ideas could be correct or incorrect on their own merits. The danger arose when those ideas became stamped with the approval of a church authority. The personal correspondence in which Huxley disparaged his enemies is often revealing on this point. The Duke of Argyll was dismissed as only "having authority as a sort of papistical Scotch dominie, bred a minister" or "that spoilt Scotch minister."[8] Tyndall acknowledged how problematic clerics could be, but wondered about an enduring puzzle:

It is curious how one's feelings vary towards them: Some of them I could hew to pieces before the Lord in Gilgal: but then others are so gentle, and noble; and their gentleness and nobleness are so intertwined with their theological wrappings. The problem of the future will be to detach the one from the other without killing the good.[9]

Frankland also maintained friendships with clerics such as Dean Stanley.[10] Huxley, despite counting a number of personal friends in the ministry, concluded that "clerically-minded people cannot be accurate, even the liberals."[11]

Further, these authorities were dangerous because they spoke for particular denominations, or in Huxley's phrasing, sects. He warned the British nation in 1877 that the true dangers did not lie abroad, but instead from "the Bashi-Bazouks of ignorance and the Cossacks of sectarianism at home."[12] Even as a young man, not yet established in science, he approvingly jotted a quote in his notebook: "I hate all people who want to form sects. It is not error but sectarian error. Nay, and even sectarian truth, which causes the unhappiness of mankind."[13] Many years later, he attributed the root of his difficulties on the London School Board to the influence of sectarian educators persisting in teaching their "tweedle-dum and tweedle-dee theological idiocies."[14]

The danger of sects, put simply, was that they thought they were correct. Their ideas, right or wrong, valuable or worthless, became dogmas that could not be questioned. The representatives and teachers of these doctrines, then, became unquestionable authorities as well. Huxley hit this point—that clerics claimed unjustified authority over laypeople— many times through the years. One eloquent example was in his "Automata" lecture:

Making a man a Bishop, or entrusting him with the office of ministering to even the largest of Presbyterian congregations, or setting him up to lecture to a Church congress, really does not in the smallest degree augment such title to respect as his opinions may intrinsically possess.

And it was in the nature of ecclesiastical authority that clerics would not only pronounce on matters incomprehensible to them, but also place moral judgment on the conclusions reached by others:

But that, if a man elect to become a judge of these grave questions; still more, if he assume the responsibility of attaching praise or blame to his fellow-men for the conclusions at which they arrive touching them, he will commit a sin more grievous than most breaches of the Decalogue.[15]

The original sin of ecclesiasticism was the reflexive assumption of authority, and the use of that authority in all circumstances.

As an educator, Huxley argued that one of the worst effects of the assumption of universal authority was the "manner in which a man's mind might be poisoned by early instruction." He blamed his own difficulty with accepting evolution on how in his "early childhood he was indoctrinated with the reasonings of a great divine [Paley]."[16] He accepted that those church authorities who controlled his education sincerely believed they were doing right, while at the same time:

> impressing upon my childish mind the necessity, on pain of reprobation in this world and damnation in the next, of accepting, in the strict and literal sense, every statement contained in the protestant Bible. I was told to believe, and I did believe, that doubt about any of them was a sin, not less reprehensible than a moral delict. I suppose that, out of a thousand of my contemporaries, nine hundred, at least, had their minds systematically warped and poisoned, in the name of the God of truth, by like discipline.[17]

To Huxley it seemed a terrible abuse of power to force children into a belief system of which they could have no comprehension, and about which they were given no choice. He said that exact strategy was also used to keep the working classes under control, constantly telling them that they must live in a certain way solely "because fellows in black with white tie tell them so."[18] Even further, he was reluctant to allow women into science because they were too susceptible to these "ignorant parsonese superstitions."[19]

And the worst, Huxley said, was that the ideas forced into helpless students' minds were simply absurd. He dismissed many theological dogmas as contradictory to our moral sense and the facts of nature, and stated that they were perpetuated with reasoning along the lines of: it is "very pleasant and very useful, therefore it is true."[20] He was frustrated that there were so many "ready to believe in any miracle so long as it is guaranteed by ecclesiastical authority."[21] This was precisely why it was so dangerous for clerics to wield such authority. Huge numbers of people would believe anything if it came from a pulpit, allowing great deception and injustice. Huxley's repeated public attacks on the Christian miracles should be seen in this light. He was not so interested in disproving the miracles per se (though, as discussed in chapter 2, he certainly disbelieved them), but rather in showing how unjustified it was to trust *the clerics who vouched for the stories' veracity.*

Huxley's seemingly endless attacks on the New Testament story of the Gadarene swine are a useful example of this. By showing that "demonology" was implicit in the story, he hoped to show the tenuousness of the doctrines promoted by Anglicans such as Gladstone.

> It was with the purpose of bringing this great fact into prominence; of getting people to open both their eyes when they look at Ecclesiasticism; that I devoted so much space to that miraculous story which happens to be one of the best types of its class. And I could not wish for a better justification of the course I have adopted, than the fact that my heroically consistent adversary has declared his implicit belief in the Gadarene story and (by necessary consequence) in the Christian demonology as a whole.

The story itself was not so important. What was important was how an analysis of the story, and the clerical authorities' subsequent insistence on its accuracy, showed how unreliable they were as authorities, even on their own territory. Huxley gloated that Gladstone's sectarian doctrines had hoisted him on his own petard:

> I may express my great satisfaction at finding that there is one spot of common ground on which both he and I stand. So far as I can judge, we are agreed to state one of the broad issues between the consequences of agnostic principles (as I draw them), and the consequences of ecclesiastical dogmatism (as he accepts it), as follows.
>
> Ecclesiasticism says: The demonology of the Gospels is an essential part of that account of that spiritual world, the truth of which it declares to be certified by Jesus.
>
> Agnosticism (*me judice*) says: There is no good evidence of the existence of a demoniac spiritual world, and much reason for doubting it.[22]

Ecclesiasticism, he insinuated, carried the seeds of its own destruction within it, by claiming unearned authority. Men of science, on the other hand, justified every claim of authority with empirical evidence and logical reasoning.

Huxley delighted in demonstrating the shifting sources of ecclesiastical authority, especially if he could find disagreement among the members of a group.[23] One of the major issues of theological authority during his period was the question of whether scripture or institution should be supreme—the varying currents of Tractarianism, evangelicalism, and the newly declared papal infallibility gave him plenty of ammunition.

> An uneasy sense of the weakness of the dogma of Biblical infallibility seems to be at the bottom of a prevailing tendency once more to substitute the authority of the "Church" for that of the Bible. In my old age, it has happened to me to be taken to task for regarding Christianity as a "religion of a book" as gravely as, in my youth, I should have been reprehended for doubting that proposition. It is a no less interesting symptom that the state Church seems more and more anxious to repudiate all complicity with the principles of the Protestant Reformation and to call itself "Anglo-Catholic."[24]

He noted that it was hard to imagine infallible scriptures without an infallible church to select them. "But it is no less true that the Hebrew and the Septuagint versions of most, if not all, of the Old Testament books existed before the birth of Jesus of Nazareth; and that their divine authority is presupposed by, and therefore can hardly depend upon, the religious body constituted by his disciples."[25] He seemed pleased to leave ecclesiastical authority in this ouroboros, with no solid rock on which the clerics could rest. By linking institutional and textual authority, he positioned himself to attack wherever an enemy sought shelter.

One of the inevitable results of sectarian authority was something close to Huxley's own heart: the naming of infidels. The very nature of sects, he said, required the naming of *others*. There could be no orthodoxy without heretics. Describing his childhood preacher:

> Ignorant alike of literature, of history, of science, and even of theology, outside that patronized by his own narrow school, [he] poured forth, from the safe intrenchment of the pulpit, invectives against those who deviated from his notion of orthodoxy. From dark allusions to "skeptics" and "infidels," . . . and, from the horror of the tones in which they were mentioned, I should have been justified in drawing the conclusion that these rash men belonged to the criminal classes.[26]

Of course this look back was filtered by Huxley's adulthood lived as one of these outcasts. Describing himself to Kingsley, he embraced that role: "Nevertheless, I know that I am, in spite of myself, exactly what the Christian world call, and, so far as I can see, are justified in calling, atheist and infidel."[27] In later years he and his friends acknowledged this with a somewhat more playful spirit.[28] But the essence of the problem remained: they were marked as lesser simply because they could not bring themselves to accept the dogmas offered by the ecclesiastics. The demeaning of those who believed differently became one Huxley's major targets for demolition.

The Perversion of Christianity

As he did with science, Huxley placed his views on ecclesiasticism into a historical framework. It was not simply that clerics wielded their power foolishly and dangerously; the ecclesiastical mind-set had actually corrupted true religion in order to further its own interests. The historical change and development of Christianity became a powerful weapon in Huxley's battle against orthodoxy, and spurred substantial amounts of his work (particularly later in life).

The core version of Huxley's narrative, around which several variants were formed, was this: a pure religion emerges and provides ethical guidance, but later figures encrust that religion with dogma and doctrines that corrupt it. Sometimes that pure religion was named as "prophetic Judaism," more often "the Nazarenism of Jesus."[29] He acknowledged that, despite his efforts, it was virtually impossible to recover these original religions and religious figures. He described the difficulty in a wonderful image of church renovation:

> The ecclesiastical "Moses" proved to be a mere traditional mask, behind which, no doubt, lay the features of the historical Moses—just as many a mediæval fresco has been hidden by the whitewash of Georgian churchwardens. And as the æsthetic rector too often scrapes away the defacement, only to find blurred, parti-colored patches, in which the original design is no longer to be traced; so, when the successive layers of Jewish and Christian traditional pigment, laid on, at intervals, for near three thousand years, had been removed, by even the tenderest critical operations, there was not much to be discerned of the leader of the Exodus.[30]

But even if Huxley could not recover the original fresco, he was sure he could show the process of defacement. He was confident that the "Church founded by Jesus has *not* made its way; has *not* permeated the world— but *did* become extinct in the country of its birth."[31] The simple faith left behind by Jesus was "overlaid and transmuted by Hellenic speculation into the huge and complex dogmatic fabric of Ecclesiastical Christianity."[32] Modern Christianity was the encrusted wall, completely forgetting that its own church fathers would now be considered heretics.[33]

Huxley often included a cyclical process as well, in which old theological errors continued to reappear. Thus, Christianity was "smothered under the old clothes of Paganism," and "polytheism comes back under the disguise of Mariolatry and the adoration of saints; image-worship becomes as rampant as in old Egypt; adoration of relics takes the place of the old fetish-worship . . . until the Christians of the twelfth century after our era were sunk in more debased and brutal superstitions than are recorded of the Israelites in the twelfth century before it."[34] Even England suffered the "Phariseeism of the Puritans, and the Sadduceeism of the Church" in the seventeenth century.[35]

It should come as no surprise that this narrative usually ended with the development of "the scientific spirit" and "the contest between Science and Ecclesiasticism" in the Renaissance.[36] The Protestants were described as making the first thrusts against the church, but "almost all of which, as soon as they were strong enough, began to persecute those who carried criticism beyond their own limit."[37] The fight was taken up by Galileo, Hobbes, Descartes, Spinoza, Rousseau, culminating, of course, in the modern science of the Victorian age.

Huxley argued that the doctrines that had become attached to the core religion could, in addition to simply being religiously corrupting, be quite dangerous: "History proves that there is no social crime that man can commit which has not been dictated by theology and committed on theological grounds."[38] He pointed out that it would be trivial "to prove that rape, murder and arson are positively enjoined in Exodus."[39] This was not a problem with the scripture itself, nor even the authors of it. Rather, "This and innumerable other passages in both the Old and New Testaments, gave rise, through the special influence of Christian ecclesiastics, to the most horrible persecutions and judicial murders of thousands upon thousands of innocent men, women, and children."[40] An otherwise model religion, he said, was perverted through the domination of clerics to create a force for great evil.

He certainly did not think this process was confined to the past, and his personal crusade against General William Booth's Salvation Army showed exactly this concern. He warned that the creation of an authoritarian religion, seen so often in the past, was repeating itself:

> The Salvation Army—a body of devotees, drilled and disciplined as a military organization, and provided with a numerous hierarchy of officers, every one of whom is pledged to blind and unhesitating obedience to the "General," who frankly tells us that the first condition of the service is "implicit, unquestioning obedience."

General Booth's demand for reflexive obeisance was exactly what allowed dogma to flourish, and for that dogma to be turned to terrible ends.[41] He repeatedly compared the Salvation Army to the development of the Franciscan order in the Middle Ages. Huxley held the Franciscans up as the exemplar of how a group founded on the noblest of ideals quickly descended into wealth, corruption, and dogmatism.[42] Again, this was not a failing of the Franciscan leaders; it was the inevitable result of combining theology and central authority. Unsurprisingly, he placed agnosticism as the complete opposite of this. No authorities, no dogmas. "I am not aware that there is any sect of Agnostics; and if there be, I am not its acknowledged prophet or pope. I desire to leave to the Comtists the entire monopoly of the manufacture of imitation ecclesiasticism."[43]

In seeking a life free from unjust authority, Huxley regarded as the last redoubt the right to think and believe as one wished. If liberty did not mean an unbound mind, it meant nothing. He said, tongue in cheek, that he was willing to submit to any authority, but that the

> pretensions of the ecclesiastical "Moses" to exercise a control over the operations of the reasoning faculty in the search after truth, thirty centuries after his age, might be justifiable; but, assuredly, the credentials produced in justification of claims so large required careful scrutiny.[44]

The affront of dictating to others what they must think was said to be a central factor of ecclesiasticism. Priests and bishops would never be content with the temporal power and earthly riches their authority could seize for them. They would instead force everyone to worship as they did.

Huxley described this effort to force belief as the result of Christian

Rome. Pagans, he said, permitted no offense against the gods, while at the same time not requiring anyone to participate in their polytheism. This state of things was brought to an end by the Christians, by disrupting Roman life to such a degree that the political authorities were forced to move against them: "Pagan Rome, therefore, systematically persecuted Christianity with the intention of averting a political catastrophe of the gravest character."[45] Once the Christians came to power themselves, they took over this practice.

> As soon as the church was strong enough, it began to persecute with a vigour and consistency which the Empire never attained. In the ages of faith, Christian ecclesiasticism raged against freedom of thought, as such, and compelled the State to punish religious dissidence as a criminal offence of the worst description. The ingenuity of pagan persecutors failed to reach the shameful level of that of the Christian inventors of the Holy Office.[46]

The early church was thus doubly damned as being both the source of the need to persecute, and the first to persecute on a large scale. Intolerance and the "extermination of heterodoxy" became part of the flesh and blood of Christianity, waxing and waning with the power of the papacy. Even the Protestant reformers, he said, inherited this noxious tradition.

As part of this persecution, the "cleric asserts that it is morally wrong not to believe certain propositions," regardless of the evidence for or against them.[47] It was this offense that made ecclesiasticism the mortal enemy of right thinking and scientific truth.

> When Ecclesiasticism declares that we ought to believe this, that, and the other, and are very wicked if we don't, it is impossible for us to give any answer but this: We have not the slightest objection to believe anything you like, if you will give us good grounds for belief; but, if you cannot, we must respectfully refuse, even if that refusal should wreck morality and insure our own damnation several times over.[48]

The requirement of holding one particular view was to Huxley fundamentally offensive. To his credit, he extended the right to worship and practice as one wished to his opponents as well. He declined to push for George Elliot to be buried in Westminster Abbey, saying that the Anglicans should not be forced to honor someone who lived her life fighting their doctrines. In a letter to Spencer:

George Eliot is known not only as a great writer, but as a person whose life and opinions were in notorious antagonism to Christian practice in regard to marriage, and Christian theory in regard to dogma. How am I to tell the Dean that I think he ought to read over the body of a person who did not repent of what the Church considers mortal sin, a service not one solitary proposition in which she would have accepted for truth while she was alive? How am I to urge him to do that which, if I were in his place, I should most emphatically refuse to do?[49]

He similarly refused to defend the aggressive atheist George Foote's belligerent writings. Huxley named himself as one of the great defenders of freedom of speech, but Foote had crossed the line by "coarsely and brutally insulting his neighbours' honest beliefs."[50] Attempting to beat down someone else's beliefs was an act that should be associated with clerics, not with freethinkers.

Institutional Theology

All of the facets of ecclesiasticism that incensed Huxley—sectarianism, authority, doctrinal corruption, enforced belief—met in a perfect storm known as the Church of England. Huxley's attacks on the established Church and its educational, social, and political privilege have been discussed at length elsewhere. I will here discuss precisely what aspects of establishment were problematic for him, so we can better understand why he chose to work with certain religious groups and not others. Our exploration of this will be assisted by looking at another theological institution against which he railed: the Roman Catholic Church.

Since the papacy was virtually the embodiment of what Huxley saw as unacceptable in religion, it is worth some investigation here. Many of his positions were simply what we would expect of an Englishman of his generation.[51] However, he often went well beyond that—for example, calling "Papistry" the "damnable perverter of mankind."[52] When visiting Rome, he wrote to his son, describing the Catholic rituals he saw there:

I have been to several great papistical functions—among others to the festa of the Cathedra Petri in St. Peter's last Sunday and I confess I am unable to understand how grown men can lend themselves to such elaborate tomfooleries—nothing but mere fetish worship—in forms of execrably bad taste, devised, one would think, by a college of ecclesiastical man-milliners for the delectation of school-girls.

His disgust filled pages. Even beyond the offensive rituals, the popes' theological errors could be seen clearly in their architectural errors:

> It is curious to note that intellectual and æsthetic degradation go hand in hand. You have only to go from the Pantheon to St. Peter's to understand the great abyss which lies between the Roman of paganism and the Roman of the papacy. I have seen nothing grander than Agrippa's work—the popes have stripped it to adorn their own petrified lies, but in its nakedness it has a dignity with which there is nothing to compare in the ill-proportioned, worse decorated tawdry stone mountain on the Vatican.[53]

As bad as the Anglicans were, at least they had good taste.

His attacks on Catholicism started much earlier, such as in 1869, when he explicitly listed the Catholic clergy as more dangerous than the Protestant. The Protestant clergy were derided as overwhelmingly ignorant, though clumsy and bungling. This was quite different from their more ancient colleagues: "Our great antagonist—I speak as a man of science—the Roman Catholic Church, the one great spiritual organisation which is able to resist, and must, as a matter of life and death, resist, the progress of science and modern civilisation, manages her affairs much better."[54] Huxley liked to speak of the Romanists' competence at identifying heresies and systematically deciding how to attack them. He suggested that slightly more skill among other ecclesiastics might provide more effective sport for the freethinkers.[55]

According to him, the Catholic Church had centuries upon centuries of practice at restricting freedom of thought and enforcing orthodoxy. The fate of one of Huxley's heroes, René Descartes, provided an object lesson in what happened to those who thought for themselves in Catholic lands:

> Descartes lived and died a good Catholic, and prided himself upon having demonstrated the existence of God and of the soul of man. As a reward for his exertions, his old friends the Jesuits put his works upon the "Index," and called him an Atheist; . . . and the misfortunes of Galileo so alarmed him, that he well-nigh renounced the pursuits by which the world has so greatly

benefited, and was driven into subterfuges and evasions which were not worthy of him.[56]

Here we can see many of Huxley's complaints about clerics: abuse of power, the labeling of the other, the anger at truth seeking, the use of fear. He labeled the whole affair "very cowardly." The climax of the tale, as he told it, was Descartes's descent into "subterfuges and evasions" instead of holding true to his ideas. He imagined that "Descartes was a man to care more about being worried and disturbed, than about being burned outright; and, like many other men, sacrificed for the sake of peace and quietness, what he would have stubbornly maintained against downright violence."[57]

Huxley's vicious expulsion of St. George Jackson Mivart from the Darwinian circle can be better understood through this lens of antiecclesiasticism. It was not sufficient that Mivart, a devout Catholic, proposed theistic interpretations of evolution (Kingsley was forgiven for this). But once Mivart brought in institutional theology—*Catholic* theology—he had moved into the realm of anathema to Huxley. A close look at Huxley's 1871 blast at Mivart makes this issue quite clear.

Huxley began by staking out his ground. On Mivart's claim that there need be no opposition between evolution and religion, he pounced, asking what precisely was meant by "religion."

> Mr. Mivart, on the contrary, is perfectly explicit, and the whole tenor of his remarks leaves no doubt that by "religion" he means theology; and by theology, that particular variety of the great Proteus, which is expounded by the doctors of the Roman Catholic Church, and held by the members of that religious community to be the sole form of absolute truth and of saving faith.[58]

He immediately placed Mivart on the ground of theology (as Huxley understood it), declaring that he was simply a stand-in for the Roman church. Mivart's claim of the harmony of evolution and Catholic doctrine stepped directly into Huxley's minefield: ecclesiastical dogma touching science. Dryly, Huxley commented:

> I confess that this bold assertion interested me more than anything else in Mr. Mivart's book. What little knowledge I possessed of Catholic doctrine, and of the influence exerted by Catholic authority in former times, had not led me to expect that modern science was likely to find a warm welcome within the pale of the greatest and most consistent of theological organisations.

Mivart's chief source for his claim was the theologian Francisco Suarez, whose works Huxley marched to the British Library to read and "acquaint myself with the true teachings of the infallible Church."[59] One can imagine that Huxley would have forgiven Mivart if he had made an argument from his own personal beliefs and experience, but the invocation of a Catholic cleric was too much.

He then dripped Suarez in all of his most despicable language: "doctor of the Catholic Church," "unchallenged authority," "unspotted orthodoxy," "Catholic doctrine," "Catholic faith requires the believer," "sanctioned by Infallible Authority, as represented by the Holy Father and the Catholic Church."[60] He argued that Suarez believed in a literal six-day creation, and therefore that Mivart had no choice but to do so as well. After thus disposing of Mivart's attempt to let Catholics believe evolution, he claimed the opposite was true as well: "But, for those who hold the doctrine of evolution, all the Catholic verities about the creation of living beings must be no less false." He clearly stated that scientific truth inevitably destroyed any ecclesiastical theology it came in contact with. Even further, he offhandedly commented that one of the reasons he liked the theory of evolution was because "it occupies a position of complete and irreconcilable antagonism to that vigorous and consistent enemy of the highest intellectual, moral, and social life of mankind—the Catholic Church."[61] Both Mivart's ideas and Mivart himself were being rejected because Huxley saw them as mindless products of clerical authority, who could offer nothing besides regurgitation of dogma. His vision of the Roman Catholic Church as being a perversion of the true original Christianity was powerfully evoked in the famous passage that exiled Mivart forever:

> Elijah's great question, "Will you serve God or Baal? Choose ye," is uttered audibly enough in the ears of every one of us as we come to manhood. Let every man who tries to answer it seriously ask himself whether he can be satisfied with the Baal of authority, and with all the good things his worshippers are promised in this world and the next. If he can, let him, if he be so inclined, amuse himself with such scientific implements as authority tells him are safe and will not cut his fingers; but let him not imagine he is, or can be, both a true son of the Church and a loyal soldier of science.[62]

While that final sentence is sometimes read as Huxley rejecting any religious identity for men of science, I suggest that we take the capital C

quite seriously. What he was proscribing was a man of science being subject to institutional, clerical, authority against which he could offer no
contest. Mivart, in seeking solace from a Catholic theologian, could not
have done better in raising Huxley's ire.

Huxley's placement of the Catholic Church as the distillation of all
that was wrong with theology also helps explain his rejection of positivism. Some of the philosophical and methodological aspects of
positivism were appealing to Huxley, but he could never accept the
"religion of humanity" cloak in which Auguste Comte had wrapped
them. It was no better to worship philosophers than to worship saints.
As with Mivart, Huxley was quite clear about the reasons he rejected
positivism.

> [It] embraces not only a philosophy and a system of politics, but culminates
> in an elaborate scheme of religious organization, equipped with priesthood,
> liturgy, and calendar. And, lest anything should be wanting to complete the
> resemblance to the Papistical model, on which primitive Positivism was ex-
> pressly founded, the Atheocracy of the new faith was to establish as thorough
> a spiritual despotism, and to exhibit as complete a contempt for liberty of
> conscience, as the Theocracy it aimed to supplant.[63]

Comte was as bad as the popes, and for exactly the same reasons. He had
somehow managed to preserve all the worst (as Huxley saw them) aspects of ecclesiasticism without even the tiny core of true religion that
Catholicism retained. Huxley declared positivism to be "as thoroughly
antagonistic to the very essence of science as anything in ultramontane
Catholicism" and famously dismissed it as "Catholicism *minus* Christianity."[64] Comte even managed to envision a system of forced belief based
on science, a "modern spiritual power" that would control religious and
temporal activity. Huxley was sure this "would be the establishment of
something corresponding with that eminently Catholic, but admittedly
anti-scientific, institution—the Holy Office."[65]

Science and Liberty

The negative space defined by Huxley's objections to ecclesiasticism and
Catholicism helps reveal the positive space that he thought necessary
for the advancement of science and learning. That was independence of

thought (scientific or otherwise). Writing to Michael Foster, he declared his lifetime goal to be the defense of "the principle of religious and philosophical freedom."[66] It was this quality that he most closely associated with scientific work.

As he often did, Huxley looked to history for heroes with which to extol his values. Erasmus was celebrated as the real core of the Reformation because he was "at heart, neither Protestant nor Papist, but an 'Independent Christian.'"[67] His refusal to abandon freedom of thought even under pressure marked him as a great father of modernity. Locke was noted for his advocacy of toleration in religious matters, though Huxley noted that Locke did not extend that toleration to either atheists or papists.[68]

In addition to intellectual freedom being necessary for science, Huxley argued that men of science were the great defenders of liberty. Only through training, and the experience of humbling themselves before natural phenomena, would men truly learn the need for such ideals. Only the practice of science could bring someone to this state of enlightenment, which burdened men of science with certain duties. This idea helped spur the scientific naturalists to become involved with politics against Gladstone's plans for Irish Home Rule. Huxley and Tyndall were particularly concerned about what they saw as a capitulation to the Roman Catholics. Tyndall wrote: "Men of science have, for centuries, been leaders in the practice and defense of true liberty of thought and speech, but they cannot, without protest, surrender to the modern demagogue."[69] They saw themselves, as men of science, as a critical bulwark against tyranny.

An essential part of Huxley's vision of intellectual freedom was the right to doubt, to criticize, and be criticized. Vigorous, even fierce debate was held to be essential to science. Anything went, so long as it allowed for freedom. Tyndall announced that the "only thing out of place . . . is dogmatism."[70] In marking the twenty-first anniversary of the *Origin of Species*, Huxley celebrated the importance of attacking an idea:

> Now the essence of the scientific spirit is criticism. It tells us that whenever a doctrine claims our assent we should reply, Take it if you can compel it. The struggle for existence holds as much in the intellectual as in the physical world. A theory is a species of thinking, and its right to exist is coextensive with its power of resisting extinction by its rivals.[71]

Given Huxley's famous joy in debate, it comes as no surprise that he declared his constant willingness to be criticized. More interesting was his use of that willingness to disparage his enemies. When attacking the Duke of Argyll and Canton Liddon, he tried to shame them by implying they were unwilling to take as good as they gave.

> I decline to assume that the standard of morality, in these matters, is lower among the clergy than it is among scientific men. I refuse to think that the priest who stands up before a congregation, as the minister and interpreter of the Divinity, is less careful in his utterances, less ready to meet adverse comment, than the layman who comes before his audience, as the minister and interpreter of nature.[72]

We here have echoes of his earlier argument that ecclesiastics were less moral than their enemies, because they force their beliefs on others rather than relying on genuine intellectual contest. Huxley positioned science as the only method of inquiry that focused on discernment rather than doctrines, and thus inculcated freedom rather than belief. He even claimed that he would prefer that evolution not be taught to children, and rather that school pupils spend time gaining "a firm foundation for the further knowledge which is needed for the critical examination of the dogmas, whether scientific or anti-scientific, which are presented to the adult mind."[73]

Huxley found an opportunity to expound on his views on the importance of liberty in science when he was invited to Birmingham to speak on Joseph Priestley. Writing to Tyndall about the invitation, he admitted no particular enthusiasm for Priestley, but revealed his plans to use the lecture for a broader agenda:

> Satan whispered that it would be a good opportunity for a little ventilation of wickedness. I cannot say, however, that I can work myself up into much enthusiasm for the dry old Unitarian who did not go very deep into anything. But I think I may make him a good peg whereon to hang a discourse on the tendencies of modern thought.[74]

The discourse he delivered used Priestley as an icon of liberty in both science and religion, and is a wonderful source for seeing what values Huxley held in highest esteem. The lecture leapt directly into praise—

but praise for Priestley as a brave freethinker, not as someone who devoted his life to religion:

> But I am sure that I speak not only for myself, but for all this assemblage,
> when I say that our purpose to-day is to do honour, not to Priestley, the Uni-
> tarian divine, but to Priestley, the fearless defender of rational freedom in
> thought and in action: to Priestley, the philosophic thinker; to that Priestley
> who held a foremost place among "the swift runners who hand over the lamp
> of life," and transmit from one generation to another the fire kindled, in the
> childhood of the world, at the Promethean altar of Science.[75]

Huxley told Priestley's story as that of a boy who was raised in orthodoxy,
expected to become a priest, but who rebelled with the help of heterodox
teachers. These teachers "encouraged the discussion of every imaginable
proposition with complete freedom, the leading professors taking opposite sides; a discipline which, admirable as it may be from a purely scientific point of view, would seem to be calculated to make acute, rather
than sound, divines." His developing religious views were described as a
continuous series of casting-offs of doctrines and dogmas.[76]

Priestley's early scientific career was said to be interrupted by ecclesiasticism when "certain clergymen" on the Board of Longitude prevented
him from joining Captain Cook's second voyage. His suffering at the
hands of a pious "Church-and-King" mob was vividly described. He and
other Dissenters fled, leaving, significantly, his papers, library, and apparatus to the flames.[77] Huxley faintly praised Priestley's scientific work
as "admirable" while declaring that his political works were "full of insight and replete with the spirit of freedom." His attacks on the clerics
were cheered:

> And while all these sparks flew off from his anvil, the controversial hammer
> rained a hail of blows on orthodox priest and bishop. While thus engaged,
> the kindly, cheerful doctor felt no more wrath or uncharitableness towards
> his opponents than a smith does towards his iron. But if the iron could only
> speak!—and the priests and bishops took the point of view of the iron.[78]

Huxley gleefully contrasted Priestley's equanimity and purity of thought
with his political and theological enemies. The orthodox were painted as
bigots who hurled accusations of "moral defects," all fired by their "sectarian hatred."[79]

Huxley showed similar pleasure in demonstrating that the Christian doctrines Priestley was persecuted for disbelieving (e.g., the details of natural immortality) had since been espoused by several prelates of the Church of England. This led inevitably to commentary on the hypocrisy of the state Church, where a Dissenter's diabolical claim became a fine idea once held by an Anglican.[80] Huxley pointed out that Priestley's then-dangerous views on ecclesiastical reform and the Test Acts would, in the reign of Queen Victoria, seem downright conservative. And for those views, he reminded his listeners, Priestley "could neither teach nor preach, and was liable to ruinous fines and long imprisonment. In those days the guns that were pointed by the Church against the Dissenters were shotted. The law was a cesspool of iniquity and cruelty."[81] Huxley was glad his own time was much improved, when even high offices of state were open to Jews and secularists. But his applause of liberal reform in the established Church was riddled through with how this demonstrated the pointlessness of its orthodoxy.

Huxley's praise for the open-mindedness of himself and his friends did not, of course, go unchallenged. Some of his students reported that his teaching fell far short of encouraging students to think for themselves.[82] Critics declared that men of science had come to "constitute in our day a sort of lay-priesthood, as narrow, and intolerant, and tyrannous in temper as the priesthood of the Church ever was in the days of its darkest supremacy."[83] Some attacked the scientific naturalists' idolization of Darwin as exactly the sort of argument from authority that they claimed to despise. Mivart wrote to Richard Owen in 1871 complaining that Huxley, who spoke so vehemently against authority in science, was critiquing a writer in the *Quarterly Review* for "not observing sufficient reverence for Mr. Darwin."[84] Argyll invoked this frequently, warning his readers to be "awake to the retarding effect of a superstitious dependence on the authority of great men."[85] He accused Huxley of "dogmatism" by "the very simple and natural mistake of confounding 'geology' with himself."[86] He even speculated about a "Reign of Terror" in science in which the Darwinians would persecute all unbelievers. In response, Huxley wondered whether

a guillotine is erected in the courtyard of Burlington House for the benefit of all anti-Darwinian Fellows of the Royal Society? Where are the secret conspirators against this tyranny, whom I am supposed to favour, and yet not have the courage to join openly?[87]

Huxley scoffed at the idea of a scientific persecution that, to him, was so opposed to the spirit of the investigation of nature that it was simply impossible. Men of science did not have the dogma, thirst for power, or desire for conformity that made ecclesiastic infidel-making possible. He knew what it was like to be outcast by the powers that be, and it was an affront that his enemies would accuse him of the same. He admitted that "as is the case with all new doctrines, so with that of Evolution, the enthusiasm of advocates has sometimes tended to degenerate into fanaticism; and mere speculation has, at times, threatened to shoot beyond its legitimate bounds."[88] The younger men of science, he said, had been warned to keep their spirits even.

"Religious" Education

This discussion of the details of Huxley's antiorthodoxy has helped show the boundaries of his willing cooperation with religious and theological groups. Where were the edges of acceptable cooperation, and what issues could be compromised on?

It could be phrased this way: for Huxley, *religious* education was acceptable, but not *sectarian* or *theological* education. His battles on the London School Board focused on just these issues. Many of the precise details are laid out in his 1870 essay "The School Boards: What They Can Do, and What They May Do." Across Britain there was rising concern about exactly what relationship the new state school system would have with the established Church. The Nonconformists and their allies (Huxley among them) were deeply worried that the new schools would serve to indoctrinate their children in Anglican dogmas. Huxley suspected exactly such a plan when it was announced that a minister of education was to be appointed, who, "in accordance with our method of making Ministers, will necessarily be a political partisan, and who may be a strong theological sectary into the bargain."[89]

In Huxley's argument against this development, we find virtually all of the concerns that have emerged in this section. He warned that teachers would not have "freedom of action," and that the minister would have "despotic" powers regarding religious education. He saw what can only be described as an ecclesiastic conspiracy, in which theologians maneuvered great powers for the minister and sought to place picked devotees onto the boards themselves:

An interest, almost amounting to pathos, attaches itself, in my mind, to the frantic exertions which are at present going on in almost every school division, to elect certain candidates whose names have never before been heard of in connection with education, and who are either sectarian partisans, or nothing.

Huxley here heard echoes of what he described as having happened innumerable times in history—clerics seizing unearned authority in areas about which they knew nothing. He thought he understood precisely what was happening, and was moving to prevent it. The language of the act, he said, was explicit that no "religious catechism or religious formulary which is distinctive of any particular denomination" could be taught in the schools. And yet, he wondered sarcastically, "Can it be that these zealous sectaries mean to evade the solemn pledge given in the Act?"[90]

Correspondence with newspapers, including the *Guardian*, suggested to Huxley that sectarians were explicitly planning to take over the school boards and vote in doctrinal education despite the clear text of the act. He suggested that these doctrines would probably be of the sort acceptable to several denominations, such as the Trinity. This would lead to a situation "offensive to every Unitarian and to every Jew in the House of Commons, besides creating a precedent which will afterwards be used to the injury of every Nonconformist."[91] Huxley suggested that there were legal structures in place to prevent this, but nonetheless we can see his fear of exactly the ecclesiastical developments that he described in his historical narratives.

Huxley found himself frustrated at the hard lines all the parties involved were drawing. He proposed that everyone move toward the middle:

We are divided into two parties—the advocates of so-called "religious" teaching on the one hand, and those of so-called "secular" teaching on the other. And both parties seem to me to be not only hopelessly wrong, but in such a position that if either succeeded completely, it would discover, before many years were over, that it had made a great mistake and done serious evil to the cause of education.

As was often the case, Huxley accused his enemies of confusing religion and theology, and even worse, his allies seemed to be accepting that conflation.

> For, leaving aside the more far-seeing minority on each side, what the "religious" party is crying for is mere theology, under the name of religion; while the "secularists" have unwisely and wrongfully admitted the assumption of their opponents, and demand the abolition of all "religious" teaching, when they only want to be free of theology—Burning your ship to get rid of the cockroaches![92]

He could not be more clear. Religious teaching was valuable, a ship to help one advance, while theological teaching was nothing but cockroaches. The religion he saw as necessary in education was "the love of some ethical ideal." Quite shockingly to his secularist allies, he even mused that if he were "compelled to choose for one of my own children, between a school in which real religious instruction is given, and one without it, I should prefer the former, even though the child might have to take a good deal of theology with it."[93] As long as the education was free of the dangerous elements of ecclesiasticism—sectarianism, unearned power, enforced belief—the poison of the theology would be outweighed by the nutrition of the religion. He had no plan for a purely secular education that involved no religious ideals, individuals, or values.

The great surprise came when Huxley approved of the reading of the Bible in the schools. While many of his followers felt this was a great betrayal, it was quite in line with the values we have seen here. He denied that there was any inconsistency in his position, clarifying that he had been in favor of secular education in the specific

> sense of education without theology; but I must confess I have been no less seriously perplexed to know by what practical measures the religious feeling, which is the essential basis of conduct, was to be kept up, in the present utterly chaotic state of opinion on these matters, without the use of the Bible.

The Greek and Roman writers, he thought, were too refined and abstract for use by children, and some clear standard was needed. Of course a "lay teacher" should use all the knowledge provided by the higher criticism, and not use passages inappropriate for children. Even after these trimmings there would still remain a "vast residuum of moral beauty and grandeur."[94]

And beyond this moral value, Huxley posited that the Bible was so interwoven with English culture and life that it would be a great crime to ignore it. It was the foundation of the stories that the English told about themselves and, also important, it did so across class boundaries. Whether poor or highborn, every Englishman had that same grounding in one text. It was also a hoard of historical and geographical information. The key for him was to ensure that the teachers avoided adding any theological glosses. His advice for this was a curious level of textual literalism:

> The teacher would do well not to go beyond the precise words of the Bible; for if he does, he will, in the first place, undertake a task beyond his strength, seeing that all the Jewish and Christian sects have been at work upon that subject for more than two thousand years, and have not yet arrived, and are not in the least likely to arrive, at an agreement; and, in the second place, he will certainly begin to teach something distinctively denominational.[95]

As always, Huxley's bugbear was sectarianism. His solution here was to invest authority in the book itself, thus preempting any claims to authority by the clerics. The same strategy used by the evangelicals to reach across sectarian lines was here used to refute the sectarians. He said the only way the Bible could be problematic was if it was read with "ecclesiastical spectacles." Indeed, it was the key to undoing the damage caused by the ecclesiastics:

> Its teachings are so infinitely superior to those of the sects, who are just as busy now as the Pharisees were eighteen hundred years ago, in smothering them under "the precepts of men"; it is so certain, to my mind, that the Bible contains within itself the refutation of nine-tenths of the mixture of sophistical metaphysics and old-world superstition which has been piled round it by the so-called Christians of later times.

As an antidote to theological poison, he recommended deep drinks from "the undefiled spring." He emphasized that it was the right and duty of every man to address the scriptures with his own judgment and without any doctrinal filter. In this way he turned the basic assumptions of Protestant missionaries back on themselves: does not a missionary assume that any potential converts can draw the necessary truths themselves? Otherwise, to what end giving Bibles to the Zulus? Huxley skillfully tied this appeal for an inherent religious judgment to the assumptions of British colonialism: "I trust that I may, without immodesty, claim to be put on the same footing as a Zulu."[96]

Huxley's praise for the Bible as a pedagogical text with a clear meaning harkened back to his own religious upbringing. He pleasantly recalled hours spent as a child with his grandmother's Bible, absorbing the moral lessons of Joseph, David, and Abraham. He testified that his own experience was sufficient evidence that a young child, "left to his own devices," could extract useful moral guidance. His *unassisted* reading was key:

> For if I had had some theological "explainer" at my side, he might have tried, as such do, to lessen my indignation against Jacob, and thereby have warped my moral sense for ever; while the great apocalyptic spectacle of the ultimate triumph of right and justice might have been turned to the base purposes of a pious lampooner of the Papacy.[97]

One of the critical moral lessons he expected students to draw from scripture was the need to battle authority. He recommended the Bible for his democratic age, naming it as one of "the great instigators of revolt against the worst forms of clerical and political despotism." Both the Jewish and Christian scriptures were said to be the best guarantors of the rights of the poor and oppressed. The scriptures' repeated assertions of the rights of humanity and the rise of liberty against injustice were critical lessons. Huxley was keen to emphasize that this rebelliousness applied to both temporal and ecclesiastic power—he expected Jesus to rail as vehemently against the pope as against the czar.[98]

Huxley went so far as to call the Bible "the most democratic book in the world." He asserted that it helped undermine the "clerico-political despotism" of the Middle Ages as well as the Catholic powers during the Reformation. In a remarkable move, he directly linked belief in an authoritative scripture to the defense of liberty: "The Protestant sects have favoured political freedom in proportion to the degree in which they have refused to acknowledge any ultimate authority save that of the Bible." He was quick to add that this had nothing to do with any of the "cosmogonies, demonologies, and miraculous interferences" found in the texts. Rather, it was simply that the Bible was a particularly clear source of ethical instruction. It certainly needed other lessons taught alongside it (science, for example), but he maintained that it was an essential part of modern education.[99] Reflecting on this controversy decades later, he mused that in the hands of lay teachers, the Bible had "gradually become modified into harmony with common sense." This was just as he had predicted, and precisely what he thought the orthodox feared. The small amount of doctrinal instruction that might have slipped through was regrettable, but necessary: "Twenty years of reasonably good primary education is 'worth a mass.'"[100]

We can now see the values that underlay Huxley's alliance with the Working Men's College. Despite Maurice's strong feelings for the Church of England, he, too, felt that sectarianism was counterproductive and would keep apart individuals who could do important work together. He certainly had as significant difficulties with theological authorities as Huxley ever did. The Working Men's College was therefore set up as being both Christian and nonsectarian, religious and antiauthoritarian. The spirit of the freedom of inquiry that Huxley saw as essential to science was cultivated deeply there, and he acknowledged this even as he admitted that he did not understand how it could grow from religious roots. Writing to Charles Kingsley, he said these sorts of endeavors were the only way to save the Church of England. All the younger men of science, he wrote, were similarly antiecclesiastic: "I know not a scoffer or an irreligious man among them, but they all regard orthodoxy as you do Brahmanism." They would never submit to men in "shovel hats" and would always demand that nature be the final authority. A typical Huxleyian screed, but he also offered that men like Kingsley who "see your way to the combination of the practice of the Church with the spirit of science" will keep the Church from being destroyed by "the advancing tide of science." So he certainly was willing to ally with Christians, so long as they shared the same values of freedom and inquiry. This is not to say that he was one of them, of course: "I don't profess to understand the logic of yourself, Maurice, and the rest of your school, but I have always said I would swear by your truthfulness and sincerity, and that good must come of your efforts."[101] But he could work with them.

Maxwell

Huxley's ability to find shared values with a Christian educational system was not merely a fluke. Similar processes were at work on the theistic side of the spectrum. Maxwell's alliance with the Christian Socialists might not seem to need any closer examination at all—he was a fairly typical recruit. However, I would like to delve deeper and show specifically that he shared many of the same antisectarian values that attracted

Huxley to the Working Men's College, and that they were similarly tied to his identity as a natural philosopher and a Christian.

At first glance, it might seem that Maxwell, a Cambridge man in the days of the Test Acts, might be exactly the sort of institutionally religious figure that Huxley despised. However, this was not the case, and Maxwell provides an important lesson in the variety of religious belief and practice that was sometimes mistaken for orthodoxy. As we saw in chapter 1, Maxwell's early religious upbringing in both the Episcopalian and Presbyterian traditions meant that it would have been nearly impossible to ally himself with a single sect. His later evangelical conversion made that even more difficult. Evangelicalism was self-consciously interdenominational, placing authority in individual religious experience and scripture rather than in institutions and their representatives. What Huxley would call antisectarianism, Maxwell would call ecumenicism.

Maxwell had no interest in quashing heterodoxy or labeling heretics. When he heard that Maurice was going to be tried for heresy, he wrote to his friend Lewis Campbell anticipating that the controversy would draw in more and more people. He noted that Maurice had made no "firm or dogmatic statements" and that this was simply the revival of old doctrinal arguments. While Maxwell was clear that he did not agree with Maurice's theology on all points, he praised the spirit of his inquiry, looking for the common elements of all Christian systems. He said the issue was indeed "whether the 'variables' of such a system ought to remain constant, as they were at some arbitrary epoch (that of sect-founders, Fathers, General Councils, Reformers, etc.), and not rather to be trusted to the true and approved Christians of every age."[102] That is, he saw Maurice's prosecution as simply the attempt of one Christian sect to control the beliefs and practices of another. He implied that Christianity should be expected to change and develop over time, and that the effect of sects was to ossify and control what should be a living, growing faith. The idea of labeling and restricting infidels would be deeply contrary to this perspective. Maxwell was known to say "I have no nose for heresy" and to look for points of agreement and cooperation with those of different positions.[103]

The correspondence with Campbell is helpful for unpacking Maxwell's thoughts on sectarianism and the clerisy—Campbell became an Anglican priest in 1857. Maxwell praised his friend's decision ("May your life and doctrine set forth God's glory, and be the means of setting forward the salvation of all men!") but voiced some concern about the role of professional clergy:

> Some of my friends think that the separation to a "holy function" puts a man into an artificial position with respect to the conduct of his thoughts, words, and acts, and that he is immersed in a professional atmosphere,—a *world*, in fact, differing from the world of business or of fashion only in the general colouring of its scenery.[104]

The professional duties of a priest, he suggested, actually pull a man away from the Christian life and toward sundry matters. Maxwell distinguished between the clerical world and the world of the true church (meaning those actually in communion with Christ):

> It has always seemed to me that men who have fallen into this "religious world" have completely failed in getting into the Church, seeing that the Church professes to be an escape from the world, and the only escape. . . . So far my theory of the Church not being a clerical world. Now I believe it not only as a theory, but as a fact, that a man will find the thing so if he will try it himself.[105]

Maxwell drew a clear distinction between the ecclesiastic life and a genuinely Christian life. He was certainly not disparaging his friend for choosing this path, only warning that the clerical life alone was not enough to be truly godly. This was characteristically evangelical, emphasizing that only one's inner state was relevant to one's relationship with the divine. Along these lines, Maxwell cautioned that becoming an institutional professional could lead to "restraints and professional stiffness of sentiment" that were not appropriate for true Christians, only for "those whom the truth has not yet made entirely free."[106] The freedom referred to here was of course freedom from sin, the crucial lesson for any evangelical. He gave Campbell advice on the core message to deliver to his congregants:

> You have to say that what men are and the nature of their actions depends on the state of their wills, and that by God's grace, through union with Christ, the contradictions and false action of those wills may be settled and solved, so that one way lies perfect freedom, and the other way bondage under the

devil, the world, and the flesh, and therefore you entreat them to give heed to the things which they have heard.[107]

Note that these suggestions have nothing to do with an established Church, any institutional system, or any appeals to religious authority. Personal decision and responsibility were the keys. Maxwell argued, from a deeply religious position, against thinking of clerics as having any special authority and for the need of individuals to come to their own conclusions—points that would seem very familiar to Huxley.

Maxwell's own relationship to religious institutions was somewhat more complicated. He apparently had no trouble subscribing to the Thirty-Nine Articles at Cambridge, though it is unclear how seriously these sorts of testaments were taken by that time: William Thomson did not find them to be very significant, whereas G. G. Stokes placed much more importance on them.[108] Maxwell also for many years served as an elder at the Presbyterian church close to his estate in rural Scotland, where his name appears in the records sporadically. His duties seem to have consisted of fund-raising, helping deliver punishments to local troublemakers such as fornicators and adulterers, and once representing the church at the Presbytery of Kircudbright and the Synod of Galloway.[109] It does not appear that he was involved in any actual theological matters, and his role as elder was more as a community leader than as a religious guide. Certainly, he was willing to be affiliated with a religious institution of this sort, but the essence of his religious life was largely internal and had little relationship to it.

There are suggestions that Maxwell was concerned about the tendency of institutions to prematurely reify theological ideas. As a young man, he read John Henry Newman's *Essay on the Development of Christian Doctrine*, and commented on its demonstration of an idea he called "Envelopment or Self-Involution." He said this was the process whereby the mind prematurely takes hold of an incomplete theory, and criticized Newman for doing this with Christianity. This transformation of a theory into a doctrine replaced useful pursuits ("the extension of knowledge, the dominion over Nature, and the welfare of mankind") with theory and argument for their own sake.[110] Similarly, one of Maxwell's objections to Maurice's theology was the latter's attempt to define and dictate what should be believed. Maxwell's biographers said that he "sometimes felt that the new teacher was apt to travesty the Popular Theology in trying to delineate it."[111]

We can see the details of his reluctance to link his religion to an institution or an authority in his reaction to the Victoria Institute. The head of the Institute, W. H. Petrie, wrote to Maxwell in the spring of 1875 inviting him to join. Petrie described the group as including "the Archbishop of Canterbury and other prelates and leading ministers, several professors of Oxford and Cambridge and other universities, and many literary and scientific men."[112] Among the distinguished members was Stokes, Maxwell's friend and colleague.

It is not known whether Maxwell sent the surviving manuscript letter declining membership. His response used an interesting choice of words: "I do not think it is my duty to become a candidate for admission into the Victoria Institute." He admitted that some of the goals of the group were ones of which he thought highly, and wished to see pursued. In particular, he wrote:

> I think men of science as well as other men need to learn from Christ and I think Christians whose minds are scientific are bound to study science ~~and so to widen their views of the glory of God~~ that their view of the glory of God may be as extensive as their being is capable of. [strikeout in the original]

So he did think it was worthwhile for men of science to consider religious issues, and for Christians to grapple with science. What, then, was the source of his reluctance to join? He made it clear that his concern was associating these practices with a society or with institutional authority:

> But I think that the results which each man arrives at in his attempts to harmonize his science with his Christianity ought not to be regarded as having any significance except to the man himself and to him only for a time, and should not receive the stamp of a Society. For it is of the nature of Science, especially of those branches of Science which are spreading into unknown regions to be continually . . . [Manuscript ends.][113]

Again we see that Maxwell's reasoning was typical for an evangelical. Religious insight and meaning was internal and individual, without any need for ecclesiastic approval. And even beyond not needing approval, it appeared that Maxwell wanted no part of it. While the manuscript letter is incomplete, context suggests that his concern was the continuous change of science being incompatible with an institution's necessary resistance to change. He was certainly tolerant of institutional authority (given his status as an elder at home), but he does not seem to have welcomed it, or considered it particularly valuable for questions regarding science and religion.

Maxwell's evangelical views on institutionalized power were not unusual, and were commonly held among theistic scientists. Many other theistic scientists were Dissenters (Carpenter) or latitudinarians (Thomson) and thus had no particular affection for an established Church. It was Protestants, not secularists, who ended the Anglican dominance at Oxbridge.[114] Those scientists who held similar attitudes but were Anglicans were often evangelical, as with Stokes, and had a looser sense of institutional authority. Even Richard Owen stressed the dangers of dogmatism and sectarian narrowness:

> It is the human element mingling with the divine one, or meddling with it, which the discoveries of Science expose; it is the fence set up about some narrow and exclusive view which they break down. Beware, therefore, of logically precise and definite summaries of Doctrine, accounting from its point of view, for all things and cases natural and preternatural, claiming to be final and all sufficient. Systems of Christianity, Schemes of Doctrine, are of human construction, the works of man's brains.

He warned against all "isms," declaring them lacking in "Christian charity." Instead, he said, follow the evidence that both science and religion provide without prejudice. Speaking to the YMCA, he invited his listeners to emancipate themselves from the things they thought were true and to read the scriptures with an open mind, "as little children."[115] Victorian Protestants saw themselves as guardians of freedom as much as political radicals did. However, they saw the source of liberty to be God, and argued that only within God and his laws could one find the strength to overcome repression both from within (the fallen nature of humanity) and without (unjust authorities).[116]

An incident in the 1860s helps illustrate how widespread these values of freedom were among theistic scientists. A group of London chemists authored a "Scientists' Declaration," which proclaimed the harmony of science and Christian doctrine.[117] Men of science overwhelmingly refused to sign it (including 90 percent of Fellows of the Royal Society). Even theistic figures, such as Faraday, Whewell, and Owen, declined, though they surely agreed with the spirit of the document. They rejected it because they saw it as an imposition on their religious freedom. John Herschel famously announced his opposition in the *Athenaeum*:

> I consider the act of calling on me, publically to avow or disavow, to approve or disapprove, in writing, any religious doctrine or statement however carefully or cautiously drawn up (in other words to append my name to a religious manifesto) to be an infringement of that social forbearance which guards the freedom of religious opinion in this country with especial sanctity.

That is, not talking about the relationship of science and religion was a matter of religious freedom, even for believers. He considered these efforts to enforce certain views to be "mischievous, having a direct tendency (by putting forward a new Shibboleth, a new verbal test of religious partisanship) to add a fresh element of discord to the already too discordant relations of the Christian world."[118] William Stanley Jevons wrote to Herschel in support of his statement, particularly incensed that "freedom of inquiry [was] being interpreted as a tendency to Irreligion. Is it worthy of Religion to assume that it must be discarded by all who freely seek after the Truth?"[119] It was the attempts to codify and enforce belief that led to the rejection of the declaration among theistic men of science.

As with Huxley, Maxwell had many concerns about religious authority and forced belief. He saw the right to doubt and question as sacrosanct—without it, how could one genuinely choose a godly life? These values were deeply ingrained in him, and as a young man he wrote to Lewis Campbell about his agenda for personal spiritual development:

> The Rule of the Plan is to let nothing be willfully left unexamined. Nothing is to be *holy ground* consecrated to Stationary Title, whether positive or negative. All fallow land is to be ploughed up, and a regular system of rotation followed.

Maxwell hoped to grapple with all issues before him, with nothing set apart. However, it would surely be a mistake to regard Maxwell as a freethinker. He was sure that Christianity best provided the opportunity to pursue this questioning. His sense of precisely what liberty meant was highly colored by the attitudes of the day, but we should not doubt that he really did care about liberty and the right to explore on one's own—

and that he valued Christianity because he thought it embodied those ideals.

He was sure that everyone, including Christians, had blind spots that they did not wish to challenge. It was human nature that was reinforced by sectarianism:

> But there are extensive and important tracts in the territory of the Scoffer, the Pantheist, the Quietist, Formalist, Dogmatist, Sensualist, and the rest, which are openly and solemnly *Tabooed*, as the Polynesians say, and are not to be spoken of without sacrilege.

All the "ists" prevented meaningful and honest questioning of beliefs. Subscriptions to schemes and doctrines hobbled any seeker. Dogmas passed down by sects restrained inquiry and forced conclusions. Maxwell, in language quite similar to Huxley's, argued that the solution to these restraints was reading the Bible:

> Christianity—that is, the religion of the Bible—is the only scheme or form of belief which disavows any possessions on such a tenure. Here alone all is free. You may fly to the ends of the world and find no God but the Author of Salvation. You may search the Scriptures and not find a text to stop you in your explorations.[120]

Close attention to the text itself (a hallmark of evangelicalism) would break away any false ideas or dogmas that someone may have picked up. The deity that Maxwell took away from scripture was surely the God of Abraham, Creator of the Universe, something quite different from the ethical guidance Huxley desired. Nonetheless, the practice of Bible reading was advocated by both for similar purposes: an opportunity to grapple with vast ideas and truths, and to come to a personally justified conclusion.

Maxwell's approach to this issue of deciding for oneself overlapped with Huxley's in another way, too: anti-Catholicism. Like Huxley, he saw Roman Catholicism as a terrible institution that functioned by compelling belief and practice. It was an affront to all the values he saw as critical to religion. Writing to R. B. Litchfield (the friend who recruited him to the Working Men's College), he explained his ideas on why Catholicism was sometimes appealing:

As to the Roman Catholic question, it is another piece of the doctrine of Liberty. People get tired of being able to do as they like and having to choose their own steps and so they put themselves under holy men who, no doubt are really wiser than themselves. But it is not only wrong but impossible to transfer either will or responsibility to another, and after the formulae have been gone through the patient has just as much responsibility as before, and feels it, too. But it is a sad thing for any one to lose sight of their work and to have to seek some conventional, arbitrary treadmill occupation prescribed by sanitary jailers.[121]

This passage touched on many of the values discussed so far: the importance of religious liberty, the suspicion of clerical authority, and the importance of personal decision making. Maxwell even provided an evocative image of idolatrous papists as madmen forced to work blindly and ignorantly by careless supervisors.

Working with Scientific Naturalists

This shared rejection of compulsion of belief and mutual values of tolerance and individual decision making created areas of common ground between theistic and naturalistic scientists that were critical for the functioning of the Victorian scientific community. These areas were subtle but deeply necessary activities such as refereeing papers, organizing lectures and demonstrations, and critiquing each other's scientific ideas. Without shared values of how to interact with people with whom one disagrees, Victorian science would have experienced severe strains that were not in evidence.

Theistic and naturalistic scientists refereed each other's papers for major journals on a regular basis. Maxwell, Tyndall, Thomson, Stokes, and William Spottiswoode all had overlapping expertise and frequently commented on each other's submitted manuscripts.[122] These evaluations were usually positive, and even when they were critical, the objections were always on clear technical matters. Despite their well-known and profound religious disagreements, such issues never intruded on the day-to-day process of scientific research and publication.

Melinda Baldwin has closely examined the Stokes-Tyndall correspondence, and concluded that their significant differences in religious perspective did not affect their genuine admiration for each other's work. They were very concerned not to offend each other when writing about religion, and Stokes once wrote to ensure that Tyndall did not change something simply because he disagreed with it. Tyndall emphasized his esteem for Stokes's tendency to "justice" and relied heavily on his judgment.[123] And despite Tyndall's frequent clashes with Thomson on various matters, he reiterated his high respect for the Scot's intellect and always sought to bury the hatchet.[124]

The theists and naturalists' shared values provided a solid foundation on which they could work productively. They wrote letters of recommendation for each other for academic positions despite any worries about placing someone with opposing beliefs. Maxwell wrote a strong letter for W. K. Clifford even after the latter's aggressive materialism became known. Clifford was praised for his "elucidation of scientific ideas by the concentration upon them of clear and steady thought."[125] Huxley and Tyndall tried to get Maxwell to contribute to their projects.[126]

Even outside the strictures of formal refereeing, the two groups crossed boundaries to seek advice on various scientific issues. In a remarkable set of letters, Herbert Spencer sought advice on his evolutionary model of the solar system from Maxwell, a firm believer in the hand of God in creation. Apparently, Maxwell had been discussing the nebular theory with Clifford, who passed along a critique to Spencer. Spencer then wrote asking for clarification, to which Maxwell responded with a long letter. His reply included detailed descriptions of theories of electricity and molecular motions that complicated Spencer's system. The physicist was clearly skeptical, and had no sympathy for Spencer's cosmological system, but nonetheless devoted a significant amount of time and energy to providing specific feedback. Maxwell even offered kind comments about Spencer's willingness to accept criticism and modify his ideas instead of reifying them.[127] Their correspondence continued for over a year, with Maxwell explaining the dynamics of heat and entropy. He made it clear that he did not support Spencer's system, and did not "quite understand the principal features of your hypothesis." He even offered some suggestions on scientific nomenclature.[128]

Additionally, Maxwell and Tyndall corresponded regarding Faraday's lines of force.[129] While that topic might not seem likely to inflame ideological differences, the presence of oxygen on a young Earth certainly does. Nonetheless, Thomson discussed exactly that with Joseph Hooker and William Turner Thiselton-Dyer (an assistant of Huxley's).[130] Again, while they no doubt disagreed on details, they shared enough common ground to discuss the issues in a meaningful and useful way. Thiselton-Dyer also contacted Maxwell to discuss how to organize teaching laboratories. Here we have one of Huxley's protégés seeking advice from a Christian aristocrat on laboratory teaching—Huxley's great weapon against orthodoxy.[131] Maxwell offered advice on how to set up a lab to benefit both highly gifted students and those without much promise. The naturalistic scientists clearly did not feel that Maxwell would indoctrinate students or that his theism would be a gateway into orthodoxy. Their seeking of his advice on science education shows that they credited his ability to create a safe space for inquiry.

Beyond this intellectual and educational intercourse, the theists and the naturalists maintained "easy social relations." While living in London, Maxwell met the X-Club's T. A. Hirst at the Unitarian William Carpenter's house, dined at Spottiswoode's, and was commonly present at Tyndall's Royal Institution.[132] Maxwell joined the Athenaeum Club with the support of both Tyndall and Hirst.[133] He never became close friends with anyone in Huxley's circle, but they were clearly affectionate and respectful. Tyndall sent a disappointed note when he missed Maxwell after a lecture; Maxwell complimented Tyndall's conversational abilities.[134] When preparing a presentation for the RI, Maxwell quipped that he had been "Tyndallizing my imagination up to the lecture point."[135] He even stood up for Tyndall during the Tait-Tyndall controversy over glaciers, when the former had attacked the latter for his popular lectures and writings. Despite his many years of friendship with Tait, Maxwell defended Tyndall's right to popularize, as long as it was done well: "Can a man do *good* service in popularising certain parts of science and *thereby* lose his claim to scientific authority? If a man has a claim to scientific authority the only way he can lose it is by writing bosch."[136] Thomson and Huxley were elected to the Royal Society at the same time, and kept on good terms—Thomson invited the biologist along on a cruise to the Hebrides.[137] E. Ray Lankester praised Thomson and Maxwell as being part of a "wonderful group of men . . . [who] possess an ingenuity and delicacy in appropriate experiment which must fill all who even partially follow their triumphant handling of Nature with reverence and admiration."[138] Maxwell and Clifford got along famously well, to the point where Maxwell was uncomfortable giving his book a poor review: "There were many things in the book that wanted trouncing, and yet the trouncing had to be done with extreme care and gentleness, Clifford was such a nice fellow."[139]

After Maxwell's death, Spottiswoode presented a eulogy at the Philosophical Club of the Royal Society, where attendees mixed theists and naturalists: Stokes, Huxley, Hooker.[140] After Huxley's death, Thomson (by then Lord Kelvin) spoke generously about him despite their past exchanges. He praised Huxley as "a resolute and untiring searcher after truth, and an enthusiastically devoted teacher of what he learned from others and what he discovered by his own work in biological science." While he could have simply offered pleasantries and ignored the controversies, he did not. Instead, he addressed the issue directly, commenting on exactly those aspects of Huxley's worldview that the theist scientists could admire:

> When he introduced the word agnostic to describe his own feeling with reference to the origin and continuance of life, he confessed himself to be in the presence of mysteries on which science had not been strong enough to enlighten us; and he chose the word wisely and well. It is a word, which, even though negative in character, may be helpful to all philosophers and theologians. If religion means strenuousness in doing right and trying to do right, who has earned the title of a religious man better than Huxley?[141]

One might read this religious label as one last jab at Thomson's old opponent, but I think we should take it seriously, and as we have seen in this chapter, Huxley probably would not have minded. Certainly, Stokes took it at face value and applauded the sentiment: "As I listened to your address, I liked very much what you said of religious opinions in speaking about Huxley."[142]

The theistic and naturalistic scientists were, generally speaking, not close confidants. They were colleagues. They respected each other and acknowledged the fundamental shared values of freedom of inquiry and antidogmatism that made basic activities such as publication refereeing possible. Charles Kingsley satirized this situation in *The Water Babies*:

> Whereon a certain great divine, and a very clever divine was he, called him a regular Sadducee; and probably he was quite right. Whereon the professor, in return, called him a regular Pharisee; and probably he was quite right too. But they did not quarrel in the least; for when men are men of the world, hard words run off them like water off a duck's back. So the professor and the divine met at dinner that evening, and sat together on the sofa afterwards for an hour, and talked over the state of female labor on the antarctic [*sic*] continent (for nobody talks shop after his claret), and each vowed that the other

was the best company he ever met in his life. What an advantage it is to be men of the world![143]

Anger at Belfast

This shared respect and expectation of liberty was, surprisingly, one of the causes of the angry reception of Tyndall's Belfast Address among theistic scientists.[144] The sentiments offered by Tyndall were not particularly new. However, I suggest that it appeared to the theists that Tyndall was trying to use his position as president of the BAAS to *enforce* his naturalism: precisely the sort of institution-based coercion of belief that both parties had agreed was antithetical to science. Whether that was an accurate reading of the address is another question, but Tyndall's aggressive phrasing, coming from the president's pulpit, struck all the wrong notes. Their comfortable, day-to-day interactions in the community of science would not be possible if the naturalists were willing to use their positions of authority to suppress those who disagreed. The year 1873 had marked a watershed for the X-Club in which its members had many of the leadership positions of the Royal Society, and perhaps Belfast had marked a new strategy of using those positions to their advantage.[145]

Maxwell recorded his reactions to the address in two comic poems, one sent to Tait, and one published anonymously in *Blackwood's Edinburgh Magazine*.[146] While Maxwell certainly used these to state his disagreement with Tyndall's molecular theories, these poems also show his visceral feeling of being attacked and coerced by the scientific naturalists. The letter he wrote to Tait immediately upon his return from Belfast presents the poem, then rather sharply notes that Spencer was also at the meeting presenting on evolution. He even uncharacteristically mocked Clifford's paper on chemical equations. The letter was from start to finish a complaint about the X-Club, with the poem at its core:

> I KNOW not what this may betoken,
> That I feel so wondrous wise;
> My dream of existence is broken
> Since science has opened my eyes.
> At the British Association
> I heard the President's speech,
> And the methods and facts of creation

Seemed suddenly placed in my reach.
My life's undivided devotion
To Science I solemnly vowed,
I'd dredge up the bed of the ocean,
I'd draw down the spark from the cloud.
To follow my thoughts as they go on,
Electrodes I'd place in my brain;
Nay, I'd swallow a live entozöon,
New feelings of life to obtain.
O where are those high feasts of Science?
O where are those words of the wise?
I hear but the roar of Red Lions,
I eat what their Jackal supplies.
I meant to lie so scientific,
But science seems turned into fun;
And this, with his roaring terrific,
That old red lion bath done.

The verse was instructed to be sung to the tune of Heinrich Heine's "Loreley." The poem began "I heard the President's speech / And the methods and facts of creation / Seemed suddenly placed in my reach." Note that Tyndall was immediately named as the president of the BAAS—this was not an abstract objection, but aimed at the particulars of the speech. Maxwell imagined a listener's reaction to the address: "My life's undivided devotion / To Science I solemnly vowed"—Tyndall had suddenly pulled the listener into an existence solely dictated by science. And Maxwell made it clear that this was not just science being pressed onto the listeners, but naturalistic science: "I'd dredge up the bed of the ocean . . . / To follow my thoughts as they go on / Electrodes I'd place in my brain," referring to Huxley's pet projects of the *Challenger* expedition and humans as automata, respectively.

He asked, "O where are those high feasts of Science? / O where are those words of the wise? / I hear but the roar of Red Lions / I eat what their Jackal supplies." The Red Lions was a social club founded by Edward Forbes, which Huxley and Tyndall joined in 1851. Its membership overlapped significantly with the X-Club, and Maxwell's use of that term certainly evoked the scientific naturalists. The image of the roaring lions drowning out the words of the wise is a powerful one—volume, strength, and savagery crushing more refined considerations.[147] Whom Maxwell felt offended by at Belfast was clear (the scientific naturalists as a group, not solely Tyndall), as was his concern (being forced to eat the carrion left behind by their scientism).

The poem he sent to Tait was written quickly and emotionally, while the poem he published later in *Blackwood's* was significantly longer, more detailed, and more precisely phrased. Titled "Notes of the President's Address," it mimics the historical structure of Tyndall's presentation:

> IN the very beginnings of science, the parsons, who managed things then,
> Being handy with hammer and chisel, made gods in the likeness of men;
> Till Commerce arose, and at length some men of exceptional power
> Supplanted both demons and gods by the atoms, which last to this hour.

The first line clearly jabs at the Huxley-Tyndall historical narrative of modern progress away from theological domination. Then those men of power dictated determinism (see chapter 6) and made impossible religious belief and practice:

> From nothing comes nothing, they told us, nought happens by chance, but by fate;
> There is nothing but atoms and void, all else is mere whims out of date!
> Then why should a man curry favour with beings who cannot exist.

Prayer and heaven both were excluded.

And what tools were used to crush these practices? Not the light of knowledge: "But not by the rays of the sun, nor the glittering shafts of the day, / Must the fear of the gods be dispelled, but by words, and their wonderful play." Tyndall's speech was criticized as just that, with no deeper substance. Calling him a "poet-philosopher," Maxwell mocked the molecular creation story that depended on nothing but incompressible spheres and force, particularly the idea that such stories could explain emotion and will. "Let us damn with faint praise Bishop Butler, in whom many atoms combined / To form that remarkable structure, it pleased him to call—his mind."

Maxwell again located his complaint specifically at the Belfast meeting, and specifically at these ideas being pushed through the institution of the BAAS:

> There is nobody here, I should say, has felt true indignation at all,
> Till an indignation meeting is held in the Ulster Hall;
> Then gathers the wave of emotion, then noble feelings arise,
> Till you all pass a resolution which takes every man by surprise.

The image that Maxwell chose to represent what he thought the Association has become was quite striking, and visibly demonstrated the roots of his anxiety: "The British Association—like Leviathan worshipped by Hobbes / The incarnation of wisdom, built up of our witless nobs."[148] He worried that the scientific naturalists were claiming the complete monopoly on power and political absolutism of Hobbes. And as the citizens ruled by Leviathan gave up their individual political activity in favor of the monarch, Maxwell feared that the diversity of individual views within British science might be quashed by the naturalistic ideology.

Maxwell's poems provide important evidence as to the concerns of theistic scientists raised by the Belfast Address. It was not solely that the ideas and claims presented were disagreeable; it was that the pulpit from which they were presented might signal the breaking down of the freedom of investigation and thought that were critical to science. The apparent linking of naturalistic values to the institution of the BAAS was seen as a danger. This was not the first time such concerns were articulated: during the Forbes-Tyndall glacier controversy in 1859, Thomson wrote that Tyndall and Huxley had been *"most improper"*—because they were trying to use the Royal Society to leverage the argument.[149] There had also been anxiety that Huxley, when he was president of the BA in 1870, would cause just such trouble. His presidential address was actually quite calm, which no doubt led to increased outrage at Tyndall—even *Huxley* had been nonpartisan.[150] Of course Huxley and the X-Club had leveled many such accusations against their enemies over the years, and it was certainly not the case that one group or the other was particularly guilty of this. Rather, it is important to note the values that underlay these objections, and how each group saw them as essential to science.

Conclusion

It should come as no surprise that Huxley's antiorthodoxy was closely associated with freedom of thought and scientific investigation. Perhaps more surprising was that he found those values mirrored in Maurice's

Christian educational mission, which created a space in which he was willing to work. Those same values were found often in Victorian Christianity, particularly among working theistic scientists such as Maxwell. We think of intellectual freedom and antisectarianism as being associated with the scientific naturalists because Huxley and friends explicitly made that case, not because it was necessarily so. Religious values provided the foundation for these ideals quite effectively. They were constantly invoked in scientific practice, and neither group could imagine science being done without them. However, these ideals also provided a focus for anger when it seemed that they were not being observed. Huxley's numerous screeds against orthodox power were classic examples of this, but the same reactions appeared on the other side after the Belfast Address. These values, usually transparent among scientists, become visible here only because of concerns on both sides about their violation. These concerns were largely rhetorical or hypothetical (at least with respect to scientists themselves), but the concerns loomed so large precisely because of the shared values. Everyone was keenly aware of what was necessary for the community of science to function, and the use of military imagery and aggressive speeches seemed to threaten those shared values. The theism-naturalism axis here was almost secondary—anyone trying to use institutional authority to compel belief would have triggered outrage. But as with uniformity, members of each side tried to seize intellectual freedom as theirs, and accused their opposites of endangering it.

CHAPTER SIX

Free Will and Natural Laws

The question was whether you could decide what to have for dinner. Did you have the freedom to choose? Victorian society's base assumption was that the soul and will could act freely, whether to select a meal or to accept divine grace. Being divinely created and endowed, the soul was qualitatively different from the crude matter around it and was thus exempt from having all its future states already determined as a rolling billiard ball would.[1] The mystery of how the soul drove the body gave rise to detailed, largely fruitless analyses, but was widely accepted as obviously true.[2]

But there was another option.

Perhaps the mind was not so different from the billiard ball. Perhaps the natural laws that allowed exact prediction of the motion of the planets also constrained the thoughts in our heads, making our choice of dinner an inevitable result of hidden processes. Conceptually, this was a straightforward move. A scientific investigator simply had to extend the uniformity of nature (as discussed in chapter 2) to the mind. There were many precedents for a science of mind, though confusion remained about precisely what it meant to have the mind as a scientific object.[3]

Huxley and the scientific naturalists took this even further. Applying the uniformity of nature to the mind, they said, demanded that animals and humans be considered as *automata*.[4] The original Greek term meant a self-moving object, but in the eighteenth century it came to refer to an entity incapable of free will, a soulless machine. The stupid, the oppressed, the conforming aristocrat, the inflexible tyrant were all automata.[5] Huxley's choice of this term thus explicitly invoked the political, social, and theological stakes that the Enlightenment had invested in human volition. To most contemporaries, moral order seemed impossible without an immortal soul able to affect its environment. Roger Smith argues that, even beyond moral order, the freedom of the will seemed essential for the universe to be understood in human terms at all. Without volition, there could be no purpose to life.[6]

It was on this issue—freedom of the will—that we can see the formation of the deepest fractures between theism and naturalism in Victorian science. This chapter will not survey the voluminous Victorian literature on free will. Rather, it will focus on the will as an expanding boundary between naturalistic and theistic approaches to science. The point of contest was whether it was acceptable and meaningful to include the mind in the uniformity of nature. Was consciousness an object governed by natural laws? Or was it of a different order entirely? Theistic and naturalistic scientists had been able to find common ground in a lawful nature, the role of hypotheses, educational systems, and intellectual freedom. But free will was the fault line from which they began to diverge profoundly.

The Victorian Mind-Body

The first half of the nineteenth century saw remarkable shifts in the understanding of the relationship between the human body and mind. The growing acceptance that the study of the mind should focus on the material structures of the brain and nervous system can be seen in (and, it has been argued, was driven by) contemporary frameworks such as phrenology.[7] An important corollary to an embodied mind was the application of experimental methods to physiology and psychology. Johannes Mueller's research in this area was enormously important, demonstrating that controlled laboratory work could yield important insights into human functions. He also taught many influential students, including Rudolf Virchow, F. G. J. Henle, Du Bois-Reymond, and Helmholtz, who all evangelized for his experimental approach.[8] Du Bois-Reymond's 1848 essay "Über die Lebenskraft" demanded a physiology founded on physics and chemistry of the organism, not vague ideas of organic exceptionalism.[9] These laboratory explorations were highly fruitful, but often came at the cost of devaluing introspection as a tool for studying the mind.[10] Measurement had become king and the brain was considered as only a physical entity. Physiological psychology became a decisive tool for the naturalistic understanding of humanity.[11]

A critical step was the establishment of the reflex as a fundamental aspect of the nervous system.[12] In 1832 Marshall Hall coined the term *reflex arc* to suggest how a sensory nerve could send a signal that "reflected" from the limb to the spine, and then back to the motor nerves, thus allowing for action independent of sensation or volition. The young physiologist W. B. Carpenter took up Hall's ideas and integrated Thomas Laycock's work as well. He fleshed out the idea of reflex as a basic function of the body and began extending it to psychology.[13] One of the major questions was whether reflex action existed above the brain stem—that is, did it appear in the areas where higher intellectual functions were localized? Carpenter became convinced that phenomena such as mesmerism, electrobiology, and hypnosis showed that "mental reflexes" must exist.[14] He developed categories in between reflex and volitional acts, such as his "ideo-motor" actions.[15] These unconscious behaviors raised difficult issues about the nature of the human will. Through the 1840s and 1850s, the will became a subject of increasing interest among men of science studying the human body.

Many philosophers interested in association psychology (the general assumption that all experience could be accounted for by combinations of sensations and perceptions) sought to fuse their systems with this new physiological approach. Alexander Bain's *The Senses and the Intellect* (1855) and *The Emotions and the Will* (1859), and Herbert Spencer's *Principles of Psychology* (1855) were highly influential efforts to build sophisticated systems for understanding the human mind.[16] Carpenter's *Principles of Human Physiology* was more restricted in scope, but became read enormously widely as the standard text in British physiology for decades. The goal of all of these systematizers was to present a more law-like portrait of human nature, both physical and mental. The incorporation of the human mind into the universe of natural law was a profound move, with great cultural significance.[17]

Into this environment came the young Huxley, literally fresh off the boat. After his years on the *Rattlesnake* he returned to England looking to make his name in the life sciences. Along with his interest in comparative physiology and jellyfish he brought a long-standing concern with the relationship between mind and body. Even his teenage journal records speculation about the interaction of matter and soul.[18] As his career developed, he increasingly addressed this problem through the lens of the body as a machine. Norton Wise has noted that Huxley originally wanted to be a mechanical engineer, and he called physiology "the mechanical engineering of living machines."[19] Carpenter's reflex actions made animals look more like machines than ever, and Huxley thought of himself as following in his footsteps. Huxley reviewed Carpenter's *Comparative Physiology* for the *Westminster Review* in 1855.[20] His anonymous essay praised the book as "among the most advanced works on the subject" and as displaying "the methods and criteria of all sound Physiology and Biology." Carpenter's great contribution, he wrote, was to bring together individual laws of nature into "a harmonious body of statutes— the Institutes of Nature."[21] He was even willing to forgive the author's error about the tail structure of prehistoric fish, since the error still argued against the progressionist theory of animal development.

Huxley embraced Carpenter's *Principles of Human Physiology* as his standard text. It was, he said, the first book that laid the groundwork for "a rational, that is to say, a physiological psychology."[22] The book stressed that the human body was composed of the same substances found in the inorganic universe, devoting an entire chapter to the topic.[23] Carpenter frequently referred to "vital force" but cautioned that it was simply one more physical force added to the suite of gravity, electricity, and magnetism, not an entity divorced from the material world.[24] The core for Huxley's "physiological psychology," however, was the extensive description of reflex actions and the law-like operation of the nervous system: "Thus we see that the nervous force, itself excited by impressions of a physical nature, can determine mental changes; whilst, conversely, certain states of mind, by exciting the nervous force, can effect changes in the bodily fabric, and, through this, upon the external objects within its reach."[25] Carpenter's text was so canonical to Huxley that, when he needed a medical authority to refute the resurrection of Jesus, he quoted it verbatim.[26]

Carpenter's work provided critical resources for two foundations of Huxley's approach to the problem of human volition. First, that the human body was subject to all the same forces and laws as the inorganic world. Second, that the human *mind* was subject to all the same forces and laws as the inorganic world. The following sections will examine his approach to these issues of uniformity, as the details were critical to his conclusions about the nature of the free will problem.

Some of Huxley's early writings, such as the 1854 "On the Educational Value of the Natural History Sciences," tried to draw a strong contrast between living and nonliving matter, in terms similar to the "vital force" described by Carpenter.[27] He later disavowed such ideas, instead stressing "one of the great tendencies of modern biological thought," the inclusion of living things in the uniformity of nature.[28] His classic argument for this was presented in an 1863 lecture to workingmen as "On Our Knowledge of the Causes of the Phenomena of Organic Nature."

As he often did, Huxley focused his lecture on a subject completely familiar to his audience: a horse. After discussing the details of equine anatomy, he wanted to explain how the parts worked together.

> A horse is not a mere dead structure: it is an active, living, working machine. Hitherto we have, as it were, been looking at a steam-engine with the fires out, and nothing in the boiler; but the body of the living animal is a beautifully-formed active machine, and every part has its different work to do in the working of that machine, which is what we call its life.[29]

To understand the nature of life, he said, you need think of the horse as a machine. The horse that seemed to be eating was really a mill grinding corn into a chemical digester, which pushed fuel into a series of pipes. The substance of the beast was protein, itself a compound of familiar carbon, hydrogen, and oxygen. These inorganic materials were unchanged when they became part of a living thing. The organic and inorganic worlds were one: "There is thus a constant circulation from one to the other, a continual formation of organic life from inorganic matters, and as constant a return of the matter of living bodies to the inorganic world."[30] The materials of life were nothing special. They simply happened to be part of the horse rather than its shoes.

More important than establishing a unity of substance, though, was showing that the *forces* of life were

> either identical with those which exist in the inorganic world, or they are convertible into them; I mean in just the same sense as the researches of physical philosophers have shown that heat is convertible into electricity, that electricity is convertible into magnetism, magnetism into mechanical force or chemical force, and any one of them with the other, each being measurable in terms of the other,—even so, I say, that great law is applicable to the living world.[31]

His description of organic forces as simply one more converted force was very similar to Carpenter's proposal. Huxley pointed to Du Bois-Reymond's researches showing electrical states in active nerves as even further evidence. Higher life and primitive life, and even life and non-life, were different only in terms of structure and complexity. A few years later, Tyndall made the case to the BA that if one could accept that the structure of salt crystals came from natural forces, then it was a simple matter of continuity for the same to be true of the structure of corn.[32] Living bodies were not exempt from the uniformity of natural laws.

A particularly important natural law for physiology and psychology was that of the conservation of energy. As Frank Turner showed, that principle became one of the pillars of the naturalistic worldview, not least because of its enormous impact on questions of mind-body interaction.[33] It gave rise to several new research programs examining the thermodynamics of life in varying levels of detail, from "black-boxed" organisms to cellular processes.[34] Efforts to apply energy concepts to animal physiology began in Germany, with J. Robert Mayer and Hermann Helmholtz being particularly influential.[35]

Measuring energy processes inside bodies turned out to be quite challenging, and a host of new measurement techniques and monitoring equipment was developed. Some investigators, such as the Alsatian engineer Gustave Hirn, took the straightforwardly crude approach of building a calorimeter big enough to hold a person. The subject worked a treadwheel and had his oxygen input and carbonic acid output measured to see if his work obeyed the rules of engines.[36] Experiments of this type often returned confusing results, particularly once the actual physiology of nutrition and energy production was considered. Huxley's friend Edwin Lankester was a pioneer in Britain of considering nutrition in terms of energy: "Man, in fact, exists in consequence of the physical and chemical changes that go on in his body as the result of taking food."[37] He argued strongly for the identity of the chemical reactions that produced light in a candle and heat in an organism. Even the function of the mind, he said, must be considered to be results of transformed food:

> Nervous matter consists of about 7 per cent. of albumen, not a very large quantity, but still this matter must be regarded by us as an intensely interesting product, because it is the material by which we are put in relation with the external world. It is this which enables us to see, to hear, to taste, to smell, and to feel. It is this which enables us to think, to feel, and to be conscious of our existence. All this depends upon the condition of the albumen in our system. Although we may sit at our breakfast partaking of the daily egg, thinking of other things, yet the laws by which the egg becomes the source of our thought is worth a thought.[38]

Edward Frankland, the chemist and X-Club member, conceived and executed a number of critical experiments to determine the actual energy conversion processes at work in the human body.[39] His experiments, often carried out on the top of mountains, were complicated but valuable. While not demonstrating exact energy conservation, they did dispute continuing attempts to show nonenergetic processes (typically vitalistic) in the human body. Some physiologists and doctors tried to work around the uncertainties of these measurements by experimenting on populations for whom they could control nutritional input and energy expenditures exactly: prisoners. Samuel Houghton set up prisoners conducting "shot drill" (a military exercise involving the lifting and moving of standard weights) under controlled conditions.[40] Helmholtz's estimate that the human body functioned with only one-eighth efficiency was based on the measurements of Dr. Edward Smith, conducted on the treadwheel of a London prison.[41]

Near the end of his life, in his royal jubilee essay, Huxley placed energy physics as one of the foundations of science: "The doctrine of the conservation of energy which I have endeavoured to illustrate is thus defined by the late Clerk Maxwell: 'The total energy of any body or system of bodies is a quantity which can neither be increased nor diminished by any mutual action of such bodies, though it may be transformed into any one of the forms of which energy is susceptible.'"[42] The breadth and scope of this doctrine meant that

> living beings, in so far as they are material, are all molar or molecular motions, these are included under the general law. A living body is a machine by which energy is transformed in the same sense as a steam-engine is so, and all its movements, molar and molecular, are to be accounted for by the energy which is supplied to it.[43]

Looking back on the previous generation of physiology, Huxley triumphantly declared that humans, just as much as the horse, were fuel-consuming, energy-limited machines. Was there anything left to set humans apart from the uniform course of nature?

Uniformity and Mind

The breach in the defenses of the sui generis human mind was the brain. As Huxley noted in his 1863 lecture, Helmholtz and Du Bois-Reymond had correlated electrical impulses in the nervous system with muscular movements of the sort normally associated with thought and intention. Herbert Spencer had insisted on the need to correlate measurable nerve phenomena with states of consciousness in his *Principles of Psychology*.[44] The hypothesis that thought had its origin in physical processes was certainly not new, and Huxley had many resources to drawn on to support it—he even enlisted David Hume. That philosopher, he wrote, "fully adopted the conclusion to which all that we know of psychological physiology tends, that the origin of the elements of consciousness, no less than that of all its other states, is to be sought in bodily changes, the seat of which can only be placed in the brain."[45] He certainly did not claim that all mental processes were understood; rather, he continually argued that the physiology of the nervous system simply showed that the mind could be treated as a scientific object: "And thus from the region of disorderly mystery, which is the domain of ignorance, another vast province has been added to science, the realm of orderly mystery."[46] Once physiologists could measure nerve force the way they measured the length of a limb, the mind could be treated as wholly within the uniformity of nature.

He did not deny that the mind existed, or that consciousness was available to introspection. Rather, he wanted to emphasize the dependence of those on the physical matter of the nervous system:

> Unless the nerve-elements of the retina, of the optic nerve, of the brain, of the spinal cord, and of the nerves of the arms, went through certain physical changes in due order and correlation, the various states of consciousness which have been enumerated would not make their appearance. So that in this, as in all other intellectual operations, we have to distinguish two sets of successive changes—one in the physical basis of consciousness, and the other in consciousness itself; one set which may, and doubtless will, in course of time, be followed through all their complexities by the anatomist and the physicist, and one of which only the man himself can have immediate knowledge.[47]

The dependence of mind on matter became a serious issue for Huxley in defending Darwin's theory, particularly around the publication of *The Descent of Man*, as some critics tried to object that human mental capacity could not have evolved by physical means. He maintained that a proper physiological perspective inevitably led to a certain conclusion:

> I am not aware that there is any one who doubts that, in the proper physiolog-
> ical sense of the word function, consciousness, in certain forms at any rate, is
> a cerebral function . . . it is the function of muscle to give rise to motion . . .
> why is the production of a state of consciousness in the other case not to be
> called a function of the cerebral substance?

He acknowledged that some objected to this position as materialistic.
With his typical caginess, Huxley toyed with the meaning of the term
until only "rhetorical sciolists" could object to its use.[48]

An important consequence of this association of mind and brain
came through comparative physiology. The human brain, it seemed, was
nothing special:

> Structure for structure, down to the minutest microscopical details, the eye,
> the ear, the olfactory organs, the nerves, the spinal cord, the brain of an ape,
> or of a dog, correspond with the same organs in the human subject. Cut a
> nerve, and the evidence of paralysis, or of insensibility, is the same in the two
> cases; apply pressure to the brain, or administer a narcotic, and the signs of
> intelligence disappear in the one as in the other. Whatever reason we have for
> believing that the changes which take place in the normal cerebral substance
> of man give rise to states of consciousness, the same reason exists for the be-
> lief that the modes of motion of the cerebral substance of an ape, or of a dog,
> produce like effects.

That is, a dog had all the equipment necessary for the higher functions
normally reserved for man. Dogs seemed to have sensations, dreams,
memories, and expectations.[49] Animals regularly performed actions
that, if executed by a human, would be marked as signs of consciousness.
Huxley used the continuity of both anatomy and behavior to extend results from animal behavior to human minds.

This was the foundation of his famous 1870 essay for the Metaphysical Society on the nature of a frog's soul. He described a series of experiments on a vivisected frog, whose nervous system was severed in various places. Even when the brain was disconnected from the body, the
frog's leg still moved to avoid irritation. The persistence of reflex actions
despite the injury made it "clear that they are effected independently of
any sensation or volition in the rest of the body." Huxley maintained that
the same purposive movements could be observed in a human with a spinal injury—even if the leg had no contact with "his consciousness and

his volition."[50] Such actions, he said, were not anomalous; they were the standard. "Every one will discover, if he considers his own actions, that he is constantly performing operations directed towards special ends of which he has no consciousness whatever. And therefore it must be granted that it is possible that all the far less complex actions of the frog *may* be equally devoid of consciousness. Whether they are so or not, is a point on which no positive evidence is attainable, or even conceivable."[51]

Such automatic actions were well known in physiology by this point, having been extensively documented by Carpenter in various editions of *Principles of Human Physiology.* Carpenter noted that most movements of the body, such as respiration, were executed without any intervention of consciousness.[52] Movements triggered by certain kinds of mental states, such as emotions, occur without mental volition. And even voluntarily controlled movements clearly did not directly move the muscles of the body: "Although it has been customary to regard the Will as directly operating on the muscular system, yet we shall hereafter find reason to consider it as exerting its power through the medium of the Automatic apparatus, to which its determinations are transmitted, and by which they are carried into execution."[53] We might decide to walk forward, but we certainly do not command the individual muscles needed. Otherwise, detailed anatomical knowledge would be necessary to do anything from picking up a cup to speaking a word.

These unconscious movements were used by Huxley as the foundation for far-reaching claims about the nature of animals and humans: his theory of automatism. The various elements of the theory had appeared many times in his lectures and writing, but it was not until the 1874 Belfast meeting of the BA that he presented it with full rhetorical vigor. Tyndall had been planning his controversial presidential address, and asked Huxley to compose a similar barn burner. Huxley was not entirely sold on his friend's plan, but declared himself "at your [Tyndall's] disposition for whatever you want me to do."[54] The result was the infamous "On the Hypothesis That Animals Are Automata, and Its History." This was classic Huxley: a verbose, pointed historical narrative about the triumph of naturalism. He chose Descartes as his avatar (in previous lectures he had named Hume and Harvey as the pioneers of automatism), declaring that the French philosopher had been a physiologist of the first rank: "I shall now endeavour to show that a series of propositions, which constitute the foundation and essence of the modern physiology of the nervous system, are fully expressed and illustrated in the works of Descartes."

The first proposition: *"The brain is the organ of sensation, thought, and emotion; that is to say, some change in the condition of the matter of this organ is the invariable antecedent of the state of consciousness to which each of these terms is applied."*[55]

He presented Descartes's well-known mechanical analysis of animals as a prescient distillation of all that Victorian physiology had discovered. The Frenchman had found that the brain was the seat of consciousness, that it was excited by sensory nerves, and that muscles were triggered by signals from the brain. His argument, Huxley said, was perfectly clear (and remarkably similar to his own):

> He starts from reflex action in man, from the unquestionable fact that, in ourselves, co-ordinate, purposive, actions may take place, without the intervention of consciousness or volition, or even contrary to the latter. As actions of a certain degree of complexity are brought about by mere mechanism, why may not actions of still greater complexity be the result of a more refined mechanism? What proof is there that brutes are other than a superior race of marionettes, which eat without pleasure, cry without pain, desire nothing, know nothing, and only simulate intelligence as a bee simulates a mathematician?[56]

The mutilated but reactive frog made a reappearance, though it was quickly replaced by a similar figure more personally relevant to his audience. This Sergeant "F—" had been wounded during fierce fighting at the Battle of Bazeilles. He had received a bullet to the brain and managed to gain revenge on the Prussian who shot him before collapsing. After years of recovery, he now had only a slight weakness in the right side of his body. However, he was also subject to "periodical disturbances of the functions of the brain."[57]

Outside these disturbances, the sergeant led a normal, healthy life. When they appeared, however, his behavior became strange. After some discomfort of the head

> he walks about as usual; but, if he is in a new place, or if obstacles are intentionally placed in his way, he stumbles gently against them, stops, and then, feeling over the objects with his hands, passes on one side of them. He offers no resistance to any change of direction which may be impressed upon him, or to the forcible acceleration or retardation of his movements. He eats,

drinks, smokes, walks about, dresses and undresses himself, rises and goes to bed at the accustomed hours.

Somewhat peculiar. Upon medical examination, even stranger behavior was displayed:

> Nevertheless, pins may be run into his body, or strong electric shocks sent through it, without causing the least indication of pain; no odorous substance, pleasant or unpleasant, makes the least impression; he eats and drinks with avidity whatever is offered, and takes asafœtida, or vinegar, or quinine, as readily as water; no noise affects him; and light influences him only under certain conditions.

The sergeant seemed to be trapped in some liminal state. He could carry out complex movements, even those typically associated with consciousness such as singing and writing letters. But he displayed no reaction to unpleasant stimuli or having his writing ink replaced by water. Was he truly conscious, or "is consciousness utterly absent, the man being reduced to an insensible mechanism?"[58]

The actions of Sergeant F were perhaps more understandable in their similarity to the actions of animals. As Descartes suggested, both were automata. Animals "may be more or less conscious, sensitive, automata; and the view that they are such conscious machines is that which is implicitly, or explicitly, adopted by most persons." What was meant by calling them automata was that their behavior was determined solely by the physical organization of their bodies.

> We believe, in short, that they are machines, one part of which (the nervous system) not only sets the rest in motion, and co-ordinates its movements in relation with changes in surrounding bodies, but is provided with special apparatus, the function of which is the calling into existence of those states of consciousness which are termed sensations, emotions, and ideas.

Huxley reminded his listeners that if they doubted that a physical state was the immediate antecedent to a state of consciousness, they were welcome to run a pin into themselves.[59]

It was his conclusion that drew the most attention, and the most alarm. His statement that changes in the brains of animals were responsible for changes in their consciousness could be accepted widely, but the inverse raised more troubling issues:

> Is there any evidence that these states of consciousness may, conversely, cause those molecular changes which give rise to muscular motion? I see no such evidence. The frog walks, hops, swims, and goes through his gymnastic performances quite as well without consciousness, and consequently without volition, as with it; and, if a frog, in his natural state, possesses anything corresponding with what we call volition, there is no reason to think that it is anything but a concomitant of the molecular changes in the brain which form part of the series involved in the production of motion.

Huxley had just spent a great deal of time explaining how the motions of the frog were fundamentally identical to the actions of Sergeant F. If the frog could function automatically, and the sergeant functioned as the frog, did human volition mean anything? Still declining to refer directly to the human mind, Huxley answered his own question:

> The consciousness of brutes would appear to be related to the mechanism of their body simply as a collateral product of its working, and to be as completely without any power of modifying that working as the steam-whistle which accompanies the work of a locomotive engine is without influence upon its machinery. Their volition, if they have any, is an emotion indicative of physical changes, not a cause of such changes.[60]

The mind was not only downgraded to an impotent shadow of the body; it was of no more significance than the whistle of a train. An indicator of activity, certainly, but of no importance to the functioning of the machine.

Huxley finally moved to explicitly connect animal automatism to humans. Before he delivered, however, he casually wondered if he would be attacked by clerical authorities for his views on animal minds.

> The question is, I believe, a perfectly open one, and I feel happy in running no risk of either Papal or Presbyterian condemnation for the views which I have ventured to put forward. And there are so very few interesting questions which one is, at present, allowed to think out scientifically—to go as far as reason leads, and stop where evidence comes to an end—without speedily being deafened by the tattoo of "the drum ecclesiastic"—that I have luxuriated in my rare freedom.[61]

Huxley's portrayal of himself as a humble investigator hoping to avoid orthodox censure was, of course, simply the lead-in to his divisive conclusion:

> It is quite true that, to the best of my judgment, the argumentation which applies to brutes holds equally good of men; and, therefore, that all states of consciousness in us, as in them, are immediately caused by molecular changes of the brain-substance. It seems to me that in men, as in brutes, there is no proof that any state of consciousness is the cause of change in the motion of the matter of the organism . . . it follows that our mental conditions are simply the symbols in consciousness of the changes which take place automatically in the organism; and that, to take an extreme illustration, the feeling we call volition is not the cause of a voluntary act, but the symbol of that state of the brain which is the immediate cause of that act.

Human volition was, indeed, illusory. There was no way in which consciousness could be considered the source of action or choice. Humans were "conscious automata," aware but unable to interfere. The mind was just one unremarkable part of "the great series of causes and effects which, in unbroken continuity, composes that which is, and has been, and shall be—the sum of existence."[62]

And how did the mind behave? Did consciousness have a life of its own? Huxley asserted that the behaviors of the mind were just as comprehensible and predictable as any physical phenomena. Mental events were subject to a "definite order," and the study of the laws implicit in that order was psychology. That there was such an order, he maintained, "is acknowledged by every sane man."[63] He refused to distinguish between the material and the mental: "All the phenomena of nature are either material or immaterial, physical or mental; and there is no science, except such as consists in the knowledge of one or other of these groups of natural objects, and of the relations which obtain between them."[64] The uniformity of nature applied equally to both. On whatever grounds physiology claimed to be a science, psychology had the same grounds.[65] As there was an anatomy of the body, so there was an anatomy of the mind.

The scientific naturalists were relentless in claiming the human consciousness for the uniformity of nature. They were unwilling to accept that the mind functioned differently from the material world. That said, they were quick to demur that they did not have all the answers—mys-

teries remained. Psychology was yet a young science, and could not be
expected to have results as mature as, say, astronomy.[66] Tyndall warned
that, in particular, the mechanism of connection between mind and body
remained obscure: "The utmost [the philosopher] can affirm is the association of two classes of phenomena, of whose real bond of union he is in
absolute ignorance. The problem of the connection of body and soul is as
insoluble, in its modern form, as it was in the pre-scientific ages. . . . Let
us lower our heads, and acknowledge our ignorance, priest and philosopher, one and all."[67] Huxley described Hume as having discarded that
problem as unintelligible, and that even if the soul exists, "we can by no
possibility know anything about it."[68] Du Bois-Reymond's famous invocation of *ignoramus et ignorabimus* was one of the strongest statements
of this—not only did he *not* know; he could *never* know.

The Will

Despite this professed ignorance, the relationship between mind and
matter was a critical inflection point in naturalistic understandings of the
mind. This crystallized in the question of the will and its alleged freedom. To establish the terms of debate for exactly what "free will" meant,
Huxley went to Hume. He listed the various attempted solutions to the
problem of how the soul could move matter: occasionalism, Leibnizian
preestablished harmony, Berkeley's idealism.[69] He preferred the position
that the question was itself unintelligible. Hume's own definition of the
will became the start of the argument:

> I desire it may be observed, that, by the *will*, I mean nothing but *the internal
> impression we feel, and are conscious of, when we knowingly give rise to any
> new motion of our body, or new perception of our mind*. This impression, like
> the preceding ones of pride and humility, love and hatred, 'tis impossible to
> define, and needless to describe any further.[70]

Huxley suggested that this needed to be expanded to include both the
idea of an action, and the desire for the occurrence of that action. His
amended version:

> Volition is the impression which arises when the idea of a bodily or mental ac-
> tion is accompanied by the desire that the action should be accomplished. It

differs from other desires simply in the fact, that we regard ourselves as possible causes of the action desired.[71]

This gave rise to two further questions. Did volition have a cause? And did volition have any effect?

In the automata lecture, Huxley toyed with exactly what could be meant by causality and desire in this context. A good definition of "free" will, he suggested, would be that there was nothing to prevent an agent from doing what it desired to do. In that sense, surely a dog chasing a rabbit was a free agent? Once released from the leash, nothing prevented the dog from carrying out its desire. And since we could account for all the dog's actions purely on mechanical and physiological principles, in what way does free will enter into the discussion?

> But if, as is here suggested, the voluntary acts of brutes—or, in other words, the acts which they desire to perform—are as purely mechanical as the rest of their actions, and are simply accompanied by the state of consciousness called volition, the inquiry, so far as they are concerned, becomes superfluous. Their volitions do not enter into the chain of causation of their actions at all.[72]

Huxley's move here brought two gains: free will should be ascribed to animals if we ascribe it to humans, and that will was of no real significance for understanding action. Mechanical principles could give a complete explanation.

Further, the standard version of volitional self-control was demonstrably illusory. The most basic verification of free will was the control of our own bodies, to walk where we want and sit where we choose. But the machinery of our body, warned Huxley, was much more complex than it seemed.

> Suppose one wills to raise one's arm and whirl it round. Nothing is easier. But the majority of us do not know that nerves and muscles are concerned in this process; and the best anatomist among us would be amazingly perplexed, if he were called upon to direct the succession, and the relative strength, of the multitudinous nerve-changes, which are the actual causes of this very simple operation.[73]

Speaking was even worse. No one, given perfect control over the nerves regulating the mouth and larynx, could actually direct the muscles to produce a word. Thus, our sense that we direct our limbs to move and our voices to speak was simply incorrect. The actual elevation of an arm was *not* due to the simple direction of the mind. We were completely reliant on the automatic systems of the body. These examples were taken directly from Carpenter's *Principles*, and Huxley used them again and again over the years.

With the body and the mind pulled firmly within the uniformity of nature, and the will defined away, Huxley arrived at a controversial position of long standing. Commonly called determinism, sometimes necessitarianism, it was usually phrased negatively: there was no room for freedom of action in the world. The laws of nature allowed no exceptions, bringing rigid causality even to the living world. The modern sense of determinism was usually attributed to Laplace:

> Given for one instant an intelligence which could comprehend all the forces by which nature is animated and the respective situation of the beings who compose it—an intelligence sufficiently vast to submit these data to analysis—it would embrace in the same formula the movements of the greatest bodies of the universe and those of the lightest atom; for it, nothing would be uncertain and the future, as the past, would be present to its eyes.[74]

Huxley and the X-Club embraced it as a chief component of scientific naturalism. Tyndall in his Belfast Address famously declared that conservation of energy was

> that doctrine which "binds nature fast in fate," to an extent not hitherto recognised, extracting from every antecedent its equivalent consequent, from every consequent its equivalent antecedent, and bringing vital as well as physical phenomena under the dominion of that law of causal connection which, so far as the human understanding has yet pierced, asserts itself everywhere in nature.[75]

Huxley's understanding of causality was somewhat more subtle, befitting his study of Hume. The conclusions he arrived at, though, were just as strong. He placed determinism as a wholly necessary consequence of natural law:

> Let us suppose, further, that we do know more of cause and effect than a certain definite order of succession among facts, and that we have a knowl-

edge of the necessity of that succession—and hence, of necessary laws—and I, for my part, do not see what escape there is from utter materialism and necessarianism.

He argued that it was not at all clear how one might prove that a given event was not the result of a necessary cause. A noncaused action would be one that was exempt from the uniformity of nature, which was a concept that had no philosophical meaning to him. And while these questions were not subject to philosophical proof, he suggested that the history of science had shown the inevitable "extension of the province of what we call matter and causation, and the concomitant gradual banishment from all regions of human thought of what we call spirit and spontaneity."[76]

An essential part of this vision was the assumption that the future was fixed—that is, that the outcome of all future events was already determined. Even though humans might be ignorant of the course of the future, nature itself had already set it through an unbroken chain of causality. Just as past events could not be altered, neither could future ones. In the last document Huxley ever wrote, he described this:

> In thinking of the future we imagine it to be indefinite and uncertain, simply so far as we know nothing about it. Yet a little consideration should produce the conviction that the future is as definite and fixed as the past. At this moment I am writing at a certain table in a certain room at 9.15 A.M. Yesterday this was part of the future, tomorrow it will be part of the fixed and unalterable past—becomes such in fact even as I write.

One can only imagine him, on his deathbed, pondering the march of the future and the retreat of the past. Huxley placed the sound prediction of future events as a crucial element of this worldview:

> Consequently any one who possessed the power of foreseeing the future yesterday or a thousand or a million years ago, must have seen me doing this exact thing at this very time and place. In fact it is not really in our power to conceive of futurity, however remote, as other than a definite series of events a, b, c, and no other. Only we do not know what a, b, and c are. Again, if the law of causation is absolute—that is to say, if nothing comes into being by chance—future events are the consequences of present events and therefore predetermined by the latter.[77]

Successful prediction was not just of metaphysical interest to Huxley; he saw it as enormously important for the authority of science. The ability to predict, even in the abstract, was the strongest tool for extending scientific reasoning to new domains. Katharine Anderson, writing in the context of Victorian meteorology, establishes that "prediction summarized what the uniformity of nature meant for the reach of the scientific observer."[78] Huxley's lecture on Zadig placed successful prediction and retrodiction as the markers of true science, and also what made it so threatening to the orthodox.[79]

Tyndall had a particular gift for colorful language describing all the consequences of this: "With the necessary molecular data, and the chick might be deduced as rigorously and as logically from the egg, as the existence of Neptune from the disturbances of Uranus."[80] Organic matter clearly fell under the domain of determinism, but he was willing to extend it even to the mind and culture. Huxley wrote to Spencer that "a favorite problem of [Tyndall's] is—Given the molecular forces in a mutton chop, deduce Hamlet or Faust therefrom. He is confident that the Physics of the Future will solve this easily."[81]

Huxley was aware that the most difficult defense of free will to stamp out would be that based on direct experience—the unbreakable sense that one can choose what to eat for dinner, therefore free will must be real. Balfour declared it "ludicrous" to think it was illusory.[82] This subjective sense of will was impossible to observe, but equally impossible to debunk. Opponents of the "doctrine of necessity," Huxley wrote, claimed this as their impregnable fortress against uniformity. He denied that it was actually relevant to the question:

> For they rest upon the absurd presumption that the proposition, "I can do as I like," is contradictory to the doctrine of necessity. The answer is; nobody doubts that, at any rate within certain limits, you can do as you like. But what determines your likings and dislikings? Did you make your own constitution? Is it your contrivance that one thing is pleasant and another is painful? And even if it were, why did you prefer to make it after the one fashion rather than the other?[83]

People could do as they like, but not like as they like. The constraints placed by the material mind perfectly controlled what someone would want to do. There could be no chance or spontaneity involved; it was all the result of the natural laws of the mind. This created a situation,

Huxley argued, that was a perfect illusion of free will: the past history of a person's brain created conditions wherein, for example, that individual would always prefer roast beef to chicken. Then, when presented with the choice between the two entrées, the person's consciousness declares that it *chooses* the roast beef, when in fact it has simply followed the deterministic forces of its history. Consciousness, that epiphenomenal steam whistle that emerged from the steam engine of our brain, had fooled itself into imagining that it was in control of its future, when in fact it was trapped by its past.

W. K. Clifford laid particularly vigorous attacks against this argument from direct experience of the will. From that subjectivist perspective, the will was simply a state of consciousness. And Clifford denied that an interior state of consciousness could be a valid source of scientific information. "The state of a man's brain and the actions which go along with it are things which every other man can perceive, observe, measure, and tabulate." Therefore, they could be part of a scientific theory such as automatism.

> But the state of a man's own consciousness is known to him only, and not to any other person. Things which appear to us and which we can observe are called *objects* or *phenomena*. Facts in a man's consciousness are not objects or phenomena to any other man; they are capable of being observed only by him. We have no possible ground, therefore, for speaking of another man's consciousness as in any sense a part of the physical world or phenomena.[84]

Anything that existed solely in consciousness could not be part of science. Thus, a statement such as "I can verify that I have free will" was not a matter of truth or untruth. Instead, it was simply nonsense. With no way to perceive someone else's volition, it could not be a subject for scientific discussion.

An even greater riposte waited for orthodox Christian objections to determinism. In his retrospective on the reception of the *Origin*, Huxley denied that science had created any neccessitarian difficulties for theology. Rather, it was a problem of theologians' own invention:

> In theological science, as a matter of fact, it has created none. Not a solitary problem presents itself to the philosophical Theist, at the present day, which has not existed from the time that philosophers began to think out the logical grounds and the logical consequences of Theism. All the real or imaginary

perplexities which flow from the conception of the universe as a determinate mechanism, are equally involved in the assumption of an Eternal, Omnipotent and Omniscient Deity.

Providence was no different from the universality of causation. They both assumed that events resulted from foreknowledge. The only difference was that men of science refused to accept the traditional misleading solutions.

> The angels in "Paradise Lost" would have found the task of enlightening Adam upon the mysteries of "Fate, Foreknowledge, and Free-will," not a whit more difficult, if their pupil had been educated in a "Real-schule" and trained in every laboratory of a modern university. In respect of the great problems of Philosophy, the post-Darwinian generation is, in one sense, exactly where the præ-Darwinian generations were. They remain insoluble. But the present generation has the advantage of being better provided with the means of freeing itself from the tyranny of certain sham solutions.[85]

Natural science did not "not dictate sundry passages in the Epistle to the Romans, nor whisper in the ear of Augustine of Hippo." Protestants did not seem particularly concerned that Luther declared man to be a beast of burden who went wherever his rider directed (be it God or the devil). And did Calvin or Jonathan Edwards understand the conservation of energy? Certainly not, but "who has ever put the case of Determinism better or more unanswerably?"[86]

As always, Huxley delighted in turning theologians against their own. He could then paint attacks on him as simple prejudice—if Balfour truly objected to determinism, why was he not attacking Luther? In truth, was this not simply one more example of orthodoxy gone awry? Augustine and Calvin were happy to see man as a conscious automaton. Given that, "Is it not just possible that smaller folk may be wrong?" Huxley noted that a "large share of this clamour is raised by the clergy of one denomination or another," and "that it really would be well if ecclesiastical persons would reflect that ordination, whatever deep-seated graces it may confer, has never been observed to be followed by any visible increase in the learning or the logic of its subject."[87] The clergy were so ignorant of even their own subject that it was sheer foolishness for them to speak on the meaning of science.

This strategy of appropriating historical figures as protonaturalists (or, less charitably, as mouthpieces for himself) was one Huxley often displayed when attacking orthodoxy. As on other issues, he found Joseph Priestley to be a particularly effective figure. Priestley's "Disquisitions Relating to Matter and Spirit" and his "Doctrine of Philosophical Necessity Illustrated" were described as "among the most powerful, clear, and unflinching expositions of materialism and necessarianism which exist in the English language, and are still well worth reading." Huxley painted eighteenth-century opinions on freedom of the will and existence of the soul as essentially identical to Victorian ones, and presented himself as persecuted on the same grounds as Priestley. In particular, he protested that both had been tarred with scurrilous labels:

> If a man is a materialist; or, if good authorities say he is and must be so, in
> spite of his assertion to the contrary; or, if he acknowledge himself unable to
> see good reasons for believing in the natural immortality of man, respectable
> folks look upon him as an unsafe neighbour of a cash-box, as an actual or po-
> tential sensualist, the more virtuous in outward seeming, the more certainly
> loaded with secret "grave personal sins."[88]

Calling names, he suggested, was the last desperate move of orthodoxy unwilling to face truth.

Consequences

Huxley's automaton theory stirred deep controversy. It was one thing for Huxley to tell people they were animals; it was something else entirely for him to tell people they were machines. Even beyond Darwin, the steam-whistle model of deterministic consciousness seemed to annihilate the last vestiges of human uniqueness. With the destruction of the possibility of an efficacious soul came a host of psychological and social threats.[89] These were listed particularly passionately by W. S. Lilly in his essay on materialism and morality in the *Fortnightly Review*. He explained how a wave of materialism had come across Europe under the diverse names of positivism, determinism, and agnosticism. Huxley's declaration that consciousness was just a result of material causes was described as the sharp edge of this attack.[90] The philosophy of determinism, Lilly argued, had led to a moral crisis. Without the concept of human freedom, the very foundations of social stability, morality, and religion were destroyed. For example, without free will, crime and justice could have no significance:

> This conception of human freedom underlies the notion of crime. Yes; the sense of crime is bound up with the belief in man's power of choice, and in his obligation to choose rightly. Where there is no faculty to judge of acts, as right or wrong, and to elect between them, as in a young child or a lunatic, there is no criminal responsibility, for there are no persons.

Right and wrong could mean nothing if there was not a sense of being able to choose between them. A will must be able to choose between two alternatives or there could be no moral accountability. *"Ought* is a meaningless word without *Can*."[91] How could people be praised for following rules, or punished for breaking them, if they had no choice in the matter? If they were directed to steal only by deterministic electrical impulses in their brain, what would be the point of trying and imprisoning them? They were no more blameworthy than a dog chasing a rabbit. How could any moral strictures remain meaningful without free will?

Lilly noted that someone might object to his concerns by finding a materialist who behaved morally. Huxley, he said, might be a perfectly nice person. But even if this were true, it said nothing about the moral results of materialism. Whatever morals that man of science held most certainly did not come from his deterministic worldview: "I do not myself know anything of the early history of this illustrious man. But I suppose that, like the rest of us, he was brought up upon the Catechism. At all events, I am quite sure that he is the product of many generations of Christian progenitors." Huxley and his materialist friends had grown up in Christian civilization. Their status as moral men was due to their upbringing, not their disruptive ideas. However, if their ideas went unchallenged, that Christian civilization would come to an end. Huxley might be moral, but would his grandchildren be?[92]

The roots of modern civilization came, Lilly argued, from a recognition of the truth of the human soul. He denied that he was a Christian dogmatist, though he could not allow that humans were wholly material beings. He insisted that "morality can have root only in the spiritual nature of man. If from that happy soil, watered by the river of life, and refreshed by the dews of heaven, you transplant it to the rocks and sands of Materialism, wither and die it must."[93]

Huxley was unfazed by these objections. When accused by Mivart

that "acts, unaccompanied by mental acts of conscious will directed towards the fulfilment of duty [are] absolutely destitute of the most in-
cipient degree of real or formal goodness," he denied that this conclusion followed from moral philosophy. He pointed out that deterministic
forces of the mind were often called "character"—they were those aspects of someone that influenced the person's actions. Acting in accordance with one's character surely should not be considered amoral. Huxley said that Mill and Carlyle both agreed with him. He found it

> extremely hard to reconcile Mr. Mivart's dictum with that noble summary
> of the whole duty of man—"Thou shalt love the Lord thy God with all thy
> heart, and with all thy soul, and with all thy strength; and thou shalt love
> thy neighbour as thyself." According to Mr. Mivart's definition, the man who
> loves God and his neighbour, and, out of sheer love and affection for both,
> does all he can to please them, is, nevertheless, destitute of a particle of real
> goodness.[94]

Surely, Huxley said, Mivart could not mean that acting out of love, rather
than completely free choice, was not moral.

As he often did, Huxley seized on Hume to defend his position:
"When any opinion leads to absurdity, it is certainly false; but it is not
certain that an opinion is false because it is of dangerous consequence."
He refused to accept that he should not pursue an idea only because of
its alleged danger to "religion and morality."[95] In any case, there was
no real danger of disrupting morality. Hume's insight into the relationship of determinism and morality, Huxley said, was in the very concept
of causality. Causality, the bugbear that raised the problem in the first
place, was actually necessary for moral judgment:

> If a man is found by the police busy with "jemmy" and dark lantern at a jew-
> eller's shop door over night, the magistrate before whom he is brought the
> next morning, reasons from those effects to their causes in the fellow's bur-
> glarious ideas and volitions, with perfect confidence, and punishes him ac-
> cordingly. And it is quite clear that such a proceeding would be grossly un-
> just, if the links of the logical process were other than necessarily connected
> together. The advocate who should attempt to get the man off on the plea
> that his client need not necessarily have had a felonious intent, would hardly
> waste his time more, if he tried to prove that the sum of all the angles of a tri-
> angle is not two right angles, but three.[96]

Intent, presumably a necessary part of any system of justice, required causality. And once there was causality, there was no escape from natural law and therefore automatism. Note Huxley's extraordinary equivalence that determinism was as inescapable as the basic laws of mathematics.

Huxley said that understanding the causes of someone's actions should not affect the consequences of those actions. Simply because we understood the forces that drove a man to theft and murder would not prevent punishment:

> Does any sane man imagine that any quantity of physiological analysis will lead people to think breaking their legs or putting their hands into the fire desirable? And when men really believe that breaches of the moral law involve their penalties as surely as do breaches of the physical law, is it to be supposed that even the very firmest disposal of their moral truths upon "a bare physical or physiological basis" will tempt them to incur those penalties?[97]

There was no reason to assume that justice required ignorance, which was essentially what the claim of free will entailed. Again, cause and effect was essential—certain actions result in certain punishments. In terms of the morality of his own behavior, Huxley was more than happy to be an automaton:

> I protest that if some great Power would agree to make me always think what is true and do what is right, on condition of being turned into a sort of clock and wound up every morning before I got out of bed, I should instantly close with the offer. The only freedom I care about is the freedom to do right; the freedom to do wrong I am ready to part with on the cheapest terms to any one who will take it of me.[98]

Carpenter Redux

A particularly vigorous critic of automatism was, ironically, William Carpenter. Even though his work was explicitly held up as the foundation of Huxley's claims, Carpenter himself was a devout Unitarian who strove to keep his physiological psychology from being associated with naturalism.[99] Even in his early works, such as the editions of *Principles of Human Psychology* that Huxley relied on, he argued for a human soul

with free will that was consistent with Christian doctrine. In the introduction to *Principles* he told his readers that the soul "'is that side of our
nature which is in relation with the infinite'; and it is the existence of this
relation, in whatever way we may describe it, which seems to constitute
the distinctive peculiarity of Man."[100] When discussing the impossibility of human minds communicating directly with each other, he distinguished between material life and spiritual:

> On the other hand, that in a future state of being, the communion of mind
> with mind will be more intimate, and that Man will be admitted into more
> immediate converse with the Supreme Intelligence, appears to be alike the
> teaching of the most comprehensive Philosophical inquiries, and of the most
> direct Revelation of the Divinity.[101]

Carpenter did not see these considerations as damaging the book's over-
all argument for the application of uniform natural laws to the human
body. He was a pioneer in presenting the body's reflex and automatic activities, and happily acknowledged human "tendencies to thought" and
"uniformities of mental action."[102]

He did not agree that this led to Huxley's conclusion of determinism,
however. Carpenter described that conclusion as one of two common errors in thinking about consciousness. He refuted it by stating that the reduction of the thinking man to a mere puppet "is so utterly antagonistic
to our own consciousness of possessing a self-determining power . . . that
we *feel* its essential fallacies with a certainty that renders logical proof
quite irrelevant." His subjective sense of free will could not be contested.
The other common error was the claim that the body never influences
the mind, which was shown to be untrue with every drink of alcohol.
Carpenter said that whatever resolution was found, it needed to be in

> harmony alike with the results of scientific inquiry into objective facts, and
> with those simple teachings of our own consciousness, which must, after all,
> be recognized as affording the ultimate test of the truth of all Psychological
> doctrines.[103]

This was the standard that set Carpenter apart from the naturalistic
physiologists. To him, it was intuitively obvious that scientific conclusions about the mind had to be compatible with humans' awareness of

their own consciousness. Our inward awareness was a scientific fact that had to be accounted for, just as much as any measurement of a nerve signal. Huxley felt no need to accept this standard, and therefore read Carpenter's book in a completely different light.

Carpenter's conclusion about the will was that mind and matter were completely separate entities, but that active mind moved passive matter just as physical forces moved passive matter. He postulated no mechanism for the will, only this abstract force/matter distinction. The automatic actions of the human body were not to be mistaken for an indication of our deterministic nature: "It is, in fact, in virtue of the Will, that we are *not* mere thinking automata, mere puppets to be pulled by suggesting-strings, capable of being played-upon by every one who shall have made himself master of our springs of action."[104] This internal sense of will was a fundamental axiom of mind without which humans could not function—the uniformity of nature and the moral sense were similarly axiomatic. Moral sense had no meaning without volition ("The idea of *Right* connects itself with voluntary action") and was a basic link to "the Being and Attributes of the Deity."[105]

He understood that some determinists pointed to phenomena such as hypnotism, electrobiology, and mesmerism as evidence that humans did not have free will. Carpenter countered that the unusual state induced by these practices was actually evidence *for* the will. When the will was suspended, as in mesmerism, the subject's actions were distinctly different. Thus, the observable changes induced by the suspension of the will showed the significance of the will. If volition had no real power, how could its abeyance have any effect? Man had a soul and was therefore not merely a machine. However, the soul's connections to the body could be disrupted, creating a state very much like that of animals and other true automata.

This soul, Carpenter contended, was exactly as described by Christianity. In particular, it would live on after death:

> But the Death of this Body is but the commencement of a new Life of the Soul; in which (as the religious physiologist delights to believe) all that is pure and noble in Man's nature will be refined, elevated, and progressively advanced towards perfection; whilst all that is carnal, selfish, and degrading, will be eliminated by the purifying processes to which each individual must be subjected, before Sin can be entirely subjugated, and Death can be completely "swallowed up of Victory."[106]

Carpenter made all of these points in his early career, well before Huxley and the scientific naturalists articulated their automaton theories.
They were clearly part of the extant theistic scientific tradition, and
Carpenter surely never imagined that his physiology would become the
launchpad for a naturalistic denial of the soul. After Huxley and Tyndall's 1874 addresses in Belfast, however, Carpenter felt the need to re-
claim his work for theism. One of his early attempts at this was a talk he
gave to the Sunday Lecture Society.[107] He began by reminding his listeners exactly what an automaton was. Like the organ behind his lectern, it
was "a machine which has within itself the power of motion, under conditions fixed *for* it, but not *by* it."[108] Was man a machine of this kind? He
described the extensive level of automatism in the body and reminded
them that the will did not have complete power—if a person wanted to
play a musical instrument, it could not be done without the appropriate
training.

This was not in doubt. The question was "whether the Ego is completely under the necessary domination of his original or inherited tendencies, modified by subsequent education; or whether he possesses
within himself any power of directioning and controlling these tendencies?" Some people say that since the cerebrum's state was just the result of its antecedent condition, it must be completely automatic. Even
some who do not consider the physiological side do not think people can
escape their character. Carpenter noted that J. S. Mill had changed his
mind on this, and now agreed that the will could affect the future.[109] His
central contention was again the reality of our own experience of volition: "I ask you to take as your guiding star, as it were, in the conduct
of your lives, these four words—'I am,' 'I ought,' 'I can,' 'I will.' . . . the
expression of reflection and self-consciousness, the looking-in upon our
own trains of thought."[110]

Carpenter's argument against naturalistic automatism was fully expanded in his *Principles of Mental Physiology*. He briefly reviewed the
positions of Huxley, Tyndall, and Clifford, but said he had seen no reason to change the conclusions he arrived at nearly forty years before: a
clear distinction between the automatic and volitional actions of a human.[111] Carpenter reiterated that he understood, and indeed helped for-
mulate, much of the physiology that the scientific naturalists claimed inevitably led to determinism. Against this he denied the possibility that
"any conceivable play of molecular forces" could explain how an idea
could come to dominate an entire nation. In any case,

while every one admits the existence of Uniformities in Human action which constitute the basis of our Social fabric, every one also admits that the closest observation of these Uniformities, and the most sagacious analysis of their conditions, does not justify anything more than a "forecast" of the course of action, either of individuals or communities, in any given contingency.[112]

An automatist might say this was simply a matter of complexity, but Carpenter insisted that the will needed to be explored as a phenomenon on its own. The immediate affirmation of consciousness was all that was needed to make the will real. No amount of rationalization could make it go away: "The direct testimony of Consciousness as to any one of its primal cognitions, must be held, as it seems to me, of higher account than the deductions of Reason from data afforded by other cognitions."[113]

Mental Physiology largely reiterated the points made in *Principles of Human Physiology*. He introduced few new ideas, and the debate rarely touched on matters of fact or observation. Rather, the disagreement was on the much deeper issue of how to apply the basic categories of science to the human consciousness. Specifically, should the expectation of the uniformity of nature be extended to the freedom of the will? And if so, how? It was on this issue that theistic and naturalistic men of science began to split. We have seen how Huxley appropriated Carpenter's physiology to create a deterministic view of the human mind, and Carpenter's attempts to restore his work's original theistic context. Many theistic men of science grappled with these questions in the second half of the nineteenth century, drawing on a variety of resources to decide how to fit consciousness into the growing power of science. Maxwell was particularly thoughtful on these issues, in a way that increasingly integrated his research into new scientific problems.

Maxwell

Maxwell's approach to the free will question was largely framed by his evangelicalism. Broadly speaking, evangelicals considered man to be naturally depraved via original sin, and life as the opportunity to prove one's morality through the exercise of free will to choose a godly life over a worldly one. The individual conscience was the critical element:

evangelicalism discarded Calvinist predestination in favor of an emphasis on man's free ability to accept God's offered grace. After his 1853
conversion, Maxwell described his own experience of this:

> All the evil influences that I can trace have been internal and not external,
> you know what I mean—that I have the capacity of being more wicked than
> any example that man could set me, and that if I escape, it is only by God's
> grace helping me to get rid of myself, partially in science, more completely in
> society,—but not perfectly except by committing myself to God as the instru-
> ment of His will, not doubtfully, but in the certain hope that that Will will be
> plain enough at the proper time.[114]

Maxwell's newfound evangelical stance was quite clear: a depraved human nature and a complete reliance on divine grace. The dominant
thought of this passage was the statement of Maxwell's acceptance of
the overwhelming importance of a correct understanding of God's will.
The evangelical outlook required a God who provided grace as a free
choice and humans who acknowledged that free choice through exercise
of their own will. This view of the relationship between human and divine will was found across Christian denominations in the nineteenth
century. Maurice's *Theological Essays*, which Maxwell was reading at
this same time, also stressed the importance of a correct understanding
of the will. The ability to choose to trust in God was a critical element in
Maurice's scheme, and was also an important aspect of Maxwell's own
understanding of his conversion experience.[115] This choice freed Christians from their worldly prison of "mere Fate or Necessity," giving them
energy and power to live an extraordinary life as agents of God.[116] On
this specific issue Maurice fit well with the Victorian religious mainstream, which celebrated will as a source of both human strength and
weakness, but always as a path to submitting to a higher power.[117] Carpenter expressed similar views:

> It is by the *assimilation*, rather than by the *subjugation*, of the Human Will
> to the Divine, that Man is really lifted towards God; and in proportion as this
> assimilation has been effected, does it manifest itself in the life and conduct;
> so that even the lowliest actions become holy ministrations in a temple conse-
> crated by the felt presence of the Divinity. Such was the Life of the Saviour;
> towards that standard it is for the Christian disciple to aspire.[118]

The human will was supposed to be a reflection of the divine will, where individual volition was an opportunity by which one could become part of the divine plan. Without meaningful freedom of the will, there could be no participation in God's ordained future.

An important part of Maxwell's religious development was his time as a member of the Cambridge discussion group known as the Apostles. His early essay for the Apostles on analogies was discussed in chapter 2 in the context of natural laws, but this essay also contained some of his earliest thinking on the impact of scientific developments on the Christian doctrine of free will:

> When we consider voluntary actions in general, we think we see causes acting like forces on the willing being. Some of our motions arise from physical necessity, some from irritability or organic excitement, some are performed by our machinery without our knowledge, and some evidently are due to us and our volitions. Of these, again, some are merely a repetition of a customary act, some are due to the attractions of pleasure or the pressure of constrained activity, and a few show some indications of being the results of distinct acts of the will.[119]

This passage suggests that Maxwell was likely familiar with Carpenter's work, and the reference to pleasure and repetition indicates that he may have read Alexander Bain. Bain's early work argued that volition was dependent on actual physical changes to the brain caused by a repeated activity seeking pleasure or avoiding pain.[120]

Maxwell read very widely, and there were a handful of other figures who likely stimulated his thinking on the problem of volition. He knew Helmholtz's physiological work, probably through his own research on color vision.[121] It is also well known that during this period Maxwell read Henry Buckle's *History of Civilization in England*. Buckle sought to explain all of human history by appeal to natural laws, and denied that consciousness was an exception to the uniformity of nature. While his analysis was only applied to large groups, his assumptions were quite deterministic: "If . . . I had a complete knowledge both of [a man's] disposition and of all the events by which he was surrounded, I should be able to foresee the line of conduct which, in consequence of those events, he would adopt."[122]

Maxwell had no doubts about the reality of the soul and the will, but he acknowledged that in light of contemporary developments in psychophysiology and reflex action, it was no longer tenable to claim that only the will was responsible for human behavior. "Some had supposed that in will they had found the only true cause, and that all physical causes are only apparent. I need not say that this doctrine is exploded."[123]

Near the end of the essay Maxwell cautioned that a natural philosopher must be careful not to generalize so broadly as to mistake one thing for another. His warning was one that reappeared several times in his career in different forms and in different contexts. "If we are going to study the constitution of the individual mental man, and draw all our arguments from the laws of society on the one hand, or those of the nervous tissue on the other, we may chance to convert useful helps into Wills-of-the-wisp."[124] The physiology of nerves and the behavior of societies were important topics that Maxwell thought deserved serious investigation, but they were incomplete. Without including the human will as a real and efficacious entity, one could mistake those scientific approximations for absolute truth and be led down a dangerous path.

However, Maxwell was quite familiar with the breakthroughs of experimental physiology, and had no interest in trying to circumvent the conservation of energy. This was a genuine dilemma with which he would grapple for the rest of his life. He was searching for some synthesis that would acknowledge the power of natural laws while retaining the possibility of man's free choice of God over sin. In an 1857 letter to his friend Lewis Campbell, he described how he presented this issue to his students:

> I have to tell my men that all they see, and their own bodies, are subject to laws which they cannot alter, and that if they wish to do anything they must work according to those laws, or fail, and therefore we study the laws. You have to say that what men are and the nature of their actions depends on the state of their wills, and that by God's grace, through union with Christ, the contradictions and false action of those wills may be settled and solved, so that one way lies perfect freedom, and the other way bondage under the devil, the world, and the flesh, and therefore you entreat them to give heed to the things which they have heard.[125]

Another letter to Campbell in 1862 shows Maxwell's initial attempts to deal with the problem without discarding established science. Apparently responding to an earlier inquiry from Campbell about Helmholtz, Maxwell praised the German scientist. He admitted that it was now clear that human bodies could be thought of as machines running on food for fuel (although in an extremely efficient fashion). These implications of the conservation of energy showed that "the soul is not the direct moving force of the body. If it were, it would only last till it had done a certain amount of work, like the spring of a watch, which works till it is run down. The soul is not the mere mover."[126] He was careful to distinguish the soul from a simple reservoir of energy. Once this was made clear, the concern that it could be "used up" was negated and the possibility of an eternal existence in heaven was retained.[127] William Thomson, discussing the problem with his brother James, was not completely sure that the soul obeyed the laws of thermodynamics, but he was "almost certain" that the soul did not produce energy.[128]

The problem of how a nonenergetic soul could meaningfully guide the body remained. If it could not exert force, how could it intervene in the body's actions? Having disposed of the crude notion that the soul powered the body, Maxwell argued that the solution was to be found in a more subtle model of the mind-body relationship:

> There is action and reaction between body and soul, but it is not of a kind in which energy passes from one to the other,—as when a man pulls a trigger it is the gunpowder which projects the bullet, or when a pointsman shunts a train it is the rails that bear the thrust. But the constitution of our nature is not explained by finding out what it is not. It is well that it will go, and that we remain in possession, though we do not understand it.[129]

The human will could act, not like an engine pushing a load, but as a delicate force that initiated a larger process, like a pebble starting an avalanche. Critically, both of the metaphors Maxwell used here were events initiated by a conscious actor—a man pulls the trigger, a pointsman shunts a train. Note the ending statement of his own personal experience of volition, far and away the most common Victorian defense of free will.

This metaphor of a railway pointsman (also called a signalman or switchman) was certainly not unique to Maxwell. The image often appeared in Victorian discussions of free will, to varying ends. Carpenter invoked it his explanation of the moral dangers of determinism.

> If the "wrong" movement of the self-acting points of a Railway gives such a direction to the train which passes over them as causes a terrible sacrifice of

life, we do not imply by our use of the word the moral criminality with which we charge a pointsman whose drunken carelessness has brought about a similar calamity. The machine *could not* help acting as it did; we assume that the pointsman *could*.

If the pointsman was solely an automaton, Carpenter said, he would have had no choice about whether to get drunk, and he could not be held responsible for his decision.[130] As with Maxwell's use of the figure, its essential purpose here was to show how conscious decisions had large ramifications. G. G. Stokes used a similar metaphor to explain his "directionalism." He, too, described the body as a train, placing the will as the "the intelligence of the engine-driver" and not "the coals under the boiler." More abstractly, he described the will as "a directing power, not counteracting the action of the physical forces, but guiding them into a determined channel."[131]

Maxwell was well aware that so far all he had been able to do was find out "what [free will] is not," and had not found a positive solution to exactly how the will can act. But his strategy for solving the problem was made clear: find a process that begins with consciousness but does not require a significant investment of energy. This was the pointsman, though Maxwell did not yet understand how it might work. Interestingly, this letter to Campbell in which he first formulated the pointsman metaphor also mentioned Rudolf Clausius's work on heat that had stimulated Maxwell to begin revising his kinetic theory of gases. Thus, the pointsman was in focus just as he tackled anew the problems of molecules and statistics, the context in which the pointsman would appear again later.

Possible Solutions

Maxwell's response to these developments appeared in an essay for the Eranus Club on science and free will.[132] He began the essay by stating that free will was the essential problem bridging physics and metaphysics.[133] He was clear that philosophy, religious or otherwise, must take into account the progress of science. His foundation was again the pointsman model:

As the doctrine of the conservation of matter gave a definiteness to statements regarding the immateriality of the soul, so the doctrine of the conser-

vation of energy, when applied to living beings, leads to the conclusion that
the soul of an animal is not, like the mainspring of a watch, the motive power
of the body, but that its function is rather that of a steersman of a vessel—not
to produce, but to regulate and direct the animal powers.[134]

The progress of physical science referred to in the essay's title had
caused one difficulty (humans obeyed energy physics), but that progress
might also have created the solution. The steering effect of an immaterial soul—the pointsman—was made more plausible by the innovative
concept of instability, which Maxwell credited to Balfour Stewart.[135]

Stewart argued that there were two kinds of mechanical systems, stable and unstable. Both could be considered as machines and obeyed the
laws of mechanics, but because they were regular and calculable, only
stable systems had been studied closely. However, there were also unstable systems where an infinitesimal amount of energy could set a system
in motion, such as when a balanced egg falls in one direction and not another. Unlike deterministic stable systems, here there was "freedom of
action."[136] Stewart made a connection between the ability of an unstable
system to magnify tiny forces and the problem of the will. If the human
nervous system was arranged in an unstable fashion, the will could influence the entire structure with a microscopic effort. As he put it later,
the inherent "incalculability" of unstable systems forced back the deter-
minist specter: "In truth, is there not a transparent absurdity in the very
thought that a man may become able to calculate his own movements, or
even those of his fellow?"[137] William James also emphasized the unstable nature of the human will in his attack on automatism. He argued that
the instabilities of consciousness could have Darwinian survival value,
which would suggest the reality of volition.[138]

Maxwell was delighted with the development of the concept of instability. He argued in an anonymous review that the stable/unstable division called into question many of the fundamentals of determinism,
most notably the notion of an unbroken causality that can be precisely
understood. "In unstable systems, like antecedents do not produce like
consequents; and as our knowledge is never more than an approxima-
tion to the truth, the calculation of what will take place in such a system is impossible to us."[139] Maxwell argued that determinism was thus
only plausible in processes that were stable at all times, which had been
the only systems studied to date by physics. Science had now advanced
to the point where instability could be studied and comprehended, and

it was these advances that "tend to remove that prejudice in favour of determinism."[140] This was certainly a large step toward the pointsman, but was clearly not a complete solution. While Stewart had reduced the amount of energy needed for volition to a tiny amount, *some* was still needed, thus still requiring the soul to be either energetic or impotent. Free will remained an experiential reality but its justification remained complicated.

Maxwell carried this argument into his 1873 address at the British Association. His lecture laid out what was known about molecules and the evidence for their existence.

And it is no wonder that [Lucretius] should have attempted to burst the bonds of Fate by making his atoms deviate from their courses at quite uncertain times and places, thus attributing to them a kind of irrational free will, which on his materialistic theory is the only explanation of that power of voluntary action of which we ourselves are conscious.[141]

There were two interesting points of emphasis here. First, that voluntary action was real, and that people knew it was real through their conscious experience of it. Second, that a purely materialistic theory cannot account for this reality. Attempts to do so become "irrational." As Maxwell wanted the pointsman to remind us, scientific theories that ignored consciousness and its efficacy would rapidly go down the wrong path. Molecular science could confront the materialist assumptions of contemporary physiology and refute them.[142]

One relevant advance of science came from France. Specifically, a development in French mathematical physics known as "singular states," which provided a sophisticated explanation for how a Lucretian swerve could happen without violating the laws of physics. It was found in the 1870s that for certain differential equations (the equations that govern the motion of particles) there were sometimes peculiar points where an entire family of solutions "overlapped" and it was impossible to tell which trajectory a particle would take. Many of the mathematicians involved used these results to deal with difficult issues relating to their Catholic context in France, including free will.[143] Maxwell quickly connected it to his own religious concerns. He interpreted these singular states to be the mechanism for his pointsman: at such a state, the laws of motion made no determination which track the metaphorical train might follow. No forces or energy would be required to affect the path of a particle:

> While [the particle] is on the enveloping path it may at any instant, at its own sweet will, without exerting any force or spending any energy, go off along that one of the particular paths which happens to coincide with the actual condition of the system at that instant.

He saw this as a dramatic improvement because he had removed the requirement for even the small amount of "trigger-work" that Stewart needed the will to perform.[144] At a singular state "a strictly infinitesimal force may determine the course of the system to any one of a finite number of equally possible paths, as the pointsman at a railway junction directs the train to one set of rails or another."[145]

This was truly the fulfillment of the promise of the pointsman. The motion of particles *could* be influenced by an entity not involved with the transfer of energy. Dynamical theory had now shown that entire future courses of events were only predictable "*in general*," and there was clear ontological space for conscious influences.[146] Best of all, this space fell directly out of the equations of motion and thus maintained the strict validity of physics. It was "much better than the insinuation that there is something loose about the laws of nature."[147] As with the scientific naturalists, Maxwell was very concerned that the uniformity of nature not be disrupted. Precisely *how* uniformity was to be maintained was the locus of disagreement.

It seems that by this point Maxwell's views of free will had reached a comfortable maturity, and we can now see the full message that was embodied in the pointsman. First, he reminded even his allies that the days of a completely unrestrained will were far in the past. The conservation of energy and psychophysiology had forcefully demonstrated that humans do not have unrestricted control over our bodies.[148] The pointsman did not have complete control over the train—he could only deflect it at certain times and under certain circumstances. The train really does run on rails. Nonetheless, the pointsman was needed to get the train to a particular destination.

The problem, Maxwell said, was that investigators had not been careful about applying results from one domain of knowledge to another:

> Many cultivators of the biological sciences have been impressed with the conviction that for an adequate study of their subject a thorough knowledge of dynamical science is essential. But the manner in which some of them have cut and pared at the facts in order to bring the phenomena within the range of

their dynamics has tended to throw discredit on all attempts to apply dynamical methods to biology.[149]

This was particularly dangerous in the case of investigating "sensation and voluntary motion" through purely psychological or purely neurological means. It was simply sloppy science to treat "a fact of consciousness as if it were an electric current."[150] The application of one kind of scientific idea to another could be immensely fruitful, but it could also be disastrous.

There were two extremes on which Maxwell thought one could err. The first was to try to explain the emergence of consciousness from material processes. Maxwell commented satirically on how simple a task some scientists seemed to think this was:

> I was dimly aware that somewhere in the vast System of Philosophy this question had been settled, because the Evolutionists are all so calm about it: but in a hasty search for it I never suspected in how quiet and unostentatious a manner the origin of myself would be accounted for.[151]

He mocked those, such as Du Bois-Reymond and Karl Wilhelm von Nägeli, who postulated a continuity of consciousness beginning with the pleasure felt by the simplest beings as making the error of naïve personification.[152] The problem with theories of this kind (including Herbert Spencer's) was that they explained away the soul.

The second extreme was to accept the existence of the soul, but then try to justify its properties in material terms. These sorts of "gross materialisations" of the soul were misguided attempts at objectivity and fundamentally flawed:

> Science has, indeed, made some progress in clearing away the haze of materialism which clung so long to men's notions about the soul. . . . No anatomist now looks forward to being able to demonstrate my soul by dissecting it out on my pineal gland, or to determine the quantity of it by the process of double weighing.[153]

Maxwell's targets on this end were usually other Christians, such as John Drysdale.[154] He argued that this sort of project resulted in either absurdities such as Isaac Taylor's energy-producing soul or a will trapped in a materialist prison not so different from Du Bois-Reymond's. Maxwell attacked both sides equally—anyone who argued that the soul was explainable imperiled its divine nature and role in God's plan. He satirized the claim that the human body and soul could be treated as isolated entities: "I often catch myself, when thinking about my body or my mind, supposing that I am thinking about myself."[155]

Instead, he said, we should return to our own introspective experience as the basic evidence for consciousness and volition: "I know that I exist now, and that I act, and that what I do may be right or wrong, and that whether right or wrong, it is my act, which I cannot repudiate."[156] This reminds us of the high stakes of the free will question for the Victorians—only people who could freely choose their actions could be held responsible for them (in the eyes both of God and of society). Maxwell leveled this criticism against Du Bois-Reymond's conclusion that humans have two minds, one material, deterministic, and active, and one immaterial, conscious, and impotent. "We might ask Prof. Du Bois-Reymond which of these it is that does right or wrong, and knows that it is his act, and that he is responsible for it."[157]

This all left the soul in a liminal position. It was outside the *explanatory* range of science:

> But as soon as we plunge into the abysmal depths of personality we get beyond the limits of science, for all science, and indeed, every form of human speech, is about objects capable of being known by the speaker and the hearer. . . . The progress of science . . . has rather tended to deepen the distinction between the visible part, which perishes before our eyes, and that which we are ourselves, and to shew that this personality, with respect to its nature as well as to its destiny, lies quite beyond the range of science.[158]

But this did not mean that mind and will should be *ignored* by science. Rather, the lesson of the pointsman was that considerations of the will were crucial for keeping one from making incorrect conclusions about the application of science to humans (i.e., automatism). The will was a reality about the world that changed what conclusions were valid—if you ignore the pointsman, you will not understand where the train is headed—and thus the will needed to be taken into account, but not explained away.

Maxwell's understanding of the human will was not a simple import of religious dogma into his natural philosophy. He was clearly not uncritical on these issues: he rejected scientifically unsatisfactory solutions to the free will problem, accepted that humans were subject to natural laws, and strongly condemned any naïve pairing of Christian doctrine with the science of the day.[159] Rather, the pointsman was a way of thinking about the basic experience of volition in a world of natural laws. He continually reconsidered it in light of new findings in physics and physiology, and it was modified and evolved over time. Even though it was incomplete, it was a resource he could and did draw upon to shape his understanding of the problems of the limits of science. And these problems were not confined to the mysteries of volition—they were found in the realm of pistons and engines as well.

These questions were not a matter of abstract metaphysics for Maxwell. They embodied fundamental issues regarding the practice and application of science that played significant roles in his technical work. Indeed, the pointsman and its attendant concerns also appeared in Maxwell's research on thermodynamics, where he used the metaphor to help resolve profound difficulties surrounding the scientific status of those investigations.

The Demon

Maxwell was one of the pioneers of the kinetic theory of gases, which sought to demonstrate that the observable characteristics of gases could be deduced from the hypothesis that all matter was made up of molecules in motion. Maxwell's great innovation in this field was his application of statistical methods. As Theodore Porter and Peter Harman have shown, Maxwell learned these techniques from social statistics, particularly Thomas Buckle's historical works. He explained that "the limitation of our faculties" made tracing individual molecules hopeless. In his earliest uses of statistical methods, he emphasized the incompleteness of this kind of knowledge, although it could still provide the "moral certainty" that would be accepted by reasonable persons.[160]

Maxwell achieved significant successes with his theory, notably deriving many of the observed pressure, volume, and viscosity relations of gases. Given kinetic theory's success in explaining the nature of heat, it was inevitable that Maxwell and his peers would apply the theory to the powerful macroscopic laws of energy being developed at the same time: the laws of thermodynamics. Through the association of higher temperatures with faster molecular speeds, the molecular hypothesis did provide some explanation for the interchange of heat and macroscopic movement described by the conservation of energy (also known as the first law of thermodynamics). But here I will focus on Maxwell's thinking on the second law of thermodynamics.

The second law as it was understood at the time was expressed in several different forms, such as: useful energy tended to dissipate; or in the absence of external work, heat flowed from hot to cold; or entropy always increased in a closed system. These statements were, generally speaking, based on macroscopic concepts like steam engines and steel rods, and it was not immediately clear how the microscopic perspective of the new kinetic theory could be reconciled with them. For example, the motions of molecules were governed solely by dynamical laws, which were reversible in time, but thermodynamics seemed to have a distinct temporal direction.[161]

Maxwell was unconvinced that the second law was as comprehensive as some claimed. He sought "to pick a hole" in the law with a novel thought experiment. In an 1867 letter to P. G. Tait he described two vessels placed in physical contact, each filled with gas at different temperatures. The second law would normally predict that the two vessels would gradually adjust to an equilibrium temperature. Maxwell thought he had a way around this straightforward result based on conceptual resources developed in his thinking on free will.

Maxwell first described a slight alteration of the setup. A diaphragm would be placed connecting the two vessels, able to open and close. "Now conceive a finite being who knows the paths and velocities of all the molecules by simple inspection but who can do no work, except to open and close a hole in the diaphragm, by means of a slide without mass." This being would watch the motion of individual molecules closely, and when a fast molecule approached, it would open the hole and allow the molecule into the adjacent vessel. The door would be closed to prevent the passage of slow molecules, resulting in the buildup of faster molecules on one side and slower molecules on the other. The kinetic theory of gases interpreted this asymmetry as a difference in temperature, meaning heat would have apparently flowed from cold to hot. The second law was violated with no work or energy, "only the intelligence of a very observant and neat fingered being has been employed." There was nothing qualitatively distinct about this being; it was simply very perceptive and very quick. Humans, according to Maxwell, were unable to do this only due to "not being clever enough."[162] The moral of the tale was that the second law could only be true in a general statistical sense, and that a being with access to better measurement could circumvent it casually.

This finite being became known as "Maxwell's Demon," a strange but perhaps not implausible creation.[163] The demon may seem more familiar to us in a later description of the same thought experiment, with the two vessels named A and B:

> Provide a lid or stopper for this hole and appoint a doorkeeper, very intelligent and exceedingly quick, with microscopic eyes but still an essentially finite being. . . . In this way the temperature of B may be raised and that of A lowered without any expenditure of work, but only by the intelligent action of a mere guiding agent (like a pointsman on a railway with perfectly acting switches who should send the express along one line and the goods along another).[164]

The same metaphor that Maxwell constructed to explore the human will reappeared here inside containers of heated gas. Why? It is clear that in the crudest sense his use of the pointsman was, as in his discussions of the will, meant to circumvent objections that a significant manipulation of energy would be needed to achieve the desired effects. The pointsman's tasks were thus expanded to explain both conscious volition and the flow of heat. But as with the free will issue, the pointsman metaphor also signaled much larger questions about the nature of scientific explanation.

Maxwell used metaphors repeatedly in his scientific career, and their significance has been analyzed by a number of scholars.[165] The pointsman fits well with Jordi Cat's argument that Maxwell used metaphors primarily for illustrative, not explanatory purposes.[166] So what aspects of real pointsmen was Maxwell trying to evoke, and what was he hoping to illustrate? A pointsman's job was all about information (where was the train, when would it clear the tunnel, when would the next train be arriving) and then acting immediately on it (shift the points, pull the lever). Incredible precision and unflagging attention were required, but they were still, in the end, human and finite.[167] This was exactly the image Maxwell wanted to conjure both with his demon and with volition: conscious awareness and actions based on that awareness could allow a small action (pulling a lever) to have huge consequences (the train goes south instead of west). The benefit of using a metaphor here was a straightforward one, in that it helped describe the unfamiliar in terms of the familiar. The conscious observations and actions of the pointsman were intimately familiar to any human being, providing a way for Maxwell to illustrate the highly alien processes involved in psychophysiology and thermodynamics.[168]

The pointsman was not intended to show the unrestricted force of the will. It was meant to show that the will could act even within a wide range of restrictions. The real pointsmen were restricted by the structure of the tracks, the momentum of the train, the schedule of arrivals and departures, and the rules of the rail company. But even if observers understood all of those things, they would still not understand how the train successfully got from place to place without appreciating that consciousness, observation, and volition were necessary to the process. Similarly, men of science who understood conservation of energy, the reflex action, and the dynamical theory of heat would still not be able to understand the true nature of either humans or entropy. Without an awareness of consciousness, observation, and volition, they would come to incorrect conclusions about scientific laws.

Maxwell did not mean the pointsman to be a literal explanation of what was happening in either the mind or a thermodynamic chamber. Maxwell did not think that the demon in his thought experiment was an actual human or divine intelligence. It has been claimed that the demon was a microscopic Laplacian calculator that was simply omniscient about the motions of molecules (i.e., God).[169] Maxwell anticipated this accusation and stated clearly that he was willing to dispense with the intelligent aspects of the demon and turn it into a sophisticated valve.[170] Based on these passages Peter Harman concludes that the demon had no supernatural connotations for Maxwell.[171] Harman is clearly trying to head off the claim that the demon was a direct divine agent. I agree with him on this specific point, but I believe Crosbie Smith and M. Norton Wise are also correct in arguing that there is no reason to think that Maxwell objected to *all* supernatural implications of the demon's activities.[172] That is, I think it is the case both that Maxwell did not intend the demon to be a literal description of ongoing divine actions and that he did think the results of the demon thought experiment might have implications for matters that could be called "supernatural" (such as human free will). The pointsman was not intended to provide a concrete explanation of an actual process, since there could be multiple explanations for what was going on (an intelligence or a valve). Rather, it was the illustration that was important: considering carefully what intelligence can do *shows* the errors of certain kinds of reasoning.[173] Maxwell thought that metaphors could play an important part in science, but the scientific function of the pointsman was a critical, not a constructive one.[174] Instead, its job was to warn against drawing unwarranted scientific conclusions.[175]

The demon was an elaboration of the pointsman model originally developed to shed light on human volition in a deterministic universe. The pointsman was, at root, an attempt to understand correctly the nature of the human will as something that could process information and act on it.[176] Maxwell was heavily invested in such a correct understanding due to his evangelicalism, but this was not simply a reiteration of religious dogma. Rather, he argued that a correct understanding of free will helped us understand better both the world and our conceptions of it. In the case of the demon, understanding the power of free will showed one how to question the universality of the second law. The demon itself did not have to be a supernatural being, and it was intended to solve a technical problem in thermodynamics. But its energetic function was based on a metaphor that was also part of a chain of reasoning grounded in manifestly religious premises. Thus, the pointsman was both a scientific and religious entity, and in both realms it was intended to raise difficult questions about the knowledge available from the current state of science.

More specifically, the pointsman was supposed to question what *level of knowledge* was available, and how that level affected the conclusions one could draw. Maxwell was concerned from an early point in his career with ensuring that physical laws and claims were understood properly: were they a description of a real entity, a hypothesis, a metaphor, or simply a mathematical convenience? The correct understanding of free will, as manifested in the pointsman, was one more tool for properly calibrating the level of knowledge in both science and society.

Molecular investigations appear to have particularly stimulated Maxwell's thinking on these issues. He claimed that the kinetic theory "forces on our attention the distinction between two kinds of knowledge, which we may call for convenience the Dynamical and Statistical." Dynamical knowledge could produce certainty and exact prediction, whereas statistical investigations could only address probabilities and general assertions. Maxwell accepted that if the molecular theory was true, then all knowledge of matter was simply statistical.[177] He did not wish to denigrate the powerful results of statistics, but he did feel the need to make clear that reliance on it generated certain "peculiarities" that indicated it was a "different department of knowledge from the domain of exact science." These peculiarities were responsible for such apparent anomalies as the reversibility of astronomy while thermal phenomena remained irreversible. This meant, fundamentally, that human knowledge via statistics could only be approximate, not absolutely accurate in the manner of astronomical predictions.[178]

The physical world, then, could sometimes present both puzzles and solutions that were only apparent, and that were dependent on our abilities rather than nature. The "telescope of theory" needed to be adjusted to focus on the right level of analysis in order for one to see clearly.[179] In the case of the second law, processes of the natural world sometimes appeared to show regularity that the lens of statistics revealed was only illusionary.[180] The intent of the demon was to demonstrate just such a situation. Maxwell asserted in his *Theory of Heat* that the second law was "undoubtedly true as long as we can deal with bodies only in mass, and have no power of perceiving or handling the separate molecules of which they are made up," and employed the demon to show how useful heat could be restored to a system by a being with precise, but finite, awareness.[181] He furthered the case for how human misunderstandings of people's own awareness and volition could lead to erroneous conclusions:

> It follows from [the activity of the demon] that the idea of dissipation of energy depends on the extent of our knowledge. . . . A memorandum-book does not, provided it is neatly written, appear confused to an illiterate person, or to the owner who understands it thoroughly, but to any other person able to read it appears to be inextricably confused. Similarly the notion of dissipated energy could not occur to a being who could not turn any of the energies of nature to his own account, or to one who could trace the motion of every molecule and seize it at the right moment. It is only to a being in the intermediate stage, who can lay hold of some forms of energy while others elude his grasp, that energy appears to be passing inevitably from the available to the dissipated state.[182]

The demon, as a simple application of the concept of the pointsman, showed dramatically how we can be fooled into seeing laws of nature where they do not truly exist. Similarly, Maxwell thought materialists such as Huxley had fooled themselves into seeing laws of nature in the mind where there were none. He was certainly concerned with the consequences of their ideas in the religious and social realms, but the wider application of the pointsman shows us that his critique was a deeper one. He was arguing that they had mistaken some regularities of nature (e.g., the function of the nervous system) for absolute laws (human automatism). This error came from paying too much attention to energy and motion and not enough to personality and the experience of the divine. With the pointsman he argued that understanding how volition could work even in a world of natural laws would have prevented the materialists from reifying erroneous conclusions based on their limited agnostic perspectives. Maxwell asserted that thinking of humans simply as machines was a choice: "Either be a machine and see nothing but 'phenomena,' or else try to be a man, feeling your life interwoven, as it is, with many others, and strengthened by them whether in life or death."[183] One could either accept the reality of our experiences of volition and sociability or discard it, but rejecting that reality was asserting a particular boundary to science. And to Maxwell, the materialists had picked the wrong one. They had asserted that consciousness was an object to be explained, rather than a cause to be taken into account. This neglect of the everyday experience of consciousness poisoned their analyses, and thus they became convinced of the absolute truth of biological laws (that humans were automata) that were in fact only approximately true.

The parallel use of the pointsman in his analyses of the second law shows that Maxwell thought an incorrect understanding of the will could have consequences in physics as well as in biology. The pointsman performed an analogous task here to its role in religion. It was a call to observers that they were focusing on the wrong level of analysis. For Maxwell, a correct understanding of the mind and the will affected how one saw the world. In science, that determined which laws of nature one could see. In society, it determined whether one thought man was moral and responsible for his own actions. Decision making was one of the fundamental problems of metaphysics, and misunderstanding it could have dramatic consequences in all aspects of human thought from molecules to morality. The pointsman did not just guide molecules in motion; he also guided the physicist to a better understanding of the world.

Maxwell's pointsman was a reminder of the need to draw correct boundaries, in two importantly related senses. First, the limits of science itself vis-à-vis human consciousness. Science could not encompass consciousness, or it became explained away. Solely materialist approaches to the human mind might seem self-consistent, though they were ultimately self-defeating. But science could not ignore consciousness either, or it would wander into philosophical dead ends. Thus, consciousness and free will sat precisely on the boundary of science. Science must acknowledge the existence and possible effect of consciousness, but could not seek to explain it.

This led directly to the second type of boundary, the one between different kinds of scientific explanation: statistical versus dynamical, or limited versus absolute. Some scientific conclusions were true only in an approximate sense, not a fundamental one. Humans do obey some of the same laws as machines, and entropy usually does increase. But we could be fooled into thinking those approximations were *really* true if we did not pay proper attention to human consciousness and its effects. Maxwell worried that ignorance of this issue was misleading Victorian scientists into dangerous waters, and intended the pointsman to show them a way out.

Conclusion

Theistic and naturalistic scientists alike had agreed that nature was uniform, and that apparent violations of uniformity needed to be resolved. Apparent interruptions to the natural order were to be smoothed out by appeal to deeper, or temporarily hidden, natural laws—as with Maxwell's unification of electricity, magnetism, and light. To Huxley, there was no reason to exempt the human consciousness from this precept. It should be treated exactly as any other physical puzzle, a guideline that led to the conclusion of determinism. However, to the theists, the validity of this move was not so obvious. Thoughtlessly extending the uniformity of nature to volition led to results contradictory to the internal sense of free will. The enormous stakes of this sense being in error—the existence of the soul, the possibility of a relationship with the divine, the moral order of Christian civilization—made them pause. Their position was that the subjective evidence of consciousness constituted real, meaningful evidence, which scientific investigations must take into account. Just as a physicist must include observations of falling bodies, physiologists must include observations of human will. As the English philosopher Henry Sedgwick put it: "Against the formidable array of cumulative evidence for Determinism there is to be set the immediate affirmation of consciousness in the moment of deliberate action."[184] The naturalists argued that this was *not* evidence. It was simply an illusion, like a straw appearing bent in a cup of water. It was to be explained away, not made part of the explanation.

Neither side could provide convincing physical scenarios for its preferred models. Huxley had no naturalistic scheme for how matter produced mind, and Maxwell had no laboratory experiment showing the pointsman at work. This dispute was a matter of the methodological values of science—how should investigators shape their exploration of the human mind and body? Roger Smith argues that Huxley was "concerned more with securing the authority of the principle of the uniformity of nature than with the logic of free will."[185] The evidence of uniform processes at work in the human body was agreed upon by all, but how to think about the significance of that evidence caused a profound split. The differing commitments of naturalists and theists could find no common ground on these issues. The human mind, particularly the will, became the thin end of the wedge.

We can see the consequences of this in James Leuba's famous 1916 survey of religious belief among scientists. He found that physical scientists were much more likely to believe in God than biologists, and psychologists and sociologists the least likely.[186] Following on the argument of this chapter, I would like to suggest that this is because of the difficulties presented by the human mind for the uniformity of nature at the end of the nineteenth century. The more distant a given scientist's research was from human volition, the less of a problem it presented. But psychologists and sociologists needed to grapple with human decision making regularly. In order to find the uniform laws that were necessary to make their sciences meaningful, they made moves quite similar to Huxley—consciousness must be an object capable of explanation on the same terms as the rest of the natural world. This move was the one the theistic scientists were unwilling to make, which made the human sciences an uninviting place to work.

This effect, while significant, is also clearly not sufficient to explain why the methodological values of science became almost completely naturalistic by the early twentieth century. To help understand this, the next chapter will explore how the scientific naturalists seized control of the means of scientific production. Production, that is, of the next generation of scientists.

CHAPTER SEVEN

How the Naturalists "Won"

Huxley won. Modern science is practiced naturalistically, and most scientists would be baffled to think that there was any other way—precisely what the scientific naturalists were trying to achieve. The separation of science and religion is today seen as the necessary step in making science modern, effective, and autonomous. Naturalistic science was methodologically unique. To become truly scientific, all theistic elements needed to be stripped away. Thanks to naturalism, science was freed.

But this is a story. It was written by particular people for particular purposes. Charles Taylor calls this the "subtraction" model of modernity, in which religion is washed away and leaves behind truth.[1] This is exactly how Huxley wanted one to think about science—it had always been naturalistic, just at times forced into a theistic prison that disguised it. All that needed to be done was to release it. However, as we have seen in previous chapters, this was not the case. The connections between theism and scientific values were deeply rooted, and indeed seemed completely necessary to most men of science. To overcome this, the scientific naturalists could not simply hope that theism would drop away from science. Rather, they had to actively appropriate and reimagine science as being their own. Science had to be shown to be able to exist without a theistic framework.

This outcome was due, of course, not solely to Huxley and his friends. The historical arc resulting in modern naturalism is long and complicated. Even in the Victorian period, many of the relevant ideas appeared outside science, such as the historical and literary approaches to scripture advocated in the famous *Essays and Reviews*. However, I am here interested in a precise, but critical, part of the story: how did *practitioners* of science come to embrace naturalism as essential to their work?

The shift among men of science from the nineteenth to the twentieth centuries was remarkable. There were surely many processes involved in the way naturalists came to dominate science. I will here concentrate on three possibilities. Two of these—taking control of science education in Britain, and naturalizing theistic concepts—were deliberate strategies on the part of the naturalists, which they carried out quite effectively. The third was the broader shifts in religious life in Great Britain at the turn of the twentieth century. All three built upon on the common grounds of theistic and naturalistic scientific practices to create a smooth transition instead of a disruptive revolution. This gentle shift allowed for the sense that there had been *no change* in science—it had always been thus.

Pipelines

It is well understood that one of the X-Club's major goals was to wrest control of the universities from the clerics.[2] This was certainly with the intent to secure paying jobs for themselves, but there was a deeper dimension as well. If the naturalists wanted to truly change science, as opposed to simply promoting themselves, they needed to alter the entire system by which professors of science were made and chosen. Huxley thought strategically about how best to achieve this. A major part of his strategy was to shape the next generation of science teachers, so as to start a pipeline of like-thinking practitioners.[3]

Huxley's efforts to change science education on the university level are well known, particularly his work on the Devonshire Commission.[4] He was deeply involved in the creation of biology professorships all over Britain in the 1870s and 1880s, and worked hard to influence who received those positions.[5] His goal was to place candidates who were ideologically sound (i.e., purely naturalistic) as well as scientifically talented. In this he was quite successful. Huxley's chief source for these candidates was his pool of classroom demonstrators. These were young, skilled men of science whom Huxley was able to train in both naturalistic science and naturalistic science *teaching*. As Adrian Desmond says, the laboratory was as much a training ground for the demonstrators as it was for the students.[6] Once they were suitably accomplished, Huxley maneuvered his lieutenants into science professorships around the world.

He managed to place his protégés William Rutherford at King's College London, and Louis Miall at Leeds. T. Jeffery Parker went to New

Zealand. Huxley's former demonstrators William Thiselton-Dyer became professor of botany at the Royal Horticultural Society, and Sydney Howard Vines was elected to the Sherardian professorship at Oxford. Frank Balfour would have had an Oxbridge chair had he not died tragically in the Alps. The peripatetic but devoted E. Ray Lankester was placed at University College London, Edinburgh, and Oxford (although his biggest contribution to the spread of naturalism may have been his long-lived "Science from an Easy Chair" popularizations).[7] Perhaps most important was the demonstrator Michael Foster, who went on to found a Huxleyian school of physiology in the heart of Anglican Cambridge. Foster himself trained many students who distributed the South Kensington naturalism even further.[8] Roy MacLeod declared that "an entire generation of new biologists . . . were among [Foster's] students."[9] Huxley even gained a critical foothold in the United States when H. Newell Martin, a student of both him and Foster, inaugurated the biology program at the new Johns Hopkins University.[10] Henry Fairfield Osborn reminisced about introducing the "Huxley method" to both Princeton and Columbia.[11]

In a short number of years Huxley had already managed to place his students, allies, and demonstrators at a dizzying array of universities.[12] They were noted for bringing a naturalistic perspective with them and evangelizing for the new scientific outlook. In addition to the effect of their own personalities, these protégés developed courses of study and training that had their roots with Huxley and became standard for generations to come.[13]

Beyond his efforts to pack the universities with his followers, Huxley had an insight that would be critical for the eventual victory of the scientific naturalists. He realized that only a tiny fraction of the British population would ever study science on the university level and thus be exposed to his naturalistic apostles. So he broadened his perspective and explored ways to bring his view of science to the elementary level, where it could touch virtually every student in the nation.

When the 1870 Elementary Education Act was passed, it created a huge demand for science teachers and, by implication, training for those teachers. Huxley pounced on this opportunity, writing textbooks and lab manuals, and running summer courses for these new teachers that would inculcate both teachers and students with a naturalistic, secular worldview.[14] He was not reticent to share his plans in colorful language: to one correspondent he described "a course of instruction in Biology which I

am giving to Schoolmasters—with the view of converting them into scientific missionaries to convert the Christian Heathen of these islands to
the true faith."[15] These courses trained new teachers to think naturalistically, and even to *see* naturalistically, as Graeme Gooday has shown.[16]
One of Huxley's pedagogical insights was to create the summer course as
a package of "transmittable basic science" that the teachers could trans-
port back to their own schools.[17] He taught them how to teach as he did.
He placed great importance on these teachers' courses, even declaring
that "so far as science teaching and technical education are concerned,
the most important of all things is to provide the machinery for training
proper teachers."[18]

Huxley had an even more efficient tool than teaching teachers: exams.
As James Elwick describes, many members of the X-Club began their
careers during the exam mania of the 1850s. Setting and grading exams were helpful ways for those struggling young investigators to make
ends meet, even if science questions were consistently given less importance than classics.[19] Huxley was particularly enthusiastic about exams
as a means for poor students to work their way up the scholarship ladder just as he had. He celebrated exams as "one of the greatest steps ever
made in this country towards spreading a knowledge of science among
the people."[20] Even in those early years, the scientific naturalists were interested in using the exam system to reshape science, as Hooker wrote
to Huxley:

> My own impression is that we shall make no great advance in teaching Nat.
> Science in this country, except by some joint effort of Botanists and Zoolo-
> gists who should pave the way by propounding a strictly scientific elementary
> system—were this once effected we have sufficient command over the public,
> as examiners in London, and as confidential advisors of Examiners and pro-
> fessors elsewhere, to ensure the cordial reception of such a system.[21]

Huxley, Tyndall, Frankland, and Hirst all became examiners for the Department of Science and Art, which enabled them to set the expectations
for science teachers across Britain. They all placed their own textbooks
on the exam syllabi.

Elwick argues that these exams were powerful ways to ensure epistemological conformity, even among those not taught directly by the scientific naturalists. The exams were gatekeepers to success as a teacher.
Through those exams

Huxley had the authority to determine who properly knew biology and who did not. Yet what he considered the proper understanding of biology was associated with metaphysical commitments to scientific naturalism and methodological materialism (the living body seen as a machine, for instance).[22]

The exams became a way to distribute and enforce a naturalistic catechism for science. Those hoping to become science students or teachers needed to study Huxley's syllabus, lessons, and textbooks. Elwick contends that far more people must have paid close attention to those than to any of Huxley's papers, columns, or popular books—their livelihood depended on how well they did so. The correct answers to the exam questions were those with naturalistic underpinnings: the properties of living matter, "the living body considered as a machine," "Hereditary transmission, and the modification of physical and mental characters by education, as the basis of a rational belief in the possibility of human progress." Huxley described these exams as some of the most important measures ever taken to reduce "Parsonic influence" in schools.[23] Future teachers studying for these exams would have come to associate science with naturalism, whether they realized it or not. Exams provided a way to mass-produce naturalistic science teachers by the thousands.

Along with developing exams came the need for textbooks. In addition to writing his own texts, Huxley played a key role in two major textbook-publishing projects: the International Scientific Series and Macmillan's Science Primers. His own 1877 *Physiography* was aimed at a young reader, following the principles he laid out for teaching children.[24] The International Scientific Series came from the Youmans publishers in New York, which asked Huxley, Tyndall, and Spencer to head the British component to the series. As Bernard Lightman points out, this was perfect for the overworked Huxley: he would have the power to spread scientific naturalism without having to do the writing himself.[25] The aim to influence children was clear—Tyndall's first contribution to the series was a rewrite of his juvenile lectures at the Royal Institution. Macmillan's Science Primer series was explicitly aimed at the elementary textbook market. Huxley got himself on the editorial board and wrote the series's first volume, his *Introductory Science Primer*. This text presented the fundamentals of naturalistic science in a format convenient for teaching, and at a level appropriate for young students. Lightman comments that "every school child that read this introduction to science would be trained to reject the very premises of theologies of nature."[26]

FIGURE 2. "On London's School Board." Huxley Papers.

With their allies in charge of both universities and the production of science teachers, members of the X-Club could be assured that the next generation of science students was trained in a naturalistic perspective, which those students would then pass on to their students. This strategy was carried out in clear public view, and often elicited comment. By the end of the century Huxley's methods were well entrenched, and by the end of the Great War few could imagine it being otherwise.[27] This is not to say that the scientific naturalists corrupted science education. The professors they placed were talented and skilled, but their teaching methods were not value-neutral either. Huxley designed his teaching to stand for what Adrian Desmond calls a "distinct ideological faction" that clearly marked off acceptable (naturalistic) from unacceptable (theistic) ways of thinking about science.[28]

Naturalization

In order for the scientific naturalists to dominate, they had to make their view of science seem obvious and inevitable. This goal had the major problem of two centuries of theistic science—how could science be naturalistic by definition if it had been practiced theistically for so many years? The naturalists' strategy was to rewrite the history of their discipline to erase the long tradition of theistic science. Scientists frequently reimagine their past in order to support their vision for the future, and the wave of scientific naturalists at the end of the nineteenth century did so to establish a particular way of thinking about science and religion. That is, that science as an enterprise only made sense in an areligious context.[29] Concepts like uniformity, which were both theistic and naturalistic in practice, became recast as *only* naturalistic.

Frank Turner and James Moore have written about how the social aspects of natural theology were co-opted by the scientific naturalists for their own purposes.[30] Turner argued that the social agenda of natural theology (chiefly social stability) needed to be expressible in naturalistic terms before science could seriously shift toward naturalism. Moore suggests that a critical part of this process was the development of a new naturalistic theodicy based on order and progress. Both make the case that scientific naturalism grew out of the earlier religious context in a relatively smooth and continuous fashion, taking over preexisting religious categories rather than trying to replace them.

Bernard Lightman has shown how Huxley co-opted literary strategies associated with natural theological writings to promote a naturalistic cosmology.[31] The "common object" strategy used by natural theologians to show God's contrivance in the world was clearly on display in Huxley's lectures and writings (such as "On a Piece of Chalk"). But he deftly used the same strategies to arrive at different conclusions, often subtly framed to paint a closed world of only natural phenomena with no room for God. A theist could read many of Huxley's pieces, appreciate the science, and see the argument as familiar. Huxley even managed to do this in his teaching. He successfully made the fundamentals of his scientific methods palatable to theists, even to the point where his religious enemies such as St. George Jackson Mivart and J. W. Dawson sent their children to study under him.[32]

But what about the practice, content, and meaning of science itself? How could this be appropriated solely for naturalism? It might seem that overthrowing a centuries-old tradition such as theistic science would require a dramatic revolution, but in fact it was surprisingly continuous. This was because, as I have argued in this book, the positions of the theistic scientists and the scientific naturalists were actually quite similar in terms of basic concepts such as the uniformity of nature. So the practices and methods of theistic scientists could often be imported into naturalistic work with simple relabeling, or sometimes without comment at all. Huxley was particularly skilled at this. For example, he proclaimed that William Carpenter's work was the foundation of "rational" methods for thinking about living things.[33] Huxley argued that Carpenter's goal of explaining life in terms of laws (as opposed to vital forces) was the key step in removing theological legacies from the life sciences. But this was quite different from the way Carpenter saw laws, of course—as discussed earlier, he saw natural laws as manifestations of the Creator, and thus infused with religious significance. And it was not that Huxley was unaware of Carpenter's thoughts along these lines, as in his private correspondence he was quite hostile to Carpenter's spiritual interpretations.[34] But in his published work Huxley took a different approach. He simply stressed the points on which they agreed—natural laws—then elsewhere argued that natural laws were solely naturalistic. He did not have to persuade his colleagues that natural laws were important, because everyone already agreed on that. When theists read Huxley's discussions of laws in life, they could nod along happily, thinking of those laws in theistic terms. When scientific naturalists read about such laws, they, too, were happy, thinking of those laws in naturalistic terms. So in his published research Huxley could gain support from both camps, while his naturalistic interpretation of that research was passed on through his students and teaching.

A useful example of the detailed naturalization of a theistic scientist was the treatment of Michael Faraday, a beloved figure to both theists and naturalists. Faraday was a member of the Sandemanians (sometimes called the Glasites). That tiny group was a stringent Christian sect that demanded close attention to the Bible and constant moral rectitude.

They observed the Sabbath rigorously, had no professional clergy, and generally kept high boundaries between themselves and outside society.[35] Faraday commonly used biblical metaphors and references in his lecturing and often spoke of the laws of nature as being divinely crafted. In particular, he saw the conservation and conversion of energy as being closely tied to theological principles. His faith was well known in the science community, and despite the Sandemanians' generally negative social status, he was embraced as a typical theistic scientist. The *Graphic* carried a poem praising Faraday's theistic science:

> He sees in Nature's laws a code divine,
> A living Presence he must first adore,
> Ere he the sacred mysteries explore,
> Where Cosmos is his temple, Earth his shrine.[36]

As a Dissenter, Faraday was somewhat excluded from the core of British intellectual power in the early Victorian period. But he and his work fit comfortably into the expectations of what it meant to do science at the time.

Maxwell, of course, was intimately familiar with Faraday's work. His reading of that work in theistic terms was straightforward and unsurprising. Maxwell's article on Faraday for the *Encyclopædia Britannica* showed this clearly. He discussed the technical content of Faraday's work, then segued into "another side of his character, to the cultivation of which he paid at least as much attention, and which was reserved for his friends, his family, and his church." Maxwell identified him as a member of "the very small and isolated Christian sect which is commonly called after Robert Sandeman." He noted that Faraday regularly attended its meetings and had twice been an elder of the church.[37] Nearly an eighth of the article was then given over to a lengthy quote from Faraday's 1854 lecture on mental education, which, Maxwell told the reader, explained his "opinion with respect to the relation between his science and his religion." This quote presented Faraday's denial that human reasoning could ever understand God, with an apparent rejection of natural theology. Further, Faraday said:

> It would be improper here to enter upon the subject farther than to claim an absolute distinction between religious and ordinary belief. . . . Yet even in earthly matters I believe that "the invisible things of Him from the creation

of the world are clearly seen, being understood by the things that are made, even His eternal power and Godhead."

He clearly maintained a deep theology of nature, and found (like Maxwell) Romans 1:20 to be a profound expression of the relationship between God and the creation. Maxwell also included a brief note that Faraday had added in the reprinting of the lecture: "These observations . . . are so immediately connected in their nature and origin with my own experimental life, considered either as cause or consequence, that I have thought the close of this volume not an unfit place for their reproduction." Maxwell's inclusion of this shows his interest in connecting Faraday's religion to his scientific work. The end of the article was a quote from Faraday's biographer, Dr. Bence Jones:

> "His standard of duty was supernatural. It was not founded on any intuitive ideas of right and wrong, nor was it fashioned upon any outward experiences of time and place, but it was formed entirely on what he held to be the revelation of the will of God in the written word, and throughout all his life his faith led him to act up to the very letter of it."[38]

Religion was the only nontechnical issue discussed in Maxwell's article, which would not have been surprising to his contemporaries. For Maxwell (and most theistic scientists), it made perfect sense to discuss religious matters alongside unifications of the laws of nature.

However, theists were not the only group interested in laying claim to Faraday and his work. The scientific naturalists, particularly Tyndall, had a deep reverence and affection for him. Faraday was responsible for Tyndall being hired at the Royal Institution, and they worked together as good friends for many years. The question was how to talk about Faraday in a manner that supported the naturalistic story of science. Tyndall was very successful at embracing Faraday's contributions to science in a way that made the theistic aspects of his science seem only naturalistic. Consider his review of Bence Jones's biography. The biography did not shy away from Faraday's religion, printing lectures, letters, and writings referencing theological matters (including the 1854 lecture that Maxwell quoted).[39] In his extensive review, Tyndall did not discuss any of these documents. Rather, he quoted a letter in which Faraday said, "There is no philosophy in my religion" (likely referring to his rejection of natural religion in favor of scripture). Tyndall analyzed it this way:

For his investigations so filled his mind as to leave no room for sceptical ques-
tionings, thus shielding from the assaults of philosophy the creed of his youth.
His religion was constitutional and hereditary . . . however its outward and
visible form might have changed, Faraday would still have possessed its elemental constituents—awe, reverence, truth, and love.

His description of Faraday's religion being shielded from science was
intended to clearly separate the two, allowing no overlap. The lifelong practice of Sandemanianism was dismissed as "constitutional and
hereditary"—that is, the sect's peculiarities were not Faraday's fault.
Tyndall's comments about the outward change of the sect's dogma, apart
from the core Romantic pantheistic values, were a clear invocation of the
standard naturalistic story of true religion being encrusted with theological doctrines. The implication was that what Faraday *really* believed
was unproblematic emotional religion, and the doctrinal, sectarian aspects of Sandemanianism were unimportant and accidental. Tyndall
then wondered

how so profoundly religious a mind, and so great a teacher, would be likely
to regard our present discussions on the subject of education. Faraday would
be a "secularist" were he now alive. He had no sympathy with those who con-
temn knowledge unless it be accompanied by dogma.[40]

Faraday, member and elder of a highly doctrinal Christian sect, had been
recruited to the cause of secularism. His position that "we have *no* right
to judge religious opinions," intended to protect his Nonconformist religious practice, became a strike against Christian theology.[41]

Elsewhere, Faraday was turned into an ally of Darwin. Tyndall noted
that many held Faraday up as a moral exemplar for men of science because of his devout religion. He said that Faraday was indeed morally
outstanding, but not for that reason: "Were our experience confined to
such cases, it would furnish an irresistible argument in favour of the association of dogmatic religion with moral purity and grace. But, as already intimated, our experience is not thus confined." He pointed out
that Darwin had all of the wonderful characteristics associated with Faraday, but without the same religious views.[42] He did not deny that Faraday had been religious; he denied that this fact determined his moral
character. He and Darwin were of a kind—their brilliance and greatness
had nothing to do with religion. Hooker's 1868 presidential address to

the British Association made a similar move. He presented Faraday as the predecessor to Darwin—just as the former's ideas were not understood in their time but eventually changed the world, so would Darwin change everything. Faraday was linked to the heroes of the scientific naturalists and was claimed as one of their own.[43]

Tyndall's 1868 book *Faraday as a Discoverer* set up the Sandemanian as an exemplar of scientific naturalism. Faraday was described as embodying all the naturalistic values—unified laws, provisional results, the use of hypotheses and theory, the moral value of scientific investigation, and freedom of thought. All of these were described in such a way that their theistic roots were abstracted away, and they could be read naturalistically instead. For example, when discussing Faraday's tendency to be both humble and proud, Tyndall wrote that he was "a democrat in his defiance of all authority which unfairly limited his freedom of thought, and still ready to stoop in reverence to all that was really worthy of reverence, in the customs of the world or the characters of men."[44] This drew attention to the commonalities between Dissenters like Faraday and unbelievers like Tyndall. The book continually emphasized Faraday's independence of mind. Theists would read that as a straightforward description of liberal Christian values of intellectual freedom (as discussed in chapter 5). Naturalists, however, would read that as Tyndall wanted—that Faraday rejected religious authorities of all kinds. Depending on the assumptions one brought to the book, quite different portraits of Faraday would emerge.

In telling the story of Faraday's family life and history, Tyndall made no mention of the Sandemanians.[45] Tyndall stated that, over fifteen years of "intimacy," Faraday never once mentioned religion

save when I drew him on to the subject. He then spoke to me without hesitation or reluctance; not with any apparent desire to "improve the occasion," but to give me such information as I sought . . . this faith, held in perfect tolerance of the faiths of others, strengthened and beautified his life.[46]

Tolerance, rectitude, and privacy would seem to be the distinguishing features of Faraday's religious life. Tyndall described Faraday's study of nature as leading to "a kind of spiritual exaltation which makes itself manifest here"—again, claiming that Faraday's religion was of the Romantic pantheist type welcomed by Tyndall, instead of the highly scriptural, orthodox sect it was.[47]

Tyndall's writings on Faraday did not deny that he was religious. Rather, they recast the role and meaning of religion for Faraday. Its connections to science were loosened to the point of ambiguity. Theists such as Maxwell could read these essays and recognize the Faraday they knew—seeking unity in the universe, calling for intellectual freedom, working with the highest moral character, disavowing unwarranted speculation. All of those aspects would have been read as obviously deeply connected to his Christianity. However, naturalists could read those same essays and recognize those aspects of Faraday as *their* values, completely separated from religious matters of any kind. Thus, there would be no firestorm from the theists due to one of their own being misrepresented, but the younger generation of scientists trained in a naturalistic worldview would recognize Faraday as one of *their* own. The overlap of scientific foundations discussed in this book made it possible to rewrite the history of science naturalistically, claiming that theistic values and investigators had actually been naturalistic. "True" science had never been theistic.

Tyndall's portrait of Faraday became the standard fairly quickly. In 1898 the engineer S. P. Thompson, himself a prominent Quaker, published a new biography that relied heavily on Tyndall's.[48] He dedicated an entire chapter to Faraday's religious life, but followed Tyndall's lead in denying any meaningful connection between it and his science. Thompson described Faraday as not talking about religion, and directly quoted Tyndall's statement that Faraday never brought up such issues.[49] He also quoted "one of [Faraday's] friends" (perhaps Tyndall) as saying that when the Sandemanian entered the meetinghouse, he "left his science behind."[50] Thompson insisted on the separation of Faraday's science from his religion: "It is doubtful whether Faraday ever attempted to form any connected ideas as to the nature or method of operation of the Divine government of the physical world, in which he had such a whole-souled belief." Faraday, he said, was able to "erect an absolute barrier between his science and his religion."[51] Early twentieth-century accounts of Faraday largely ignore his religion, with the notable exception of the Marxist J. G. Crowther's *British Scientists of the Nineteenth Century*. He noted that both Faraday and Maxwell were "exceptionally religious," but denied that that could have any other than "psychological significance."[52]

Opportunities to recast theistic science as naturalistic often appeared in the form of memoirs and memorials, which Huxley and friends were happy to take. Occasionally, chances to rewrite history were thrust upon them, as when Huxley was asked to write an assessment of Richard

Owen's contributions to science for the paleontologist's biography. This lengthy essay was really a Huxleyian history of biology that turned to Owen only at the end. Huxley framed the history of biology as a tension between amateurs and professionals. He emphasized that Owen worked in the era before scientific specialization, and therefore that his work should not be considered deeply scientific:

> It is no disparagement to Owen to say that he was not a physiologist in the modern sense of the term. In fact, he had done a large part of his work before modern physiology, in which no progress can be made without clear mechanical, physical, and chemical conceptions, came into existence.[53]

Huxley narrated the history of the unity of plan among animals—the concept around which Owen built his career—as the production of German thinkers. The work that contributed to Owen's fame, Huxley said, was really in support of Darwin's discoveries.[54] Owen's archetypes were dismissed as "sublimated theism" and rhetorically separated from "essentially naturalistic abstractions—'secondary causes,' 'forces,' and 'polarity'"—that he used to explain animal structures. Huxley denied that the natural laws Owen talked about had any connection to his theism. Rather, he wrote that whenever Owen discussed natural laws, he was really invoking naturalistic evolution: "The looking to 'natural laws' and 'secondary causes' for the 'progression' of 'organic phenomena' is the substantial acceptance of evolution, as set forth by Goethe, Oken, Lamarck, and Geoffroy."[55] Whatever work stemmed from the archetype itself was not really scientific:

> I believe I am right in saying that hardly any of these speculations and determinations have stood the test of investigation, or, indeed, that any of them were ever widely accepted. I am not sure that any one but the historian of anatomical science is ever likely to recur to them; and considering Owen's great capacity, extensive learning, and tireless industry, that seems a singular result of years of strenuous labour.

Huxley ended by damning Owen with faint praise:

> It would be a great mistake, however, to conclude that Owen's labours in the field of morphology were lost, because they have yielded little fruit of the kind he looked for. On the contrary, they not only did a great deal of good by

awakening attention to the higher problems of morphology in this country; but they were of much service in clarifying and improving anatomical nomenclature, especially in respect of the vertebral region.[56]

In this essay we can see another variant of the naturalistic reclaiming strategy. General matters of scientific practice, such as uniformity, were labeled naturalistic and explicitly cut off from theistic ideas. Contributions to science were connected to naturalistic science, usually Darwin. Ideas too theistic to easily relabel were dismissed as having no relationship to science, or were placed as prescientific and not worthy of discussion. A reader of this biography from a generation later would conclude that Owen made minor contributions to science when he was thinking naturalistically, but wasted much of his time thinking theistically—that is, not acting like a man of science. Indeed, Huxley's essay could serve as an excellent primer for the claim that paleontology only became scientific insofar as it was not done theistically.

The key to this naturalization strategy was for Huxley to tell a new story about the history of science. By naturalizing theistic science, he was able to argue that science had *always* been naturalistic. That is, by naturalizing the tradition of theistic science, he was able to remove it from history completely, making naturalism the obvious and solitary way to do science. This was why he was always so eager to place his arguments in the mouths of historical figures—it gave historical continuity and gravitas to those arguments. For example, in *Evolution and Ethics* he declared that "scientific naturalism took its rise among the Aryans of Ionia." Naturalism was described as appearing wherever "traces of the scientific spirit" were visible. The only debt science owed to religious thought was that the contradictions of supernaturalism often gave rise to naturalistic thinking:

> So far as physical possibilities go, the prophet Jeremiah and the oldest Ionian philosopher might have met and conversed. If they had done so, they would probably have disagreed a good deal; and it is interesting to reflect that their discussions might have embraced questions which, at the present day, are still hotly controverted.[57]

Huxley's efforts to promote naturalism were, then, simply continuities of the earliest stirrings of rational thought, which were similarly oppressed by the Bronze Age equivalent of the Church of England.

Huxley's repeated use of figures such as Descartes and Joseph Priestley was an elaboration of this strategy. Through Huxley, Priestley became converted from Unitarian divine to naturalistic scientist, providing
another point on Huxley's curve of ascending science. Even Descartes,
whose theism could hardly be questioned, was claimed as the father of
naturalistic physiology. All the arguments that that Frenchman used to
support the soul and divine truth became tools for the elimination of
free will and the construction of a naturalistic mind. Huxley's vision of
the history of science was one of expanding naturalism, beaten down occasionally by orthodoxy, but never corrupted in its purity. Theism could
be found beside science, or obscuring it, though never in it. The connections between theism and science, such as natural laws, that were so clear
to Maxwell and his contemporaries were relabeled as something quite
different. Thus, Huxley did not have to remove figures such as Faraday
and Descartes from the history of science; he only had to tweak their
stories to support his sort of science rather than their own. John Brooke
has referred to this as a kind of Trojan horse, in which the fundamental values of scientific practice (as discussed in chapters 2, 3, 4, and 5)
provided a way for the naturalists to enter the core of science: "It smug-
gled a full-blown naturalism into territory that upholders of a more conservative natural theology . . . still considered holy ground."[58] Similarly,
Charles Taylor has written that the uniformity of nature

> came to serve as grist to the mill of exclusive humanism is clearly true. That es-
> tablishing it was already a step in that direction is profoundly false. This move
> had a quite different meaning at the time, and in other circumstances might
> never have come to have the meaning that it bears for unbelievers today.[59]

He warns that telling the story of secularization as one of epistemic
shifts, as Huxley does, is a *choice*, and only one of many possibilities.[60]

The Secularization of Britain

The changes Huxley was seeking in science were supported in powerful ways by major shifts in the social role of religion in Britain at the end
of the Victorian period. These have been extensively studied by historians and sociologists of religion. Most of the discussions have been under
the aegis of the "secularization thesis." This thesis has appeared in many

forms, but generally postulates that a decline in religious belief, practice, and social significance is an inevitable result of industrialization,
modernization, and urbanization.[61] In short: Weber's disenchantment
of the world. Britain has been the major focus of secularization theory
mainly due to statistics indicating a precipitous decline in church atten-
dance in the late nineteenth century, continuing throughout the twentieth. The accuracy and utility of this data have been questioned.[62] But
more important, scholars have questioned whether institutional atten-
dance is the appropriate measurement of religious belief and practice.
The persistence of "popular religion," "minimal religion," or "diffuse
Christianity"—for example, participation in religious rituals and atten-
dance at Sunday schools—has been well documented even as the pews
became empty.[63] The basic process was one in which individuals continued to self-identify as Christian while refusing to allow clergy to define
the details of religion for them. Jose Harris suggests that this movement
of religious practice from public to private spaces was itself the result of
a critical Victorian religious value—religion should be purely a matter of
private conscience.[64]

This privatization of religion inevitably led to a decline of public professions of faith and statements regarding theological matters. Scientists, of course, were not exempt from this process, and spoke less and
less about their religious faith. Ronald Numbers has warned not to mistake this decline in "God talk" for a genuine diminution of belief. "The
deletion of religious references—by both Christians and unbelievers—
went hand in hand with the depersonalization of scientific rhetoric, such
as avoiding the first person and adopting the passive voice."[65] That is,
an important part of this change was simply epiphenomenal to the new
standards of professional scientific conversation.[66] Bernard Lightman
has shown that the venerable "clergyman-naturalist" tradition survived
the attacks of the scientific naturalists.[67] Writers such as J. G. Wood continued to invoke natural theological themes and strategies without explicit references to God or scripture. He left it to his reading audience
to fill in the missing references to the Creator.[68] The survival of theistic science was, like Edwardian Christianity in general, quiet and easily
overlooked.

Peter Bowler has examined closely the statistical evidence available
regarding the religious beliefs of scientists in the first decades of the
twentieth century. He found that most scientists seemed to maintain a
form of religious belief even while being suspicious of dogmatic theol-

ogy. No major shift away from religion was evident. Most scientists were modernist (i.e., liberal) theists with sympathy for the Christian message but not for orthodoxy.[69] Numbers has shown similar survival of religious belief among American scientists of the period.[70]

Bowler documents the efforts of these liberals to "reconcile" science and religion. But these liberals did not see themselves as continuing the Victorian tradition of theistic science. Rather, they saw themselves as beginning a *new* tradition of religious science that would sweep away the alleged materialism of the nineteenth century.[71] They accepted the story that the scientific naturalists told—that theology had never been in science. Their rhetoric about how the new science was welcoming religion only made sense if science had, in fact, been purged of religious thought. It was a tacit embrace of the narrative that Huxley and friends had created, in which science had no traces of theism.[72] The naturalistic story had become the default. British scientists spoke less about religion just as Britons in general spoke less about religion (be it for reasons of privatization or secularization). The difference was that scientists told a particular story about that: Huxley's story.

It was not until a full generation after the death of the X-Club that young scientists were typically nonreligious. Bowler points out the strong influence of teachers in this development.[73] Suggestively, this generation of teachers would be the first group born since Huxley's naturalistic science education system became the standard. Theism no longer had a home within professional science.

The Leap to America

The development and expression of X-Club naturalism was a distinctively British story. However, naturalism by no means stayed in Britain. The works of Huxley, Tyndall, and Spencer moved to America quite quickly. Their writings proved just as popular as in Britain. An observer at Dartmouth in 1873 noted that the "staple subject for reading and talk" was "English science. Herbert Spencer, John Stuart Mill, Huxley, Darwin, Tyndall."[74] They arrived just as American universities were undergoing a massive shift from ecclesiastically controlled learning institutions to disciplinary-focused research universities.[75] The key figures behind naturalistic science literature appearing in America were the Youmans brothers, Edward and William, and their connections to the New York–

based publishing house D. Appleton. They signed all of the major scientific naturalists to the International Scientific Series, which spread Darwin and Huxley across the United States.[76] Edward Youmans also edited the *Popular Science Monthly*, which proved to be a powerful force for distributing naturalistic science.[77] When Tyndall, Spencer, and Huxley arrived personally in the United States, they were welcomed with open arms and feted across the country.[78] Their naturalism was distinctly noted. When Huxley addressed the newly founded Johns Hopkins and the ceremonies closed there without prayer, Presbyterians were warned to stay away: "Honor here was refused to the Almighty."[79] Their antiorthodox arguments, aimed at the Church of England, were still intelligible to American intellectuals unhappy with religiously based colleges.

The British scientific naturalists gained a number of vigorous supporters in America who pushed their ideas widely, including Simon Newcomb, Andrew Dickson White, and Henry Draper. Newcomb was an influential spokesperson, but at the Naval Observatory he had limited influence over education.[80] White's screed *A History of the Warfare of Science with Theology in Christendom* actually had little to say about the doing of science, being mostly a list of the errors of religious thought.

By far the most important American convert to naturalism was John Dewey. That educational philosopher was a major figure in making naturalism the default mode of conversation in science education.[81] As an undergraduate, Dewey read Tyndall and Spencer, as well as widely in publications such as *Westminster Review, Fortnightly Review, Nineteenth Century*, and *Quarterly Review*.[82] When he took a physiology class, he used Huxley's *Elements of Physiology* textbook, which he credited with sparking his deeper study of knowledge: "I imagine that was the beginning of my interest in philosophy—the organic character of living creatures impressed me deeply."[83] He was particularly struck with Huxley's efforts to replace theistic understandings of the mind with naturalistic ones.[84] After reading Huxley, he began devouring philosophy, including Hume and Mill. To do his graduate work, he went to Johns Hopkins, where Huxley and Foster's student Martin had seeded naturalistic biology. One of his mentors there, G. Stanley Hall, emphasized that philosophers needed to think about science, even spending time in the laboratory. Hall was no secularist, even while arguing that psychology should take religion as a subject. Dewey studied psychophysics (including doing his own lab work) and psychological ethics with Hall, gaining understanding of naturalistic approaches to the mind and body.[85] His interest in the mind as a scientific object even drew him to read Faraday and Maxwell to help grasp the interconversion of forces.[86] He was hired for his first job at the University of Michigan precisely to teach Spencer, Lewes, and to talk about religion in "a fair, comprehensive and completely undogmatic way, and based upon data." Dewey was assigned mostly science courses, including a psychology lab, and taught an entire class on Spencer.[87]

Dewey's naturalistic ideas about science became central to American educational reform in the early twentieth century. This period saw energetic reform of science teaching on the secondary level, particularly through developing new syllabi for biology teaching.[88] These reforms relied heavily on the development of new textbooks. Huxley's *Physiography* was held up as the exemplar of what a science text should be, and both Huxley and Tyndall were frequently invoked as models of science teaching.[89] The new textbooks being developed depended heavily on Dewey's ideas, particularly his 1910 *How We Think* and "The Influence of Darwin on Philosophy."[90] When discussing Darwin, he denied that religious considerations were an issue. Dewey classified religious thought as "not creative but conservative." Anti-Darwinian ideas were not religious per se, because religion could not create new thoughts: "There is not, I think, an instance of any large idea about the world being independently generated by religion."[91] He did not accept that religion could be part of a scientific conversation at all, even as an enemy. Conversations about science had to be purely naturalistic.

How We Think presented Dewey's famous five-step scientific method. The book advocated for all the values of naturalistic science: the danger of dogma and authority, the inevitability of uncertainty, the need for hypothesis and theory, the gathering of facts under general principles. He quoted Clifford on the special powers of science to deal with the unknown, and Darwin's revelation that *"science consists in grouping facts so that general laws or conclusions may be drawn from them."*[92] John Rudolph has shown how this document became a textbook for science teachers, being widely distributed across the country. The naturalistic "scientific method" quickly became the standard. By 1918, science educators were even using Dewey's formulation to characterize historical scientists such as Newton, Descartes, and Pasteur.[93] Huxley's vision of an unbroken naturalistic science through history had come to America.

Dewey was not a ferocious evangelical naturalist like Huxley or Tyndall, and that is precisely why he is significant. He had become convinced that naturalism was the ordinary and obvious way to do science, and he expressed that in his philosophy of education. Naturalism did not need to be treated as a controversial topic. How could science be anything other than naturalistic? Naturalism became a fundamental part of American science teaching almost without comment.

Conclusion

The naturalists' own explanation of why they came to dominate science is clearly stated: naturalistic science is methodologically superior to all other possibilities, and therefore won. That is Huxley's story, and we have now seen how it came to be gospel despite naturalism's deep methodological similarities to theistic science. Even later theistic scientists, educated in Huxley's scientific pedagogy, accepted this story. It is remarkable how the naturalistic narrative came to be the standard even for religious figures, who seem to have forgotten their own intellectual ancestry.[94] It is important to note that Huxley's strategies did not make it impossible to be a religious scientist—rather, they flipped the default setting for scientists from theistic to naturalistic. Religious scientists in the twentieth century were the ones under the obligation to justify themselves, just as the young Huxley and Tyndall had been forced to do.

Huxley and colleagues managed to take over the community of science without a serious disruption in the practice of science. One sign of this was the survival and continued relevance of the same scientific societies (e.g., the British Association and the Royal Society) throughout the transition from theism to naturalism. There was no dramatic break in which naturalistic men of science had to create their own community, as Boyle and friends had to do in the seventeenth century. Despite the naturalists' claims, their methodological values were perfectly recognizable to theistic men of science and could be judged on the same terms (with the increasing exception of the sciences of the mind). That is not to say there were not controversies over specific claims (natural selection did not appear in the *Philosophical Transactions* for a long time) or matters of leadership (Huxley's ascension to the presidency of the Royal Society). But on the question of the fundamentals of science, the theists and the naturalists played on the same field. The overlapping values documented in the earlier chapters of this book were necessary for all the strategies used by the naturalists to take control of the community. Without those overlaps, they would never have been acknowledged as members of the community at all, much less been recognized as authorities worthy of leadership.

A major remaining issue is why the theistic scientists let this happen. Why were they outmaneuvered by Huxley? It is certainly the case that theists in British science can be categorized into groups with common interests and goals (such as Crosbie Smith's North British group). However, those groups were never organized into cohesive associations with coordinated actions such as the X-Club. Lightman notes that those espousing a theistic view of science in the late nineteenth century had nothing "remotely similar."[95] Maxwell's Cavendish lab certainly trained as many influential scientists as South Kensington, and his theories did form the core of a distinct group of scientific investigators. Why did they not spread Maxwell's theistic science? The answer is simply that he did not train them to do so. It never occurred to him to act as Huxley and create an easily spreadable curriculum and set of practices that would disseminate his worldview. He did not try to place his students in positions of power that would carry on his work.

To a certain degree this was simply a matter of complacency. Theistic science had been the default mode for a very long time. Proactive organization and training to protect it seemed unnecessary for the system that was already embedded in power. Theistic men of science did not seriously think that theism could be completely displaced from science, any more than Christianity could be truly displaced from the core of British life. By the time that they realized that elementary science education was in the hands of naturalists (if indeed they ever noticed), it was far too late. In this sense, theism was displaced from below—the naturalists successfully used unimportant positions (say, the School of Mines) and drudge work (the endless grading of tedious exams) to dig tunnels and spread into enemy territory. Just as women managed to enter into science through work designated as uninteresting or unworthy of men, the naturalists broke into science through the new bureaucracy of state education. Once inside, however, they convinced everyone that they had always been there.

Conclusion

In terms of doing science, not much changed. From 1850 to 1890, the most common justifications for the principles of science shifted from theistic to naturalistic, but the principles themselves remained remarkably stable. Even enormous theoretical innovations such as Darwinian evolution had little effect on scientific practice.[1] Despite the claims of Huxley and the X-Club, naturalism did not bring a dramatic revolution to how men of science conducted their research.

Rather, the scientific naturalists adopted methodological values that had been established by theistic scientists generations before. Concepts that were seen as critical to science—the uniformity of nature, limits to scientific knowledge, the moral value of science, intellectual freedom— were originally formed and justified in a profoundly religious context. The values that made the practice of science possible were deeply linked to a Christian worldview. They provided a foundation on which Victorian science thrived and grew. This book's close examination of Maxwell's work and career has shown how these values were not merely incidentally theistic, but instead that theistic men of science saw them as unique to their worldview. A nontheistic science seemed impossible.

Huxley did not contest the values that made Victorian science work (with the important exception of the scientific status of the mind). He adopted them as his own, arguing that in fact they could only be justified in his naturalistic worldview. Even as he researched alongside theistic men of science, he labored to make their identity a virtual oxymoron. By adopting a generational strategy to shift the training of science teachers in a naturalistic direction, he recast the methodological values that had supported Maxwell's work. They would no longer be theistic, but instead allied by definition with the lack of theological considerations. Huxley took science from theism to naturalism by telling a story that denied any change had ever taken place—science could never have been anything other than naturalistic. The naturalists successfully wrapped the history of theistic science in their own worldview, making any alternative seem absurd.

The transition from theism to naturalism was remarkably smooth. They functioned side by side for a generation, sharing values and methods regardless of the dramatic rhetorical attacks their practitioners leveled at each other. In terms of community and organization, there was no deep split. Since they agreed on most methodological values, the naturalists and theists did not separate themselves into different scientific societies. They published, presented, and worked together. The BAAS and the Royal Society gradually shifted from dominance by one group to the other. This is a further indication that, on the most fundamental level, they were doing the same kind of work. Neither group felt the need to create a new space for themselves, because their research was perfectly recognizable even to their enemies. In contrast, the Society for Psychical Research was founded because it was not at all clear that investigating ghosts counted as "science." Those persons interested in studying spiritualism (despite scientific excellence in other areas) failed to make the case that this work was part of mainstream science. Maxwell had no such problems with his molecular science, despite its clear theistic context. Continuity of institutions was made possible by continuity of practice.

Quite different, however, is the intelligent design community of the twenty-first century. ID scholars have not been able to participate in mainstream science journals and organizations. This is generally not because of scientific dogma or prejudice, but rather because they refuse to accept the principles of the uniformity of nature, the provisional character of science, and so forth, which have been the core methodological values of science since at least the dawn of the Victorian period. Adherents of ID have instead had to create their own parallel journals and societies such as *BIO-Complexity* and the Discovery Institute. These are necessary because they wish to create their own rules about how to do science. Phillip Johnson and Alvin Plantinga talk about simply wanting to include religious knowledge in science. If that were all they wanted, we should expect their claims to look very much like those of Maxwell and the highly successful tradition of theistic science. This book has shown how scientists can function within the scientific community while still maintaining a robust Christianity. But members of the ID community are clearly unwilling to do this, which suggests that they are less interested in participating in the scientific enterprise than in attacking the values of modernity and pluralism that they see as inherent to naturalism.

A major factor that sets ID apart from theistic science is the deep concerns of Maxwell and others about the further development of science. Despite his reverence for the Bible and divine creation, Maxwell worked hard to avoid what are today called "science stoppers." Consider the very common scenario in which an investigator encounters a mysterious phenomenon that she cannot understand. One alternative is that she assumes that through further research she *will* be able to understand it, and continues her work. Another is that she decides the mystery to be beyond human understanding, and ceases any attempts to understand it. This move is the "science stopper"—a declaration that a mysterious phenomenon *will never* be understood, and must simply be accepted as divine action. An important example of this is Michael Behe's claims that the lack of understanding of certain biochemical processes indicates that science will never understand those properties, and therefore nonnatural explanations (chiefly divine action) must be considered.[2] If this claim is accepted, then biochemistry is at an end—no further research can be done, and nothing new can ever be learned. Maxwell was well aware of this danger of design arguments, and as we saw in chapter 2, he found a productive interpretation of design that avoided exactly this problem. This was characteristic of the Victorian theistic science community— confidence that scientific exploration into current unknowns was better than simply declaring something to be an eternally unknowable mystery. Without this move, science becomes pointless, as every unknown becomes impenetrable. The philosopher Philip Kitcher emphasizes that any natural process looks mysterious—until it is understood.[3]

Plantinga accepts that science stoppers are essentially fatal to the enterprise of science, but is still willing to allow them.[4] And it is probably not a coincidence that a group willing to embrace science stoppers has also had great difficulty generating new research. The claim of the Discovery Institute is that ID is a research project that will generate new science showing the validity of its ideas. It is remarkable that despite plentiful funding and support, the ID community has essentially produced nothing compelling to anyone outside its own ranks.[5] The *Dover* decision explicitly called this out as a reason to not allow ID in the science classroom: "It is additionally important to note that ID has failed to gain acceptance in the scientific community, it has not generated peer-reviewed publications, nor has it been the subject of testing and research."[6] Barbara Forrest's historical overview of ID notes that the production of peer-reviewed research was originally an essential part of the Discovery Institute's strategic plan, making the absence of such work that much more remarkable. She notes that even Michael Behe, one of the few "real" scientists (that is, having a PhD and faculty position in the sciences) in the ID community, has not done any research based on those ideas.[7] If ID is an important framework for science, someone must demonstrate that by using it to *do* science. Huxley was able to create a new justification for science because he showed how his new worldview could generate methodologies recognizable to other scientific researchers of his day. This is precisely what ID must do if it wishes to be taken seriously as a new approach to science.

It is somewhat peculiar that these modern controversies over naturalism have appeared in America instead of the concept's original home in Great Britain. An important factor in this was the differences in educational systems between the two countries. In early twentieth-century Britain the way science was taught became more and more similar between elementary-secondary and university levels as both came under closer government control. However, during the same period American education saw a widening gap between elementary-secondary and higher education.[8] The colleges increasingly moved away from orthodoxy while lower levels of education were more exposed to the religious revivals in twentieth-century America. This resulted in what we might think of as a "naturalism gap"—professional scientists and other intellectuals are thoroughly educated in the Huxleyian views of science, while the broader public is not. To the intellectuals, naturalism seems obvious. To many others, it seems elite and dogmatic. Further, the language of naturalism that we have inherited from the Victorians does not always translate well to the American present. Huxley's attacks on the established Church were largely understood by his contemporaries for what they were—protests against institutional privilege and enforced belief. However, in modern America there is no established church and Huxley's verbiage reads as an assault on Christianity as a whole. Since naturalism's original target is not present, it seems hostile to a much broader audience than originally intended.

This suggests a wider problem with the use of the term *naturalism* by science advocates today. They use it in the same sense that Huxley did, intending to point to positive scientific values while leaving "true religion" untouched. But the term cannot seem to shake its original pejorative connotation of opposition to the supernatural. It sounds irredeemably hostile to religion, regardless of the subtleties we might want to attach to it. Modifying it to "methodological naturalism" does not help much—Plantinga is correct that it sounds like a simple cover for "provisional atheism." Certainly Maxwell would not have agreed that his work was methodologically naturalistic—he saw God and religious considerations as critical facets of his scientific methodology.

Much of the appeal of the term *naturalism* among science proponents today is that it seems to provide a clear and easy standard by which to reject ID. This is a continuation of the long-standing demarcation problem in the philosophy of science—how are we to distinguish science from nonscience? Jeffrey Koperski and Larry Laudan have suggested that demarcation is not a particularly helpful way to attack ID.[9] Laudan notes that philosophy has not been very successful at defining science, which makes accusations that ID is "unscientific" rather vague. He suggests that

> we ought to drop terms like "pseudo-science" and "unscientific" from our vocabulary; they are just hollow phrases which do only emotive work for us . . . our focus should be squarely on the empirical and conceptual credentials for claims about the world. The "scientific" status of those claims is altogether irrelevant.[10]

Using demarcation to simply declare ID unscientific eliminates the strongest arguments scientists have against it—that it is just not a very good explanation for what we can see in the world. Laudan points out that if creationists make claims, "we should confront their claims directly and in piecemeal fashion by asking what evidence and arguments can be marshaled for and against each of them."[11] If their claims are testable, they should be tested.

But if we instead declare them unscientific because they fail the test of naturalism, those claims become irrefutable. And even worse, it makes the ground rules of science seem arbitrary and dogmatic by excluding certain claims by definition. This provides ammunition to those attacking science, who do not hesitate to paint science as functioning only through oppressive authority. Refusing to acknowledge an idea because it has its roots in religion makes scientists look as though they are afraid of open debate.[12] Worse, insisting on using the term *naturalism* (regardless of the metaphysical/methodological distinction) only reifies the claims of Richard Dawkins and Steven Weinberg that science is inextricably allied to atheism. This is both historically inaccurate and quite damaging for science by making an unnecessary enemy of anyone with religious beliefs (the vast majority of Americans).

As an alternative, I suggest simply that critics of ID be more precise in their objections. Instead of dismissing a claim because it is not naturalistic (which sounds as if it is being rejected because it is religious), it should be attacked on the grounds that it does not fit into the uniformity of natural laws (which explains that it falls outside the standards set by science for the last few centuries). Plantinga should not be criticized because he wants God as part of his understanding of nature; he should be criticized because he suggests that "earthquakes, the weather, and radioactive decay" might not be subject to natural laws.[13] There is no reason to use *naturalism* as though it is synonymous with uniformity, or provisional knowledge, or empiricism, or any of the other foundations of science that, as this book has argued, are perfectly justifiable in religious terms. We have plenty of precise terms to explain the problems of ID, and we should use them.

Using language that is more precise instead of simply antireligious will also help welcome religious thinkers who seek to embrace science. These scholars pursue something that is often called "religious naturalism."[14] Their ideas are strongly reminiscent of the theistic science tradition: a God who creates through natural laws and encourages scientific exploration, a sophisticated use of theory and hypothesis, and a rejection of sectarian dogma. The scholars involved explicitly make the case that their ideas are distant from ID's approach to science.[15]

Someone pursuing such an approach must confront Ernan McMullin's challenge that theistic science cannot be universal.[16] McMullin argued that theistic science was impossible because a Christian researcher needed to be able to convince a Hindu of his results. It is surely true that scientific knowledge must be able to cross religious boundaries. However, this book has shown how the work of deeply religious, or deeply areligious, scientists *can* cross boundaries. Scientists in the end have to persuade their colleagues that their ideas are correct. But we should not lay a priori restrictions on how they get their ideas. They may get their ideas from anywhere (cold empiricism or mystical inspiration), but they need to be able to justify them in terms that make sense to other people trained and experienced in the exploration of the phenomena in question. As Koperski says, we must judge arguments by their evidence, not their intent.[17] The same problem emerges with politicians, where we cannot prevent politicians from having their views formed by their religion, but that religious formation also cannot be sufficient to support them. No one would try to make the case that Martin Luther King Jr. should not have been involved in politics because of his religious motivations. He was able to convince many people that his views were worth supporting, even if they did not share his religious beliefs—just so with scientists. If a scientist can convince other scientists that her conclusions are useful, who cares whether she was thinking about God when she did the work? Michael Ruse argues that scientists should not talk about religion in the same way that people would not want to get political advice from their doctor.[18] That may be true, but surely we would not prevent a doctor from practicing medicine simply because he wanted to do so for political reasons. Our concerns should focus on whether the individual is a good doctor. William James suggested that when thinking about the nature of religion, we should be more concerned with results than with causes: "By their fruits ye shall know them, not by their roots."[19] His advice is sound for science as well: if someone's research is persuasive to the scientific community, we should not call it unscientific only due to its origins.

Intelligent design theory is incompatible with the methodological values of science for a host of reasons, but simply being rooted in religious thought is not one of them. *Huxley's Church and Maxwell's Demon* has shown that there are many ways for religion to be part of the values of science that look nothing like ID. Theistic science was once the mainstream of science, and its successes suggest that there are a variety of ways to think about the foundations of scientific practice. Today we live in Huxley's church, and it is easy to forget that it was not always there. The demons it was designed to protect us from have largely been exiled as myth and superstition, though careful study can see their effects in the past. But the construction of that church was far from easy or obvious, and sometimes its walls can blind us to other possibilities. Science can be a house with many rooms.

Acknowledgments

Although he probably does not remember, this book began life during a lunch with Michael Clough at Iowa State University. I am grateful to him for the enduring stimulation provided by that conversation. Another colleague at Iowa State, David Wilson, provided invaluable advice during the project's initial development. These ideas would have gone nowhere if not for the support of a Mellon Fellowship at the Institute for Advanced Study. The Institute's extraordinary scholarly environment gave me both uninterrupted time to think and critical guidance when I was forming the book's foundational ideas. The argument was built there during long walks in the woods. I extend special thanks to Heinrich von Staden, Jonathan Israel, and to whoever was responsible for the snacks provided at each afternoon's coffee breaks. My research in the United Kingdom was supported by the British Academy and Frank James at the Royal Institution.

I must thank especially the Gallatin School of New York University. The school, and particularly Dean Susanne Wofford, provided the resources needed to complete this book. My colleagues here have created a wonderful interdisciplinary environment. My scholarship has matured greatly since I arrived—this book would be quite different if not for the constant exposure to their electrifying ideas.

Michael Gordin, Bernard Lightman, and James Elwick provided critical feedback and advice throughout the writing process. Gabriel Henderson, Alexander Thompson, Michael Swain, and Eric Adamson were all essential research assistants. I appreciate access and assistance provided by archivists and librarians at Iowa State University, Michigan State University, New York University, the Institute for Advanced Study, Princeton University, the New York Public Library, the British Library,

Imperial College London, Cambridge University Library, the London Natural History Museum, Royal Botanic Gardens Kew, King's College London, Aberdeen Library, the National Library of Scotland, and the Royal Society of London. Portions of chapter 2 previously appeared as "By Design: James Clerk Maxwell and the Evangelical Unification of Science," *British Journal for the History of Science* 45 (March 2012): 57–73; and "The Uniformity of Natural Laws in Victorian Britain: Naturalism, Theism, and Scientific Practice," *Zygon* 46, no. 3 (September 2011): 536–60. Parts of chapter 6 were originally presented as "The Pointsman: Maxwell's Demon, Victorian Free Will, and the Boundaries of Science," *Journal of the History of Ideas* 69, no. 3 (July 2008): 467–91.

Karen Merikangas Darling and everyone at the University of Chicago Press were wonderfully supportive throughout the process. The manuscript was notably improved by the suggestions of the Press's anonymous reviewers. This book of course relies heavily on the work of previous scholars, whom I hope I have credited fully in the notes. However, three require special comment. My thinking about scientific naturalism owes a great deal to the late Frank M. Turner. It would have been very difficult to write this book without Adrian Desmond's pioneering work on Huxley. It would have been virtually impossible to write this book without P. M. Harman's compilation and editing of Maxwell's papers and letters. I am grateful to them all.

No doubt my failing memory and incomplete notes mean that I have forgotten critical contributions from other people. I offer them my apologies in advance. As always, all mistakes are my own.

Notes

Introduction

1. Michael Ruse, "Methodological Naturalism under Attack," *South African Journal of Philosophy* 24, no. 1 (2005): 46.

2. Phillip E. Johnson, "Evolution as Dogma: The Establishment of Naturalism," in *Intelligent Design Creationism and Its Critics: Philosophical, Theological, and Scientific Perspectives*, ed. Robert Pennock (Cambridge, MA: MIT Press, 2001), 66 (hereafter cited as *IDC*).

3. *Naturalism* is a somewhat vaguely defined term, but it is useful both as an actor's category and for historical investigation. It is a convenient category of analysis but is not an intrinsically stable entity. On the use of such concepts, see Ralph O'Connor, "Reflections on Popular Science in Britain: Genres, Categories, Historians," *Isis* 100 (2009): 333–45. This use of *naturalism* should not be confused with the epistemic naturalism sometimes discussed by philosophers of science. See Joseph Rouse, *How Scientific Practices Matter: Reclaiming Philosophical Naturalism* (Chicago: University of Chicago Press, 2003). Similarly, *naturalist* will here refer to someone who holds the view of naturalism, rather than to someone who simply studies nature.

4. The 2005 *Dover* decision keeping ID out of classrooms dates this to the sixteenth and seventeenth centuries. *Tammy Kitzmiller, et al. v. Dover Area School District, et al.*, 400 F. Supp. 2d 707 (M.D. Pa. 2005), 64–65. ID philosopher Alvin Plantinga dates this to the Enlightenment. Alvin Plantinga, "Methodological Naturalism?," in *IDC*, 342–43.

5. On the fortunes of creationism and intelligent design in the courts, see Ronald L. Numbers, *The Creationists: From Scientific Creationism to Intelligent Design* (Cambridge, MA: Harvard University Press, 2006); Edward J. Larson, *Trial and Error: The American Controversy over Creation and Evolution* (Oxford: Oxford University Press, 2003); Marcel C. LaFollette, *Creationism, Science, and the Law: The Arkansas Case* (Cambridge, MA: MIT Press, 1983); and

Michael Ruse, "A Philosopher's Day in Court," in *But Is It Science? The Philosophical Question in the Creation/Evolution Controversy*, ed. Michael Ruse, 13–35 (Buffalo, NY: Prometheus Books, 1988).

6. *Kitzmiller v. Dover*, 65.

7. Michael Ruse, *Evolutionary Naturalism: Selected Essays* (London: Routledge, 1995), 2; and Robert T. Pennock, *Tower of Babel: The Evidence against the New Creationism* (Cambridge, MA: MIT Press, 1999), 194–95.

8. Ruse, "Methodological Naturalism under Attack," 54.

9. Gowan Dawson and Bernard Lightman, introduction to *Victorian Scientific Naturalism: Community, Identity, Continuity* (Chicago: University of Chicago Press, forthcoming).

10. Frank M. Turner, *Between Science and Religion: The Reaction to Scientific Naturalism in Late Victorian England* (New Haven, CT: Yale University Press, 1974), 24. The naturalists were not the only group to claim energy physics for their own. See Crosbie Smith, *The Science of Energy: A Cultural History of Energy Physics in Victorian Britain* (Chicago: University of Chicago Press, 1998).

11. The best overview is Ronald L. Numbers, "Science without God: Natural Laws and Christian Beliefs," in *When Science and Christianity Meet*, ed. David C. Lindberg and Ronald L. Numbers, 265–85 (Chicago: University of Chicago Press, 2003).

12. On this distinction, see Robert T. Pennock, "Naturalism, Evidence and Creationism: The Case of Philip Johnson," *Biology and Philosophy* 11, no. 4 (1996): 543–44; and Robert T. Pennock, *Tower of Babel*, 190. "Provisional atheism" comes from Plantinga, "When Faith and Reason Clash: Evolution and the Bible," in *IDC*, 137. The term *methodological naturalism* was first coined by Paul de Vries in 1983. See Numbers, "Science without God," 266.

13. Ruse, "Methodological Naturalism under Attack," 44.

14. Ibid., 58; Pennock, *Tower of Babel: The Evidence against the New Creationism*, 192; and Michael Ruse, *Can a Darwinian Be a Christian? The Relationship between Science and Religion* (Cambridge: Cambridge University Press, 2001), 99.

15. Barbara Herrnstein Smith, *Natural Reflections: Human Cognition at the Nexus of Science and Religion* (New Haven, CT: Yale University Press, 2009), 121–26.

16. See, for example, Johnson and Pennock's exchanges in *IDC*. This volume is a valuable collection of resources for anyone interested in these issues.

17. Bernard Lightman, "Victorian Sciences and Religions: Discordant Harmonies," in *Osiris*, vol. 16, *Science in Theistic Contexts (Cognitive Dimensions)*, ed. John Hedley Brooke, Margaret J. Osler, and Jitse van der Meer (Chicago: University of Chicago Press, 2001), 362.

18. Turner, *Between Science and Religion*, 5.

19. Matthew Stanley, *Practical Mystic: Religion, Science, and A. S. Eddington* (Chicago: University of Chicago Press, 2007), 6–7.

20. Peter Galison, *Image and Logic: A Material Culture of Microphysics* (Chicago: University of Chicago Press, 1997), 803–40.

21. Ibid., 806.

22. On the Darwin literature, see "Defining Darwinism: One Hundred and Fifty Years of Debate," *Studies in the History of Philosophy of Biology and Biomedical Science* 42 (2011): 2–4; William Brown and Andrew Fabian, eds., *Darwin (The Darwin College Lectures)* (Cambridge: Cambridge University Press, 2010); David J. Depew, "Darwinian Controversies: An Historiographical Recounting," *Science and Education* 19 (2010): 323–66; Janet Browne, *Charles Darwin: Voyaging*, vol. 1 (New York: Knopf, 1995); and Janet Browne, *Charles Darwin: The Power of Place*, vol. 2 (New York: Knopf, 2002). On Darwin's changing place in the historiography of the nineteenth century, see Gowan Dawson, "First among Equals," *Times Literary Supplement*, January 9, 2009, 7–8.

23. Pennock, *Tower of Babel*, 321–24; Plantinga, "When Faith and Reason Clash," 138; and Philip Kitcher, "Born-Again Creationism," in *IDC*, 257–87.

24. For example, see Johnson, "Evolution as Dogma"; Phillip E. Johnson, *Reason in the Balance: The Case against Naturalism in Science, Law, and Education* (Downers Grove, IL: InterVarsity Press, 1993), 7–9, 45; and Phillip E. Johnson, *Darwin on Trial* (Crowborough, UK: Monarch, 1991), 116.

Chapter One

1. See Jordi Cat, *Land, Lines, Colours and Toys: Becoming James Clerk Maxwell* (Oxford: Oxford University Press, forthcoming), for more on Maxwell's deep family history. The best resource on Maxwell's life remains the Victorian biography by Lewis Campbell and William Garnett, *The Life of James Clerk Maxwell* (London: Macmillan, 1882) (hereafter cited as *Life*). The literature on Maxwell's scientific work will be discussed in later chapters.

2. Callum G. Brown, *The Social History of Religion in Scotland since 1730* (London: Methuen, 1987), 137–39.

3. A. C. Cheyne, *Studies in Scottish Church History* (Edinburgh: T&T Clark, 1999); Keith Robbins, "Religion and Community in Scotland and Wales since 1800," in *A History of Religion in Britain*, ed. Sheridan Gilley and W. J. Sheils, 363–80 (Cambridge, MA: Blackwell, 1994); and Stewart Brown and Michael Fry, eds., *Scotland in the Age of Disruption* (Edinburgh: Edinburgh University Press, 1993).

4. Campbell and Garnett, *Life*, 420.

5. C. W. F. Everitt, "Maxwell's Scientific Creativity," in *Springs of Scientific Creativity*, ed. Rutherford Aris, Horace B. Davis, and Roger H. Stuewer, 71–141 (Minneapolis: University of Minnesota Press, 1983), 114.

6. On Maxwell's religious background, see Paul Theerman, "James Clerk Maxwell and Religion," *American Journal of Physics* 54 (1986): 312–17; and Crosbie Smith, *The Science of Energy: A Cultural History of Energy Physics in Victorian Britain* (Chicago: University of Chicago Press, 1998), chap. 11.

7. John Clerk Maxwell to James Clerk Maxwell, March 15, 1853, quoted in Campbell and Garnett, *Life*, 185.

8. Campbell and Garnett, *Life*, 27–28, 31.

9. Ibid., 49–50, 54.

10. On Maxwell's debt to Scottish thought, see Richard Olson, *Scottish Philosophy and British Physics, 1750–1880: A Study in the Foundations of the Victorian Scientific Style* (Princeton, NJ: Princeton University Press, 1975).

11. Campbell and Garnett, *Life*, 130–31.

12. Susan Faye Cannon, *Science in Culture: The Early Victorian Period* (New York: Dawson, 1978), 225.

13. Laura J. Snyder, *Reforming Philosophy: A Victorian Debate on Science and Society* (Chicago: University of Chicago Press, 2006); Laura J. Snyder, *The Philosophical Breakfast Club: Four Remarkable Friends Who Transformed Science and Changed the World* (New York: Broadway, 2011); and Richard Yeo, *Defining Science: William Whewell, Natural Knowledge and Public Debate in Early Victorian Britain* (Cambridge: Cambridge University Press, 2003).

14. Campbell and Garnett, *Life*, 107.

15. Cannon, *Science in Culture*, 268.

16. Campbell and Garnett, *Life*, 175.

17. David B. Wilson, "A Physicist's Alternative to Materialism: The Religious Thought of George Gabriel Stokes," in *Energy and Entropy*, ed. Patrick Brantlinger, 177–203 (Bloomington: Indiana University Press, 1989); and David B. Wilson, *Kelvin and Stokes: A Comparative Study in Victorian Physics* (Bristol, UK: A. Hilger, 1987).

18. James Clerk Maxwell to anonymous, September 16, 1850, quoted in Campbell and Garnett, *Life*, 144–45.

19. Crosbie Smith and M. Norton Wise, *Energy and Empire: A Biographical Study of Lord Kelvin* (Cambridge: Cambridge University Press, 1989); Wilson, *Kelvin and Stokes*; and Joe D. Burchfield, *Lord Kelvin and the Age of the Earth* (New York: Science History Publications, 1975).

20. Crosbie Smith, *Science of Energy*.

21. Campbell and Garnett, *Life*, 363n220.

22. F. J. A. Hort to L. Campbell, February 4, 1882, quoted in Campbell and Garnett, *Life*, 417.

23. Ibid., 630.

24. Andrew Warwick, *Masters of Theory: Cambridge and the Rise of Mathematical Physics* (Chicago: University of Chicago Press, 2003), 191–200.

25. William Paley, *Evidences of Christianity* (1794; repr., London: W. Clowes and Sons, 1851); and William Paley, *Natural Theology* (1802; repr., Boston: Gould and Lincoln, 1872).

26. Robert M. Young, *Darwin's Metaphor: Nature's Place in Victorian Culture* (Cambridge: Cambridge University Press, 1985), 127; and Cannon, *Science in Culture*.

27. John Hedley Brooke, *Science and Religion: Some Historical Perspectives* (Cambridge: Cambridge University Press, 1991), 198–99.

28. Ibid., 211.

29. John Hedley Brooke and Geoffrey Cantor, *Reconstructing Nature: The Engagement of Science and Religion* (Oxford: Oxford University Press, 1998), 153.

30. Ibid., 142–48.

31. Jonathan R. Topham, "An Infinite Variety of Arguments: The Bridgewater Treatises and British Natural Theology in the 1830's" (PhD diss., Lancaster University, 1993).

32. Ibid., chap. 6.

33. James Clerk Maxwell to Rev. L. Campbell, December 22, 1857, quoted in Campbell and Garnett, *Life*, 294.

34. P. M. Harman, *The Natural Philosophy of James Clerk Maxwell* (Cambridge: Cambridge University Press, 1998), 162–63; and Matthew Stanley, "A Modern Natural Theology?," *Journal of Faith and Science Exchange* 3 (1999): 105–12.

35. Young, *Darwin's Metaphor*, 139–40.

36. A useful start for understanding evangelicalism is Mark A. Noll, David W. Bebbington, and George A. Rawlyk, eds., *Evangelicalism: Comparative Studies of Popular Protestantism in North America, the British Isles, and Beyond, 1700–1990* (New York and Oxford: Oxford University Press, 1994). On its relation to science, see David N. Livingstone, D. G. Hart, and Mark A. Noll, eds., *Evangelicals and Science in Historical Perspective* (New York: Oxford, 1999); William J. Astore, *Observing God: Thomas Dick, Evangelicalism, and Popular Science in Victorian Britain and America* (London: Ashgate, 2001); and Aileen Fyfe, *Science and Salvation: Evangelical Popular Science Publishing in Victorian Britain* (Chicago: University of Chicago Press, 2004). In the context of Victorian thought generally, see Boyd Hilton, *The Age of Atonement: The Influence of Evangelicalism on Social and Economic Thought, 1785–1865* (Oxford: Oxford University Press, 1991).

37. Campbell and Garnett, *Life*, 167–71.

38. James Clerk Maxwell to Charles Benjamin Tayler, July 8, 1853, in *The*

Scientific Letters and Papers of James Clerk Maxwell, ed. P. M. Harman, 3 vols. (Cambridge: Cambridge University Press, 1990), 1:220–21 (hereafter cited as *SLP*). Maxwell had shown some inclination toward evangelical attitudes earlier that year: on February 20, 1853, he wrote to Lewis Campbell that he could not apprehend evil with rational means, which suggests a more emotional, feeling-based approach to religion. See Campbell and Garnett, *Life*, 182–83.

39. James Clerk Maxwell to R. B. Litchfield, August 23, 1853, quoted in Campbell and Garnett, *Life*, 191–92.

40. John Clerk Maxwell to James Clerk Maxwell, December 16, 1853, quoted in Campbell and Garnett, *Life*, 193–95.

41. Campbell and Garnett, *Life*, 170n64.

42. Ibid., 258–59.

43. James Clerk Maxwell to Miss K. M. Dewar, May 13, 1858, quoted in Campbell and Garnett, *Life*, 311–12.

44. There is an anecdote from Karl Pearson presenting Maxwell as a biblical literalist, quoted in Crosbie Smith, *Science of Energy*, 307. This story is quite different from the other evidence we have about Maxwell's religious beliefs and outlook. I would infer that Pearson mistook Maxwell's deep respect for scripture for a slavish literalism. Maxwell was certainly not averse to interpreting biblical passages. See Maxwell to Charles John Ellicott, bishop of Gloucester and Bristol, November 22, 1876, in *SLP*, 3:416–18.

45. James Clerk Maxwell to his wife, April 13, 1860, quoted in Campbell and Garnett, *Life*, 328.

46. James Clerk Maxwell to his wife, June 28, 1864, quoted in Campbell and Garnett, *Life*, 339.

47. James Clerk Maxwell to Rev. C. B. Tayler, February 2, 1866, quoted in Campbell and Garnett, *Life*, 345.

48. Ibid., 322–23, on 323n.

49. Simon Schaffer, "Metrology, Metrication, and Victorian Values," in *Victorian Science in Context*, ed. Bernard Lightman, 438–74 (Chicago: University of Chicago Press, 1997); and Simon Schaffer, "Accurate Measurement Is an English Science," in *The Values of Precision*, ed. M. Norton Wise, 135–72 (Princeton, NJ: Princeton University Press, 1995).

50. Campbell and Garnett, *Life*, 426.

51. On Huxley's biography, see Adrian Desmond, *Huxley: From Devil's Disciple to Evolution's High Priest* (Reading, MA: Addison-Wesley, 1997); and Paul White, *Thomas Huxley: Making the "Man of Science"* (Cambridge: Cambridge University Press, 2003).

52. Thomas Henry Huxley, "Autobiography," in *Collected Essays of T. H. Huxley*, 9 vols. (Bristol, UK: Thoemmes Press, 2001), 1:5 (hereafter cited as *CE*); and Frank M. Turner, "Victorian Scientific Naturalism and Thomas Carlyle," *Victorian Studies* 18, no. 3 (1975): 327.

53. Thomas Henry Huxley, "Thoughts and Doings," Huxley Papers, HP 31.7, Imperial College Archives, London.

54. Thomas Henry Huxley, "Autobiography," in *CE*, 1:7.

55. Desmond, *Huxley*, 11–14.

56. Adrian Desmond, "Redefining the X Axis: 'Professionals,' 'Amateurs' and the Making of Mid-Victorian Biology: A Progress Report," *Journal of the History of Biology* 34, no. 1 (2001): 36.

57. Desmond, *Huxley*, 18–26.

58. Ibid., 43–46.

59. Ibid., 61–62.

60. Ibid., 149–56.

61. Ibid., 161.

62. Willam Hodson Brock, Norman D. McMillan, and R. Charles Mollan, eds. *John Tyndall: Essays on a Natural Philosopher* (Dublin: Royal Dublin Society, 1981); and Bernard Lightman, *The Origins of Agnosticism: Victorian Unbelief and the Limits of Knowledge* (Baltimore: Johns Hopkins University Press, 1987), 97–99.

63. Colin A. Russell, *Edward Frankland: Chemistry, Controversy and Conspiracy in Victorian England* (Cambridge: Cambridge University Press, 1996).

64. On Hooker, see Jim Endersby, *Imperial Nature: Joseph Hooker and the Practices of Victorian Science* (Chicago: University of Chicago Press, 2008). On Owen, see Nicolaas A. Rupke, *Richard Owen: Biology without Darwin* (Chicago: University of Chicago Press, 2009).

65. Desmond, *Huxley*, 162.

66. Ibid., 189–90.

67. Bernard Lightman, *Victorian Popularizers of Science: Designing Nature for New Audiences* (Chicago: University of Chicago Press, 2007), 358–59; White, *"Man of Science,"* 71–72; and James A. Secord, *Victorian Sensation: The Extraordinary Publication, Reception, and Secret Authorship of "Vestiges of the Natural History of Creation"* (Chicago: University of Chicago Press, 2000), 499.

68. Desmond, *Huxley*, 388.

69. Ibid., chap. 22.

70. Frank M. Turner, "The Victorian Conflict between Science and Religion: A Professional Dimension," *Isis* 69, no. 3 (1978): 356–76. On the difficulties with the term *professional*, see Ruth Barton, "'Men of Science': Language, Identity and Professionalization in the Mid-Victorian Scientific Community," *History of Science* 41 (2003): 73–119.

71. Lightman, *Origins*, 119–20.

72. John Tyndall, "Professor Virchow and Evolution," in *Fragments of Science: A Series of Detached Essays, Addresses, and Reviews*, vol. 2 (New York: D. Appleton, 1915), 381.

73. Turner, "Victorian Scientific Naturalism and Thomas Carlyle."

74. Thomas Henry Huxley to Charles Kingsley, September 8, 1860, quoted in Leonard Huxley, *Life and Letters of Thomas Henry Huxley*, 2 vols. (1900; repr., New York: AMS Press, 1979), 1:237.

75. Quoted in Lightman, *Origins*, 130. Tyndall's position has sometimes been called pantheistic: see Ruth Barton, "John Tyndall, Pantheist: A Rereading of the Belfast Address," *Osiris* 2, no. 3 (1987): 111–34.

76. Turner, "Victorian Scientific Naturalism and Thomas Carlyle," 337–38.

77. Thomas Henry Huxley, "The Origin of Species," in *CE*, 2:52.

78. Lightman, *Origins*, 121–22.

79. Frank M. Turner, "The Victorian Crisis of Faith and the Faith That Was Lost," in *Victorian Faith in Crisis: Essays on Continuity and Change in Nineteenth-Century Religious Belief*, ed. Richard Helmstadter and Bernard Lightman (Stanford, CA: Stanford University Press, 1990), 17.

80. James R. Moore, "Theodicy and Society: The Crisis of the Intelligentsia," in *Victorian Faith in Crisis: Essays on Continuity and Change in Nineteenth-Century Religious Belief*, edited by Richard J. Helmstadter and Bernard Lightman (Stanford, CA: Stanford University Press, 1990), 169–75.

81. White, *"Man of Science,"* 106.

82. Thomas Henry Huxley, "Science and Religion," *Builder* 17 (1859): 35.

83. Thomas Henry Huxley, "Universities: Actual and Ideal," in *CE*, 3:191–92.

84. Thomas Henry Huxley to Charles Darwin, July 20, 1868. Cambridge University Library, Charles Darwin Papers, 221.4: 254. Also see Bernard Lightman, "Pope Huxley and the Church Agnostic: The Religion of Science," *Historical Papers* (1983): 150–63.

85. Thomas Henry Huxley, "Scientific Education: Notes of an after-Dinner Speech," in *CE*, 3:133.

86. Thomas Henry Huxley, "Administrative Nihilism," in *Collected Essays*, vol. 1 (New York: D. Appleton, 1902), 284.

87. Desmond, *Huxley*, 571.

88. Thomas Henry Huxley, "Science and 'Church Policy,'" *Reader* 4 (1864): 821.

89. David Lindberg and Ronald L. Numbers, eds., introduction to *God and Nature: Historical Essays on the Encounter between Christianity and Science* (Berkeley: University of California Press, 1986); and James R. Moore, *The Post-Darwinian Controversies: A Study of the Protestant Struggle to Come to Terms with Darwin in Great Britain and America* (Cambridge: Cambridge University Press, 1979), chap. 1.

90. Adrian Desmond, *Archetypes and Ancestors: Palaeontology in Victorian London, 1850–1875* (Chicago: University of Chicago Press, 1984).

91. Desmond, *Huxley*, 222–24. On Lubbock, see J. F. M. Clark, "'The Ants Were Duly Visited': Making Sense of John Lubbock, Scientific Naturalism and

the Senses of Social Insects," *British Journal for the History of Science* 30, no. 2 (1997): 151–76.

92. Ruth Barton, "Evolution: The Whitworth Gun in Huxley's War for the Liberation of Science from Theology," in *The Wider Domain of Evolutionary Thought*, ed. D. Oldroyd (Dordrecht, Netherlands: Springer, 1983), 278.

93. White, *Man of Science*, 60.

94. Thomas Henry Huxley, "Origin of Species," 23.

95. Desmond, *Huxley*, 268; and Barton, "Evolution: The Whitworth Gun," 279. On Huxley's role in the reception of the *Origin* generally, see Desmond, *Huxley*, chap. 15.

96. Exactly what happened at the debate is surprisingly difficult to assess. See David N. Livingstone, "That Huxley Defeated Wilberforce in Their Debate over Evolution and Religion," in *Galileo Goes to Jail and Other Myths about Science and Religion*, ed. Ronald L. Numbers, 152–60 (Cambridge, MA: Harvard University Press, 2009).

97. Paul White, "Ministers of Culture: Arnold, Huxley, and Liberal Anglican Reform of Learning," *History of Science* 43 (2005): 115–38.

98. Desmond, *Huxley*, 544–45.

99. Ruth Barton, "'An Influential Set of Chaps': The X-Club and Royal Society Politics, 1864–85," *British Journal for the History of Science* 23, no. 1 (1990): 53–82; J. Vernon Jensen, "The X Club: Fraternity of Victorian Scientists," *British Journal for the History of Science* 5, no. 1 (1970): 63–72; Roy M. MacLeod, "The X Club: A Social Network of Science in Late-Victorian England," *Notes and Records of the Royal Society of London* 24, no. 2 (1970): 305–22; and Desmond, "Redefining." The source of the club's name is contested.

100. Thomas Hirst, *Hirst Journals*, November 6, 1864, vol. 4, fol. 1702, quoted in Barton, "'Influential Set of Chaps': The X-Club," 57.

101. Joan L. Richards, *Mathematical Visions: The Pursuit of Geometry in Victorian England* (San Diego: Academic Press, 1988), 103–14; and William Kingdon Clifford, *Lectures and Essays*, ed. Leslie Stephen and Frederick Pollock, 2 vols. (London: Macmillan, 1879), 1:1.

102. Quoted in Brooke and Cantor, *Reconstructing Nature*, 257.

103. Lightman, *Victorian Popularizers of Science*, 353–54.

104. Desmond, *Huxley*, 368.

105. Alan Willard Brown, *The Metaphysical Society: Victorian Minds in Crisis, 1869–1880* (New York: Columbia University Press, 1947).

106. Andrew Lang, "Science and Demonology," *Illustrated London News*, January 30, 1894, 822.

Chapter Two

1. John F. W. Herschel, *A Preliminary Discourse on the Study of Natural Philosophy* (London: Longman, Rees, Orme, Brown, and Green, and J. Taylor, 1830); and Thomas Henry Huxley, "Presidential Address to the Royal Society," *Nature* 33, no. 840 (1885): 112–19.

2. Herschel, *Preliminary Discourse*, 13–14.

3. Ibid., 35.

4. Ibid., 119.

5. Ronald L. Numbers, "Science without God: Natural Laws and Christian beliefs," in *When Science and Christianity Meet*, ed. David C. Lindberg and Ronald L. Numbers, 265–85 (Chicago: University of Chicago Press, 2003); and Peter Harrison, "The Development of the Concept of the Laws of Nature," in *Creation: Law and Probability*, ed. Fraser Watts (Aldershot, UK: Ashgate, 2008). The question of whether the laws of nature had any causal force, or if they were merely descriptive, was a controversial issue throughout the century. That issue will not be thoroughly addressed here; instead, the focus will be on the uniformity of those laws.

6. Herschel, *Preliminary Discourse*, 37.

7. Baden Powell, *The Connexion of Natural and Divine Truth; or, the Study of the Inductive Philosophy, Considered as Subservient to Theology* (London: J. W. Parker, 1838). The definitive work on Powell is Pietro Corsi, *Science and Religion: Baden Powell and the Anglican Debate, 1800–1860* (Cambridge: Cambridge University Press, 1988). See also Adrian Desmond, *Archetypes and Ancestors: Palaeontology in Victorian London, 1850–1875* (Chicago: University of Chicago Press, 1984), 44–45.

8. Powell, *Connexion*, 18.

9. Ibid., 21.

10. Ibid., 106.

11. Ibid., 116–17.

12. Ibid., 168.

13. Ibid., 156.

14. Baden Powell, *Essays on the Spirit of Inductive Philosophy, the Unity of Worlds, and the Philosophy of Creation* (London: Longman, Brown, Green, and Longmans, 1855), 153–56.

15. Ibid., 155.

16. Michael Ruse characterizes Powell as a liberal and Whewell as a conservative, though both largely agreed on the importance of natural laws. Michael Ruse, "The Relationship between Science and Religion in Britain, 1830–1870," *Church History* 44, no. 4 (1975): 505–22.

17. Stephen Brush, C. W. F. Everitt, and Elizabeth Garber, eds., *Maxwell on Saturn's Rings* (Cambridge, MA: MIT Press, 1983), 73.

18. James Clerk Maxwell, "Introductory Lecture, Aberdeen," in *The Scientific Letters and Papers of James Clerk Maxwell*, ed. P. M. Harman, 3 vols. (Cambridge: Cambridge University Press, 1990), 1:542–43 (hereafter cited as *SLP*).

19. James Clerk Maxwell, "On the Dynamical Evidence of the Molecular Constitution of Bodies," in *The Scientific Papers of James Clerk Maxwell*, ed. W. D. Niven, 2 vols. (Cambridge: Cambridge University Press, 1890), 2:418 (hereafter cited as *SP*).

20. James Clerk Maxwell, "Attraction," in *SP*, 2:491.

21. James Clerk Maxwell to Charles John Ellicott, November 22, 1876, in *SLP*, 3:418.

22. Colossians 1:16, Hebrews 2:6, and Psalm 8 (New International Version).

23. James Clerk Maxwell to Charles John Ellicott, November 22, 1876, in *SLP*, 3:417.

24. Maxwell, "Introductory Lecture, Aberdeen," 1:543.

25. William Benjamin Carpenter, "Nature and Law," in *Nature and Man: Essays Scientific and Philosophical* (London: Kegan Paul, Trench, 1888), 382, 367–68.

26. William Benjamin Carpenter, *Nature and Man: Essays Scientific and Philosophical* (London: Kegan Paul, Trench, 1888), 36–37.

27. P. M. Harman, *The Natural Philosophy of James Clerk Maxwell* (Cambridge: Cambridge University Press, 1998), 98–101; John Hendry, *James Clerk Maxwell and the Theory of the Electromagnetic Field* (Bristol, UK, and Boston: A. Hilger, 1986), 143; Daniel M. Siegel, *Innovation in Maxwell's Electromagnetic Theory: Molecular Vortices, Displacement Current, and Light* (Cambridge: Cambridge University Press, 1991), 120–43; and James Clerk Maxwell, "On Physical Lines of Force," in *SP*, 1:451–513. "On Physical Lines" was originally published in several parts in *Philosophical Magazine* in 1861 and 1862.

28. Harman, *Natural Philosophy*, 102–6; and Crosbie Smith, *The Science of Energy: A Cultural History of Energy Physics in Victorian Britain* (Chicago: University of Chicago Press, 1998), 226–27.

29. Harman, *Natural Philosophy*, 106–8.

30. Maxwell, "On Physical Lines of Force," 1:499.

31. Christa Jungnickel and Russell McCormmach, *Intellectual Mastery of Nature: Theoretical Physics from Ohm to Einstein*, 2 vols. (Chicago: University of Chicago Press, 1986), 1:144–46.

32. These are the numbers Maxwell gives in October 1861 (Maxwell to Michael Faraday, October 19, 1861, in *SLP*, 1:685), but in "On Physical Lines," in *SP*, 1:499–500, Maxwell gives the speed of light as 195,647 miles per second. In Maxwell to William Thomson, December 10, 1861, in *SLP*, 1:695, he presents four different measurements, all quite close. See Harman, *Natural Philosophy*, 64–67.

33. Maxwell to Michael Faraday, October 19, 1861, in *SLP*, 1:686.

34. James Clerk Maxwell, "On the (Physical) Dynamical Explanation of Electric Phenomena," in *SLP*, 3:172.

35. James Clerk Maxwell, "On Faraday's Lines of Force," in *SP*, 1:156. Maxwell's thinking on analogies was likely influenced by Bishop Butler's famous 1736 *Analogy of Religion*, which he studied closely. See also Kevin Lambert, "The Uses of Analogy: James Clerk Maxwell's 'On Faraday's Lines of Force' and Early Victorian Analogical Argument," *British Journal for the History of Science* 44, no. 1 (2011): 61–88.

36. James Clerk Maxwell, "Essay for the Apostles on 'Analogies in Nature,' February 1856," in *SLP*, 1:376.

37. Robert Kargon, "Model and Analogy in Victorian Science: Maxwell's Critique of the French Physicists," *Journal of the History of Ideas* 30, no. 3 (1969): 423–36; and Daniel Siegel, "Thomson, Maxwell, and the Universal Ether in Victorian Physics," in *Conceptions of Ether: Studies in the History of Ether Theories, 1740–1900*, ed. G. N. Cantor and Michael Jonathan Sessions Hodge (Cambridge: Cambridge University Press, 1981), 243.

38. Richard Olson, *Scottish Philosophy and British Physics, 1750–1880: A Study in the Foundations of the Victorian Scientific Style* (Princeton, NJ: Princeton University Press, 1975), 290–92. Jordi Cat argues for a variety of sources for Maxwell's reliance on analogies. See his *Land, Lines, Colours and Toys: Becoming James Clerk Maxwell* (Oxford: Oxford University Press, forthcoming), chap. 7.

39. Maxwell, "'Analogies in Nature,'" 1:380.

40. James Clerk Maxwell, "Inaugural Lecture at Aberdeen," quoted in Reginald Victor Jones, "James Clerk Maxwell at Aberdeen, 1856–1860," *Notes and Records of the Royal Society of London* 28, no. 1 (1973): 80–81. The lecture also appears in *SLP*, 1:542–47.

41. Maxwell, "'Analogies in Nature,'" 1:377.

42. Ibid., 1:382.

43. James Clerk Maxwell, "Inaugural Lecture at King's College London, October 1860," in *SLP*, 1:670. The phrase "unsearchable Wisdom" is probably a reference to the Westminster Confession of Faith 5.4.

44. Maxwell Papers, University Library Cambridge, Add. MS 7655/Vh/1.

45. James Clerk Maxwell, "Address to the Mathematical and Physical Sections of the British Association (1870)," in *SP*, 2:224.

46. James Clerk Maxwell, "Molecules (A Lecture)," in *SP*, 2:361–77, on 2:376.

47. Ibid., 2:361–77, on 2:377.

48. Despite Paley's prominence, his version of the design argument was certainly not the only one used in the Victorian period. For examples, see David W. Bebbington, "Science and Evangelical Theology in Britain from Wesley to Orr," and Jonathan R. Topham, "Science, Natural Theology, and Evangelicalism in

Early Nineteenth-Century Scotland: Thomas Chalmers and the *Evidence* Controversy," both in *Evangelicals and Science in Historical Perspective*, ed. David N. Livingstone, D. G. Hart, and Mark A. Noll (Oxford: Oxford University Press, 1999); Richard Yeo, "The Principle of Plenitude and Natural Theology in Nineteenth Century Britain," *British Journal for the History of Science* 19 (1986): 263–82; as well as Topham's unpublished PhD diss., "An Infinite Variety of Arguments: The Bridgewater Treatises and British Natural Theology in the 1830's" (Lancaster University, 1993).

49. James Clerk Maxwell, "Appendix II: Fragments of an Apostles Essay 'What Is the Nature of Evidence of Design?' 1853," in *SLP*, 1:228.

50. Maxwell, "'Analogies in Nature,'" 1:382.

51. James Clerk Maxwell to Charles John Ellicott, November 22, 1876, in *SLP*, 3:418.

52. James Clerk Maxwell, "Draft Letter to Francis W. H. Petrie, circa March 15, 1875," in *SLP*, 3:194.

53. Different rates of change for particular elements of science are discussed in Peter Galison, *How Experiments End* (Chicago: University of Chicago Press, 1987), 246–57. In Galison's terms, Maxwell would say that science and religion should harmonize their "long-term constraints."

54. Maxwell, "Inaugural Lecture at Aberdeen," 71.

55. James Clerk Maxwell, "Review of *An Essay on the Mathematical Principles of Physics*," in *SP*, 2:338–42, on 2:338.

56. Maxwell, "Address to the Mathematical and Physical Sections of the British Association," 2:215–29, on 2:217.

57. John Hedley Brooke and Geoffrey Cantor, *Reconstructing Nature: The Engagement of Science and Religion* (Oxford: Oxford University Press, 1998), 141–246; John Hedley Brooke, *Science and Religion: Some Historical Perspectives* (Cambridge: Cambridge University Press, 1991), 192–225; and Matthew Stanley, "A Modern Natural Theology?," *Journal of Faith and Science Exchange* 3 (1999): 105–12.

58. James Clerk Maxwell to K. M. Dewar, May 9, 1858, in Lewis Campbell and William Garnett, *The Life of James Clerk Maxwell* (London: Macmillan, 1882), 311 (hereafter cited as *Life*).

59. James Clerk Maxwell to his wife, June 23, 1864, in Campbell and Garnett, *Life*, 338–39.

60. James Clerk Maxwell to his wife, June 22, 1864, in Campbell and Garnett, *Life*, 338.

61. See Livingstone, Hart, and Noll, *Evangelicals and Science in Historical Perspective*, particularly Bebbington, "Science and Evangelical Theology," 120–41; and Topham, "Science, Natural Theology, and Evangelicalism," 142–74.

62. Bebbington, "Science and Evangelical Theology," 120–22.

63. For example, William J. Astore, *Observing God: Thomas Dick, Evangel-*

icalism, and Popular Science in Victorian Britain and America (London: Ashgate, 2001); Aileen Fyfe, *Science and Salvation: Evangelical Popular Science Publishing in Victorian Britain* (Chicago: University of Chicago Press, 2004); David W. Bebbington, *Evangelicalism in Modern Britain: A History from the 1730s to the 1980s* (Abingdon, UK: Routledge, 1989), 50–60; and David B. Wilson, *Kelvin and Stokes: A Comparative Study in Victorian Physics* (Bristol, UK: A. Hilger, 1987), chap. 4.

64. John Hedley Brooke, introduction to *Evangelicals and Science in Historical Perspective*, ed. David N. Livingstone, D. G. Hart, and Mark A. Noll, 23–29 (Oxford: Oxford University Press, 1999), 24–26. He of course cautions against taking too seriously the notion of such an archetype.

65. John Hedley Brooke and Reijer Hooykaas, *New Interactions between Theology and Natural Science* (Milton Keynes, UK: Open University Press, 1974).

66. Topham, "Science, Natural Theology, and Evangelicalism," 145.

67. Ibid., 165–67.

68. Bebbington, "Science and Evangelical Theology," 128–29.

69. Crosbie Smith, *Science of Energy*, 18–21.

70. Boyd Hilton, *The Age of Atonement: The Influence of Evangelicalism on Social and Economic Thought, 1785–1865* (Oxford: Oxford University Press, 1991), 362.

71. Ibid., 21.

72. Ibid., 21–22. See also Bebbington, "Science and Evangelical Theology," 133.

73. Hilton discusses similar approaches to science in *Atonement*, 304–14.

74. Maxwell, "Inaugural Lecture at Aberdeen," 78.

75. Ibid., 80; and Maxwell, "Inaugural Lecture at King's College London," 662–74.

76. Maxwell, "Inaugural Lecture at Aberdeen," 77.

77. Ibid., 78.

78. Maxwell, "Molecules," 363.

79. Frederick Temple, *The Relations between Religion and Science: Eight Lectures Preached before the University of Oxford in the Year 1884* (London: Macmillan, 1884), 8, 27; and George Douglas Campbell Argyll (Duke of), *The Reign of Law*, 5th ed. (1867; repr., London: Strahan, 1870), 3.

80. This fusion of theism and apparently naturalistic reasoning is sometimes today called "religious naturalism." For more, see Willem B. Drees, *Religion, Science, and Naturalism* (Cambridge: Cambridge University Press, 1996); David Ray Griffin, *Religion and Scientific Naturalism: Overcoming the Conflicts* (Albany: SUNY Press, 2000); and J. A. Stone, "Varieties of Religious Naturalism," *Zygon* 38, no. 1 (2003): 89–95.

81. Hans Christian Oersted, as quoted in Powell, *Essays*, 113.

82. Thomas Henry Huxley, "Review of *Vestiges of the Natural History of Creation*," in *The Scientific Memoirs of Thomas Henry Huxley*, ed. Michael Foster and E. Ray Lankester, 5 vols. (London: Macmillan, 1902), 5:5 (hereafter cited as *SM*).

83. James Secord, *Victorian Sensation: The Extraordinary Publication, Reception, and Secret Authorship of "Vestiges of the Natural History of Creation"* (Chicago: University of Chicago Press, 2000).

84. Thomas Henry Huxley, "Review of *Vestiges of the Natural History of Creation*," 5:1; and Secord, *Victorian Sensation*, 500–504.

85. Thomas Henry Huxley, "Review of *Vestiges of the Natural History of Creation*," 5:2.

86. Ibid., 5:3.

87. Ibid., 5:6.

88. Ibid., 5:6–7.

89. Desmond, *Archetypes and Ancestors*, 43.

90. Richard Owen, "On Some Instances of the Power of God as Manifested in His Animal Creation," Owen Collection, Library and Archives, The Natural History Museum, London, OC 59.1/63–98, 76–78, November 17, 1863.

91. Ibid., 90.

92. Desmond, *Archetypes and Ancestors*, 47. See also Peter J. Bowler, *Evolution: The History of an Idea*, 3rd ed. (Berkeley: University of California Press, 2003), 124–29.

93. Thomas Henry Huxley, "On the Theory of the Vertebrate Skull," in *SM*, 1:538.

94. Thomas Henry Huxley, "On the Theory of the Vertebrate Skull," 1:571. Owen's archetypes were attacked by theistic scientists as well; see Desmond, *Archetypes and Ancestors*, 51–54.

95. Thomas Henry Huxley, "On the Theory of the Vertebrate Skull," 1:584–85.

96. Desmond, *Archetypes and Ancestors*, 85–108. Huxley's difficulty in reconciling Darwin's ideas with his own views on species change can be seen in the careful phrasing found in his work in the 1860s and 1870s.

97. Thomas Henry Huxley, "On Natural History, as Knowledge, Discipline, and Power," in *SM*, 1:307.

98. Ibid., 1:309.

99. Ibid., 1:310.

100. Mario A. Di Gregorio, *T. H. Huxley's Place in Natural Science* (New Haven, CT: Yale University Press, 1984), 402–5; and Peter J. Bowler, *Fossils and Progress: Paleontology and the Idea of Progressive Evolution in the Nineteenth Century* (New York: Science History Publications, 1976).

101. Thomas Henry Huxley, "On the Persistent Types of Animal Life," in *SM*,

2:90, 2:92. Also see Sherrie Lyons, "Thomas Huxley: Fossils, Persistence, and the Argument from Design," *Journal of the History of Biology* 26, no. 3 (1993): 545–69.

102. Thomas Henry Huxley, "On Certain Zoological Arguments Commonly Adduced in Favour of the Hypothesis of the Progressive Development of Animal Life in Time," in *SM*, 1:301.

103. Thomas Henry Huxley to Charles Darwin, November 23, 1859, quoted in Leonard Huxley, *Life and Letters of Thomas Henry Huxley*, vol. 1 (1900; repr., New York: AMS Press, 1979), 188–89 (hereafter cited as *LL*).

104. Thomas Henry Huxley, "The Darwinian Hypothesis," in *Collected Essays of T. H. Huxley*, 9 vols. (Bristol, UK: Thoemmes Press, 2001), 2:10–11 (hereafter cited as *CE*).

105. Ibid., 2:13–14.

106. Thomas Henry Huxley, extract from "The Reception of the *Origin of Species*," in *Life and Letters of Charles Darwin*, quoted in *LL*, 1:182.

107. Thomas Henry Huxley, "Time and Life: Mr. Darwin's *Origin of Species*," *Macmillan's Magazine* 1 (1859).

108. Ibid., 143.

109. Bowler, *Fossils and Progress*, 138–39.

110. See David N. Livingstone, *Darwin's Forgotten Defenders: The Encounter between Evangelical Theology and Evolutionary Thought* (Grand Rapids, MI: Eerdmans Publishing, 1987), chap. 3; Thomas Henry Huxley, "Joseph Priestley," *Macmillan's Magazine* 30 (1874): 1–37; Desmond, *Archetypes and Ancestors*; and Thomas Henry Huxley, "On the Physical Basis of Life," in *CE*, 1:130–65.

111. Temple, *Relations between Religion and Science*, 121.

112. Argyll, *Reign of Law*, 29. Some modern views on theistic evolution can be found in Robert T. Pennock, ed., *Intelligent Design Creationism and Its Critics: Philosophical, Theological, and Scientific Perspectives* (Cambridge, MA: MIT Press, 2001), 471–536.

113. Argyll to Owen, February 29, 1863, Richard Owen Papers, Argyll, 1/184–259, 18/244 (R), 230.

114. Thomas Henry Huxley, extract from "The Reception of the *Origin of Species*," in *Life and Letters of Charles Darwin*, quoted in *LL*, 1:181.

115. Joseph Hooker to Thomas Henry Huxley, March 27, 1888, Joseph Hooker Papers, Royal Botanic Gardens, Kew Library, (JDH/2/13), 172. Reproduced with the kind permission of the Board of Trustees of the Royal Botanic Gardens, Kew.

116. Thomas Henry Huxley, *Introductory Science Primer* (London: Appleton, 1887), 12–13 (hereafter cited as *ISP*).

117. Thomas Henry Huxley, "Science and Pseudo-Science," in *CE*, vol. 5.

118. Thomas Henry Huxley, *ISP*, 14.

119. Thomas Henry Huxley, "On the Animals Which Are Most Nearly In-

termediate between Birds and Reptiles," in *Notices of the Proceedings at the Meetings of Members of the Royal Institution of Great Britain with Abstracts of the Discourses Delivered at Evening Meetings*, vol. 5, 278–86 (London: William Clowes and Sons, 1869), 279.

120. Thomas Henry Huxley, *ISP*, 7.

121. Ibid., 11.

122. Thomas Henry Huxley to Sir Charles Lyell, June 25, 1859, in *LL*, 1:187.

123. Thomas Henry Huxley, "A Lobster; or, the Study of Zoology," in *CE*, 9:205–6.

124. Thomas Henry Huxley, "On Our Knowledge of the Causes of the Phenomena of Organic Nature," in *CE*, 2:316–17.

125. Thomas Henry Huxley, "The Scientific Aspects of Positivism," in *Lay Sermons, Addresses and Reviews* (New York: D. Appleton, 1872), 144–45.

126. Thomas Henry Huxley, "On the Method of Zadig," in *CE*, vol. 4.

127. Thomas Henry Huxley, "On the Method of Paleontology," in *SM*, 1:432–34.

128. Ibid., 1:434–35.

129. Ibid., 1:436; and Thomas Henry Huxley, "Review of Haeckel's *Natural History of Creation*," *Academy* 1 (1869): 12–14, 40–43, 13–14.

130. For the details of this controversy, see Crosbie Smith and M. Norton Wise, *Energy and Empire: A Biographical Study of Lord Kelvin* (Cambridge: Cambridge University Press, 1989), 579–90; and Joe D. Burchfield, *Lord Kelvin and the Age of the Earth* (New York: Science History Publications, 1975).

131. Sir William Thomson, "On Geological Time," in *Popular Lectures and Addresses*, 3 vols. (London: Macmillan, 1894), 2:44. He also provided evidence for a short time scale based on tidal retardation theory, but that is less relevant to the uniformity issues being discussed here.

132. Ibid., 2:10.

133. Ibid., 2:44.

134. Ibid., 2:45–46.

135. Thomas Henry Huxley, "Geological Reform," in *CE*, 8:307.

136. Ibid., 8:319.

137. William Thomson, "On Geological Time," 2:108.

138. Crosbie Smith and M. Norton Wise, *Energy and Empire*, 581.

139. William Thomson, "On Geological Time," 2:77, 2:75.

140. Ibid., 2:44. Hooykaas's distinction between actualism and uniformitarianism is helpful for understanding the issues in this controversy. See Reijer Hooykaas, *Principle of Uniformity in Geology, Biology and Theology* (Leiden, Netherlands: Brill, 1963).

141. Burchfield, *Lord Kelvin and the Age of the Earth*, 3, 28.

142. Crosbie Smith and M. Norton Wise, *Energy and Empire*, 544–48.

143. Quoted in Crosbie Smith and M. Norton Wise, *Energy and Empire*, 638.

144. Quoted in Burchfield, *Lord Kelvin and the Age of the Earth*, 48.

145. William Thomson, "On Geological Time," 2:45–46.

146. Quoted in Crosbie Smith and M. Norton Wise, *Energy and Empire*, 535.

147. Burchfield, *Lord Kelvin and the Age of the Earth*, 49; and Crosbie Smith and M. Norton Wise, *Energy and Empire*, 555.

148. Thomas Henry Huxley, "The Progress of Science," in *CE*, 1:127.

149. Thomas Henry Huxley, prologue to *Essays upon Some Controverted Questions*, by Thomas Henry Huxley (New York: D. Appleton, 1892), 3–4.

150. Ibid., 16, 22. Huxley noted that he did not think that science made supernaturalism per se impossible; rather, it simply disproved all existing forms of supernatural claims.

151. Thomas Henry Huxley, "Evolution in Biology," in *CE*, 2:223–24.

152. A brief overview of the military metaphor can be found in David Lindberg and Ronald L. Numbers, eds., *God and Nature: Historical Essays on the Encounter between Christianity and Science* (Berkeley: University of California Press, 1986), 1–18.

153. John Tyndall, "Address," in *Report of the Forty-Fourth Meeting of the British Association for the Advancement of Science Held at Belfast in August 1874* (London: John Murray, 1875), lxxxviii. See also Ruth Barton, "John Tyndall, Pantheist: A Rereading of the Belfast Address," *Osiris* 2, no. 3 (1987): 111–34.

154. Tyndall, "Address," lxvii.

155. Ibid., xcv.

156. John Tyndall, *Fragments of Science for Unscientific People: A Series of Detached Essays, Addresses, and Reviews*, vol. 1 (New York: D. Appleton, 1871), 67, 49.

157. For modern examples of this split, see Michael Ruse, *Can a Darwinian Be a Christian? The Relationship between Science and Religion* (Cambridge: Cambridge University Press, 2001), 95; and Michael Ruse, "Methodological Naturalism under Attack," *South African Journal of Philosophy* 24, no. 1 (2005): 44.

158. Robert Bruce Mullin, *Miracles and the Modern Religious Imagination* (New Haven, CT: Yale University Press, 1996). The early modern engagement of science with miracles is addressed in R. M. Burns, *The Great Debate on Miracles: From Joseph Glanvill to David Hume* (Lewisburg, PA: Bucknell University Press, 1981), and Peter Dear, "Miracles, Experiments, and the Ordinary Course of Nature," *Isis* 81, no. 4 (December 1990): 663–83.

159. Thomas Henry Huxley, "Hume" in *CE*, 6:132.

160. Thomas Henry Huxley, "Science and Morals," in *CE*, 9:183.

161. Thomas Henry Huxley, "Possibilities and Impossibilities," in *CE*, 5:204–6.

162. Thomas Henry Huxley, "Science and Pseudo-Science," 5:157.

163. Thomas Henry Huxley, "Agnosticism and Christianity," in *CE*, 5:309–65, on 5:359–60. This is in reference to the story found in Matthew 8:28, Luke 8:26, and Mark 5:1 in which Jesus casts demons into a herd of pigs. The location of the

story is rendered variously as "the country of the Gadarenes" or "Gerasenes," depending on which book and translation.

164. Thomas Henry Huxley, "Science and Morals," 9:180–82.

165. Ibid., 9:184–85.

166. Thomas Henry Huxley, "Agnosticism and Christianity," 5:309–65, on 5:332.

167. Thomas Henry Huxley, "An Episcopal Trilogy," in *CE*, 5:126–59, on 5:142.

168. Mullin, *Miracles*, 42.

169. Robert Bruce Mullin, "Science, Miracles, and the Prayer-Gauge Debate," in *When Science and Christianity Meet*, edited by David C. Lindberg and Ronald L. Numbers, 203–24 (Chicago: University of Chicago Press, 2003). Huxley wrote an entire book on Hume.

170. Mullin, *Miracles*, 44.

171. John Tyndall, quoted in Frank M. Turner, "Rainfall, Plagues, and the Prince of Wales: A Chapter in the Conflict of Religion and Science," *Journal of British Studies* 13, no. 2 (1974): 46–65, on 48. See also John Tyndall, "The 'Prayer for the Sick': Hints towards a Serious Attempt to Estimate Its Value," *Contemporary Review* 20 (1872): 205.

172. Tyndall, *Fragments of Science for Unscientific People*, 1:36.

173. Ibid., 1:409.

174. Temple, *Relations between Religion and Science*, 32.

175. Argyll, *Reign of Law*, 5.

176. Ibid., 12–13.

177. Temple, *Relations between Religion and Science*, 195.

178. George Douglas Campbell Argyll, "Letter to the Editor," *Times*, February 8, 1892.

179. Argyll, *Reign of Law*, 22. Also see Ruse, "Relationship between Science and Religion in Britain," 509–10.

180. Robert Grant, *Miracle and Natural Law in Graeco-Roman and Early Christian Thought* (Eugene, OR: Wipf and Stock, 1952), 217–18.

181. Ruse, "Relationship between Science and Religion in Britain," 510–11.

182. Argyll, *Reign of Law*, 17–30.

183. Temple, *Relations between Religion and Science*, 195–96.

184. Mullin, "Science, Miracles, and the Prayer-Gauge Debate," 205–6.

185. Temple, *Relations between Religion and Science*, 219–20.

186. Griffin, *Religion and Scientific Naturalism*, 38–40. The point that it is possible to retain the value of a miracle even after explaining it with natural laws is also made in Ruse, *Can a Darwinian Be a Christian?*, 96.

187. Thomas Henry Huxley, "Geological Contemporaneity and Persistent Types of Life," in *CE*, 8:287–88.

Chapter Three

1. Duke of Argyll, "Professor Huxley on the Warpath," *Ninteenth Century*, no. 167 (1891): 27.

2. Thomas Henry Huxley, "On the Method of Zadig," in *Collected Essays of T. H. Huxley*, 9 vols. (Bristol, UK: Thoemmes Press, 2001), 4:6–8 (hereafter cited as *CE*).

3. The search for a boundary between science and nonscience was characteristic of this period. For an overview of some of the relevant issues, see Martin Fichman, "Biology and Politics: Defining the Boundaries," in *Victorian Science in Context*, ed. Bernard Lightman, 94–118 (Chicago: University of Chicago Press, 1997). This chapter will not address the liminal issue of spiritualism, which has been examined thoroughly in the scholarly literature. For examples, see Janet Oppenheim, *The Other World: Spiritualism and Psychical Research in England, 1850–1914* (Cambridge: Cambridge University Press, 1985); Richard Noakes, "Spiritualism, Science, and the Supernatural in Mid-Victorian Britain," in *The Victorian Supernatural*, ed. Nicola Bown, Carolyn Burdett, and Pamela Thurschwell, 23–43 (Cambridge: Cambridge University Press, 2004); and Courtenay Grean Raia, "The Substance of Things Hoped For: Faith, Science and Psychical Research in the Victorian Fin de Siècle" (PhD diss., University of California, Los Angeles, 2005).

4. Thomas Henry Huxley, *Introductory Science Primer* (London: Appleton, 1887), 16 (hereafter cited as *ISP*). Emphasis in the original.

5. Adrian Desmond, *Huxley: From Devil's Disciple to Evolution's High Priest* (Reading, MA: Addison-Wesley, 1997), 630–32.

6. Bernard Lightman, *The Origins of Agnosticism: Victorian Unbelief and the Limits of Knowledge* (Baltimore: Johns Hopkins University Press, 1987), chap. 1.

7. Ibid., 14.

8. Ibid., 10, 93.

9. Thomas Henry Huxley, "Agnosticism," in *CE*, 5:239.

10. Bernard Lightman, "Huxley and Scientific Agnosticism: The Strange History of a Failed Rhetorical Strategy," *British Journal for the History of Science* 35, no. 3 (2002): 271–89.

11. Thomas Henry Huxley, "Hume: With Helps to the Study of Berkeley," in *CE*, 6:57–58.

12. Ibid., 6:66.

13. Ibid., 6:70–71.

14. Thomas Henry Huxley, "Agnosticism," 5:236.

15. Thomas Henry Huxley, "Agnosticism, a Symposium," *Agnostic Annual* 1 (1884): 5.

16. Thomas Henry Huxley, "Agnosticism," 5:246.

17. Ibid., 5:247.

18. Thomas Henry Huxley, "Agnosticism, a Symposium," 5–6.

19. Thomas Henry Huxley to M. Henri Gadeau de Kerville, February 1, 1887, quoted in Leonard Huxley, *Life and Letters of Thomas Henry Huxley*, vol. 2 (1900; repr., New York: AMS Press, 1979), 172 (hereafter cited as *LL*).

20. Thomas Henry Huxley, "Agnosticism, a Symposium," 5.

21. Joseph Hooker to Thomas Henry Huxley, October 21, 1886, J. D. Hooker Papers, Royal Botanic Gardens Kew Library, J. D. Hooker Letters to T. H. Huxley, c. 1851–1894 (JDH/2/13), 160–61.

22. Thomas Henry Huxley to Charles Kingsley, May 22, 1863, quoted in Leonard Huxley, *LL*, 1:262.

23. Thomas Henry Huxley, "Agnosticism and Christianity," in *CE*, 5:311–12.

24. Thomas Henry Huxley, "On the Physical Basis of Life," in *CE*, 1:156–57.

25. Ibid., 1:162.

26. Quoted in Thomas Henry Huxley, "On the Physical Basis of Life," 1:163. The quote is from Hume's *Inquiry Concerning the Human Understanding*.

27. Thomas Henry Huxley, "On the Physical Basis of Life," 1:163.

28. Ibid., 1:164.

29. Ibid., 1:165.

30. For instance, see William Samuel Lilly, "Materialism and Morality," *Fortnightly Review* 40 (October 1886): 575; David Knight, "Thomas Henry Huxley and Philosophy of Science," in *Thomas Henry Huxley's Place in Science and Letters: Centenary Essays*, ed. Alan P. Barr, 51–66 (Athens: University of Georgia Press, 1997), 54–57; and Lightman, *Origins*, 22–27. *Empiricist* at this time generally suggested that one was a disciple of J. S. Mill or Alexander Bain, rather than a general philosophical position.

31. Thomas Henry Huxley, "On the Educational Value of the Natural History Sciences," in *CE*, 3:45.

32. Thomas Henry Huxley, *ISP*, 16. Emphasis in the original.

33. Ibid., 17.

34. Lilly, "Materialism and Morality," 578.

35. Thomas Henry Huxley, "Science and Morals," in *CE*, 9:120.

36. Thomas Henry Huxley, "On Our Knowledge of the Causes of the Phenomena of Organic Nature," in *CE*, 2:361.

37. Thomas Henry Huxley, *ISP*, 16–17.

38. Ibid., 67. Emphasis in the original.

39. Ibid., 68.

40. Thomas Henry Huxley, "The Progress of Science," in *CE*, 1:62.

41. Ibid., 1:62.

42. Ibid., 1:65.

43. Thomas Henry Huxley, "On Our Knowledge of the Causes of the Phenomena of Organic Nature," 2:449.

44. Ibid., 2:466–67.

45. Ibid., 2:468–69.

46. John Tyndall, "On the Study of Physics," in *Fragments of Science: A Series of Detached Essays, Addresses, and Reviews*, vol. 1 (New York: D. Appleton, 1915), 282.

47. John Tyndall, "The Constitution of Nature," in *Fragments of Science: A Series of Detached Essays, Addresses, and Reviews*, vol. 1 (New York: D. Appleton, 1915), 4.

48. John Tyndall, "Scientific Use of the Imagination," in *Fragments of Science: A Series of Detached Essays, Addresses, and Reviews*, vol. 2 (New York: D. Appleton, 1897), 103.

49. Ibid., 2:107.

50. Ibid., 2:126–27.

51. Ibid., 2:128.

52. John Tyndall, "Apology for the Belfast Address," in *Fragments of Science: A Series of Detached Essays, Addresses, and Reviews*, vol. 2 (New York: D. Appleton, 1915), 206.

53. Ibid., 2:191, 2:194.

54. Ibid., 2:208.

55. John Tyndall, "Professor Virchow and Evolution," in *Fragments of Science: A Series of Detached Essays, Addresses, and Reviews*, vol. 2 (New York: D. Appleton, 1915), 385.

56. John Tyndall, "Matter and Force," in *Fragments of Science: A Series of Detached Essays, Addresses, and Reviews*, vol. 2 (New York: D. Appleton, 1915), 72.

57. Tyndall, "Professor Virchow and Evolution," 2:385.

58. Tyndall, "Matter and Force," 2:72.

59. Tyndall, "Professor Virchow and Evolution," 2:391; and John Tyndall, "The Rev. James Martineau and the Belfast Address," in *Fragments of Science: A Series of Detached Essays, Addresses, and Reviews*, vol. 2 (New York: D. Appleton, 1915), 233–34.

60. Tyndall, "Professor Virchow and Evolution," 2:389.

61. Tyndall, "The Belfast Address," in *Fragments of Science: A Series of Detached Essays, Addresses, and Reviews*, vol. 2 (New York: D. Appleton, 1897), 141.

62. Ibid., 2:201.

63. Ibid., 2:197–98.

64. Ibid., 2:196. Emphasis in the original.

65. Bernard Lightman, "Fighting Even with Death: Balfour, Scientific Naturalism, and Thomas Henry Huxley's Final Battle," in *Thomas Henry Huxley's Place in Science and Letters*, ed. Alan P. Barr (Atlanta: University of Georgia Press, 1997), 324–28. See also Leon Stephen Jacyna, "Science and Social Order in the Thought of A. J. Balfour," *Isis* 71, no. 1 (1980): 11–34.

66. Arthur James Balfour, *The Foundations of Belief: Being Notes Introductory to the Study of Theology* (London: Longmans, Green, 1895), 7–8.

67. Ibid., 68–69.

68. Ibid., 135.

69. Ibid., 28. Emphasis in the original.

70. Ibid., 83.

71. Ibid., 94.

72. Ibid., 115, 21–22.

73. Ibid., 134.

74. Lightman, "Fighting Even with Death," 336–37.

75. Thomas Henry Huxley, "Mr. Balfour's Attack on Agnosticism," *Nineteenth Century* 37 (1895): 527–40.

76. James Clerk Maxwell, "A Review of *Paradoxical Philosophy*," in *The Scientific Papers of James Clerk Maxwell*, ed. W. D. Niven, 2 vols. (Cambridge: Cambridge University Press, 1890), 2:759 (hereafter cited as *SP*).

77. James Clerk Maxwell, "Ether," in *SP*, 2:775.

78. Psalms 146:4, quoted in Maxwell, "Review of *Paradoxical Philosophy*," 2:759.

79. James Clerk Maxwell to Richard Buckley Litchfield, March 5, 1858, in *The Scientific Letters and Papers of James Clerk Maxwell*, ed. P. M. Harman, 3 vols. (Cambridge: Cambridge University Press, 1990), 1:587–88 (hereafter cited as *SLP*).

80. James Clerk Maxwell, "Inaugural Lecture at Marischal College, Aberdeen, November 3, 1856," in *SLP*, 1:425.

81. Book of Wisdom, 11:20. It is unclear which version of the Apocrypha Maxwell was using here. On the importance of measurement to Maxwell and contemporaries, see Simon Schaffer, "Metrology, Metrication, and Victorian Values," in *Victorian Science in Context*, ed. Bernard Lightman, 438–74 (Chicago: University of Chicago Press, 1997); and Simon Schaffer, "Accurate Measurement Is an English Science," in *The Values of Precision*, ed. M. Norton Wise, 135–72 (Princeton, NJ: Princeton University Press, 1995).

82. James Clerk Maxwell, "Draft on the Methods of Physical Science," in *SLP*, 3:75–78.

83. Maxwell, "Inaugural Lecture at Marischal College," 1:425.

84. Jacyna, "Science and Social Order in the Thought of A. J. Balfour," 24.

85. James Clerk Maxwell, "Hermann Ludwig Ferdinand Helmholtz," in *SP*, 2:593.

86. James Clerk Maxwell, "On the Theory of Colours in Relation to Colour-Blindness," in *SP*, 1:119–20.

87. Maxwell's models have been studied extensively. See P. M. Harman, *The Natural Philosophy of James Clerk Maxwell* (Cambridge: Cambridge University Press, 1998), 91–112; Robert Kargon, "Model and Analogy in Victorian Sci-

ence: Maxwell's Critique of the French Physicists," *Journal of the History of Ideas* 30, no. 3 (1969): 423–36; M. Norton Wise, "The Maxwell Literature and British Dynamical Theory," *Historical Studies in the Physical Sciences* 13, no. 1 (1982): 175–205; Thomas K. Simpson, *Maxwell on the Electromagnetic Field: A Guided Study* (New Brunswick, NJ: Rutgers University Press, 1997); and Daniel M. Siegel, *Innovation in Maxwell's Electromagnetic Theory: Molecular Vortices, Displacement Current, and Light* (Cambridge: Cambridge University Press, 1991). Also see Daniel M. Siegel, "Mechanical Image and Reality in Maxwell's Electromagnetic Theory" and Jed Z. Buchwald, "Modifying the Continuum," in *Wranglers and Physicists: Studies on Cambridge Mathematical Physics in the Nineteenth Century*, ed. P. M. Harman (Manchester, UK: Manchester University Press, 1985). On scientific models in general, see Max Black, *Models and Metaphors: Studies in Language and Philosophy* (Ithaca, NY: Cornell University Press, 1962); Mary Hesse, *Models and Analogies in Science* (Notre Dame, IN: University of Notre Dame Press, 1966); and Mary S. Morgan and Margaret Morrison, eds., *Models as Mediators: Perspectives on Natural and Social Science* (Cambridge: Cambridge University Press, 1999). The first chapter of Morgan and Morrison has an excellent bibliography and overview of the scholarship on modeling.

88. James Clerk Maxwell, "On Faraday's Lines of Force," in *SP*, 1:155. On the reliability of mathematics at this time, see Joan L. Richards, "God, Truth, and Mathematics in Nineteenth-Century England," in *The Invention of Physical Science: Intersections of Mathematics, Theology, and Natural Philosophy since the Seventeenth Century: Essays in Honor of Erwin N. Hiebert*, ed. Mary Jo Nye, Joan L. Richards, and R. H. Stuewer (Dordrecht, Netherlands: Kluwer Academic Publishers, 1992).

89. Maxwell, "On Faraday's Lines of Force," 1:156.

90. Ibid., 1:183.

91. Ibid., 1:187.

92. Ibid., 1:208. Emphasis in the original.

93. Ibid., 1:207. Emphasis in the original.

94. Ibid., 1:159.

95. See Crosbie Smith, *The Science of Energy: A Cultural History of Energy Physics in Victorian Britain* (Chicago: University of Chicago Press, 1998), chap. 11; and Harman, *Natural Philosophy of James Clerk Maxwell*, chap. 6.

96. James Clerk Maxwell, "Review of Thomson and Tait's *Natural Philosophy*," in *SP*, 2:783–84.

97. Note that Maxwell was not successful in persuading all his colleagues that his theory was within the bounds of acceptable hypothesis. In particular, William Thomson objected vigorously to the *Treatise*'s methods. Thomson's objections to Maxwell's theory are discussed in Ole Knudsen, "Mathematics and

Physical Reality in William Thomson's Electromagnetic Theory," in *Wranglers and Physicists: Studies on Cambridge Mathematical Physics in the Nineteenth Century*, ed. P. M. Harman, 149–79 (Manchester, UK: Manchester University Press, 1985); Crosbie Smith and M. Norton Wise, *Energy and Empire: A Biographical Study of Lord Kelvin* (Cambridge: Cambridge University Press, 1989), 445–90; and David B. Wilson, *Kelvin and Stokes: A Comparative Study in Victorian Physics* (Bristol, UK: A. Hilger, 1987), 9.

98. James Clerk Maxwell, *A Treatise on Electricity and Magnetism*, 2 vols. (Oxford: Clarendon Press, 1873), 1:xi–xii. A reader's guide to the treatise can be found in Thomas K. Simpson, *Figures of Thought: A Literary Appreciation of Maxwell's "Treatise on Electricity and Magnetism"* (Santa Fe: Green Lion Press, 2005).

99. Maxwell, *Treatise on Electricity and Magnetism*, 1:36.

100. Ibid., 1:59.

101. Ibid., 1:127.

102. Ibid., 2:6–7.

103. Ibid., 2:383.

104. Ibid., 2:415.

105. Ibid., 2:416–17.

106. Ibid., 2:437.

107. Ibid., 2:438.

108. Harman has commented on provisionality as a distinctive characteristic of Maxwell's models. P. M. Harman, "Edinburgh Philosophy and Cambridge Physics: The Natural Philosophy of James Clerk Maxwell," in *Wranglers and Physicists: Studies on Cambridge Mathematical Physics in the Nineteenth Century*, ed. P. M. Harman (Manchester, UK: Manchester University Press, 1985), 217.

109. James Clerk Maxwell, "On the Dynamical Evidence of the Molecular Constitution of Bodies," in *SP*, 2:419.

110. James Clerk Maxwell, "Review of Tait's Thermodynamics," in *SP*, 2:662–63.

111. James Clerk Maxwell, "Review of Thomson, *Papers on Electrostatics and Magnetism*," in *SLP*, 2:306.

112. James Clerk Maxwell, "Plateau on Soap-Bubbles," in *SP*, 2:399.

113. James Clerk Maxwell, "Address to the Mathematical and Physical Sections of the British Association (1870)," in *SP*, 2:216.

114. James Clerk Maxwell to Peter Guthrie Tait, December 23, 1867, in *SLP*, 2:335–39.

115. Maxwell, "Address to the Mathematical and Physical Sections of the British Association," 2:216.

116. Ibid., 2:228–29.

117. Ibid., 2:229.

118. Kevin Lambert, "Mind over Matter: Language, Mathematics, and Electromagnetism in Nineteenth Century Britain" (PhD diss., UCLA, 2005).

119. Richard Olson, *Scottish Philosophy and British Physics, 1750–1880: A Study in the Foundations of the Victorian Scientific Style* (Princeton, NJ: Princeton University Press, 1975), particularly chap. 12. On the Scottish natural philosophical tradition of this period more generally, see David B. Wilson, *Seeking Nature's Logic: Natural Philosophy in the Scottish Enlightenment* (University Park: Pennsylvania State University Press, 2009).

120. Olson, *Scottish Philosophy and British Physics*, 288.

121. David B. Wilson, "The Educational Matrix: Physics Education at Early-Victorian Cambridge, Edinburgh and Glasgow Universities," in *Wranglers and Physicists: Studies on Cambridge Mathematical Physics in the Nineteenth Century*, ed. P. M. Harman, 12–48 (Manchester, UK: Machester University Press, 1985); and Harman, "Edinburgh Philosophy and Cambridge Physics."

122. Harman, "Edinburgh Philosophy and Cambridge Physics," 212–13.

123. Maxwell, "Inaugural Lecture at Marischal College," 1:421.

124. See David N. Livingstone, D. G. Hart, and Mark A. Noll, eds., *Evangelicals and Science in Historical Perspective* (Oxford: Oxford University Press, 1999), particularly David W. Bebbington, "Science and Evangelical Theology in Britain from Wesley to Orr," 120–41; and Jonathan R. Topham, "Science, Natural Theology, and Evangelicalism in Early Nineteenth-Century Scotland: Thomas Chalmers and the *Evidence* Controversy," 142–74.

125. John Hedley Brooke, introduction to *Evangelicals and Science in Historical Perspective*, ed. David N. Livingstone, D. G. Hart, and Mark A. Noll, 23–29 (Oxford: Oxford University Press, 1999), 24–26.

126. Lightman, *Origins*, 31, 37.

127. Maxwell, "Address to the Mathematical and Physical Sections of the British Association," 2:225.

128. James Clerk Maxwell, "Molecules (A Lecture)" in *SP*, 2:361.

129. Ibid.

130. Ibid., 2:376.

131. James Clerk Maxwell, "Atom," in *SP*, 2:482.

132. Crosbie Smith and M. Norton Wise, *Energy and Empire*, 497–523; and Crosbie Smith, *Science of Energy*, 100–125.

133. William Thomson, "On Mechanical Antecedents of Motion, Heat, and Light," in *Report of the Annual Meeting of the British Association for the Advancement of Science*, vol. 24 (London: John Murray, 1854), 61.

134. Maxwell, "Address to the Mathematical and Physical Sections of the British Association," 2:226.

135. James Clerk Maxwell to Mark Pattison, April 7, 1868, in *SLP*, 2:358–61.

136. Ibid., 2:359.

137. Ibid., 2:361.

138. Ibid.

139. William Kingdon Clifford, *Lectures and Essays*, ed. Leslie Stephen and Frederick Pollock, 2 vols. (London: Macmillan, 1879), 1:221–22.

140. Tyndall, "Scientific Use of the Imagination," 2:133.

141. Ibid., 2:132.

142. Tyndall, "Matter and Force," 2:73.

143. Thomas Henry Huxley, "Progress of Science," 1:104.

144. Thomas Henry Huxley, "The Darwinian Hypothesis," in *CE*, 2:10.

145. John F. W. Herschel, *A Preliminary Discourse on the Study of Natural Philosophy* (London: Longman, Rees, Orme, Brown, and Green, and J. Taylor, 1830), 38.

146. Thomas Henry Huxley, "Progress of Science," 1:79–80.

Chapter Four

1. Some of the problems with the Lemon test are discussed in Stephen L. Carter, *The Culture of Disbelief: How American Law and Politics Trivialize Religious Devotion* (New York: Anchor Books, 1994), 109–15.

2. W. C. Lubenow, *The Cambridge Apostles, 1820–1914: Liberalism, Imagination, and Friendship in British Intellectual and Professional Life* (Cambridge: Cambridge University Press, 1998), 341–56; and Olive J. Brose, *Frederick Denison Maurice: Rebellious Conformist, 1805–1872* (Athens: Ohio University Press, 1971), 11–15, 19. The details of Maurice's biography can also be found in John Frederick Maurice, *The Life of Frederick Denison Maurice: Chiefly Told in His Own Letters* (New York: Charles Scribner, 1884).

3. Brose, *Rebellious Conformist*, 1–28, 65–76.

4. Ibid., 92, 78, 159–60, 64.

5. Ibid., 157, 75.

6. F. J. C. Hearnshaw, *The Centenary History of King's College London, 1828–1928* (London: G. G. Harrap, 1929), 175.

7. Frederick Denison Maurice, "Personal Explanation," *Working Men's College Magazine* 1, no. 26 (1861): 14.

8. Edward R. Norman, *The Victorian Christian Socialists* (Cambridge: Cambridge University Press, 1987), 1–13, 182–85. The term *socialism* at this time generally referred to Owenism, and was quite different from later Marxist uses. See J. F. C. Harrison, *Robert Owen and the Owenites in Britain and America: The Quest for the New Moral World* (London: Routledge and Kegan Paul, 1969), 35.

9. J. F. C. Harrison, *Learning and Living, 1790–1960: A Study in the History of the English Adult Education Movement* (Toronto: University of Toronto Press, 1961), 93.

10. John Ludlow, "A New Idea," in *Church and State in the Modern Age: A Documentary History*, ed. J. F. Maclear (Oxford: Oxford University Press, 1995), 242. The idea that education was a remedy to disruptive movements such as Chartism was fairly widespread. J. F. C. Harrison, *Learning and Living*, 77.

11. J.T., "A Few Words to the *Reasoner*," in *The Reasoner and Theological Examiner*, vol. 10 (London: James Watson, 1851), 302. Emphasis in the original.

12. Norman, *Victorian Christian Socialists*, 35–57.

13. Parson Lot, "Bible Politics; or, God Justified to the People," *Christian Socialist*, November 9, 1850, 9.

14. Charles Kingsley, "Workmen of England!" quoted in Frances Eliza Grenfell Kingsley, ed., *Charles Kingsley: His Letters and Memories of His Life*, 2 vols. (London: Henry S. King, 1877), 1:157.

15. "Our Principles," *Christian Socialist*, November 2, 1850, 2.

16. Ludlow, "New Idea," 243.

17. "Usurpation and Slavery," *Christian Socialist*, November 16, 1850, 17–18.

18. Parson Lot, "My Political Creed," *Christian Socialist*, December 14, 1850, 50.

19. "Letters on the Public Health," *Christian Socialist*, December 7, 1850, 47.

20. "The Guardian and Christian Socialism," *Christian Socialist*, March 21, 1851, 161.

21. Frederick Denison Maurice, *Working Men's College Magazine* 1, no. 12 (1859): 189.

22. "Guardian and Christian Socialism," 162.

23. Hearnshaw, *Centenary History of King's College*, 196.

24. "Our Principles," 2.

25. J. F. C. Harrison, *Learning and Living*, 163–72.

26. D. S. L. Cardwell, *The Organisation of Science in England: A Retrospect* (London: William Heinemann, 1972), 40–41; and J. F. C. Harrison, *Learning and Living*, 57–74.

27. David Vincent, *Bread, Knowledge and Freedom: A Study of Nineteenth-Century Working Class Autobiography* (London: Europa Publications, 1981), 109–19.

28. Ibid., 127, 35.

29. James Secord, *Victorian Sensation: The Extraordinary Publication, Reception, and Secret Authorship of "Vestiges of the Natural History of Creation"* (Chicago: University of Chicago Press, 2000), 348; and Cardwell, *Organisation*, 196.

30. Vincent, *Bread, Knowledge, and Freedom*, 142–43; and J. F. C. Harrison, *Learning and Living*, 52–54.

31. Gordon W. Roderick and Michael D. Stephens, *Scientific and Technical Education in Nineteenth-Century England: A Symposium* (Newton Abbot, UK: David and Charles, 1972), 109–10; and Cardwell, *Organisation*, 71.

32. Brose, *Rebellious Conformist*, 198–99.

33. R. B. Litchfield, *The Beginnings of the Working Men's College* (London: Working Men's College, 1902), 4–5.

34. Henrietta Litchfield, *Richard Buckley Litchfield: A Memoir Written for His Friends by His Wife* (Cambridge: Cambridge University Press, 1910), 7.

35. "What Use Is It?" *Working Men's College Magazine*, no. 9, September 1, 1859, 138.

36. J. F. C. Harrison, *Learning and Living*, 75, 200.

37. Frederick Denison Maurice, "Notebook," King's College London Archives, box 5199.M3 (1872); and "History of the Working Men's College," *Working Men's College Magazine* 1, no. 24 (1860): 188–92.

38. "Meeting of the Cambridge Working Men's College," *Working Men's College Magazine* 1, no. 3 (1859): 53–57.

39. Godfrey Heathcote Hamilton, *Queen Square: Its Neighbourhood & Its Institutions* (London: L. Parsons, 1926), 52.

40. *Working Men's College Magazine* 1, no. 1 (1859): 12.

41. "Report of the Manchester Working Men's College Meeting," *Working Men's College Magazine* 1, no. 2 (1859): 32.

42. J. N. Langley, "A Few Thoughts on the Franchise," *Working Men's College Magazine* 1, no. 28 (1861): 45.

43. Jonathan Rose, *The Intellectual Life of the British Working Classes* (New Haven, CT: Yale University Press, 2001), 20–23.

44. "Report of the Manchester Working Men's College Meeting," 32.

45. Langley, "A Few Thoughts on the Franchise," 46.

46. Henry Solly, "Reasons for a Working Men's College," *Working Men's College Magazine* 1, no. 18 (1860): 110.

47. Henrietta Litchfield, *Litchfield: A Memoir*, 60–61.

48. Harvey Goodwin, *Education for Working Men: An Address Delivered in the Town-Hall of Cambridge* (Cambridge: Deighton, Bell, 1855), 45.

49. Godfrey Lushington, "Shall We Learn Latin?," *Working Men's College Magazine* 1, no. 17 (1860): 71.

50. "Scheme of Mathematical Study," *Working Men's College Magazine* 1, no. 1 (1859): 10.

51. Frederick Denison Maurice, "Introductory Lecture on the Studies of the (London) Working Men's College," 5.

52. Ibid., 7–8.

53. J. F. C. Harrison, *Learning and Living*, 174.

54. Parson Lot, "Bible Politics; or, God Justified to the People II," *Christian Socialist*, November 23, 1850.

55. Lubenow, *Cambridge Apostles*.

56. Lewis Campbell and William Garnett, *The Life of James Clerk Maxwell* (London: Macmillan, 1882), 418 (hereafter cited as *Life*).

57. Ibid., 172, 218.

58. Ibid., 217.

59. Henrietta Litchfield, *Litchfield: A Memoir*, 38.

60. Ibid., 88–89.

61. Ibid., 98–100.

62. Ibid., 104.

63. Goodwin, *Education for Working Men: An Address*, 2.

64. Thomas Dinham Atkinson, *Cambridge Described and Illustrated: Being a Short History of the Town and University* (Cambridge: Macmillan, 1897), 221–22.

65. "Proceedings at a Meeting of the Working Men's College," *Cambridge Chronicle*, March 26, 1858, 4–5.

66. "Anniversary of the Working Men's College," *Cambridge Chronicle*, May 24, 1856, 7.

67. Goodwin, *Education for Working Men: An Address*, 12–13.

68. Enid Porter, *Victorian Cambridge: Josiah Chater's Diaries, 1844–1884* (London: Phillimore, 1975).

69. James Clerk Maxwell to John Clerk Maxwell, September 27, 1855, in *The Scientific Letters and Papers of James Clerk Maxwell*, ed. P. M. Harman, 3 vols. (Cambridge: Cambridge University Press, 1990), 1:327 (hereafter cited as *SLP*).

70. James Clerk Maxwell to John Clerk Maxwell, December 3, 1855, in *SLP*, 1:336.

71. Campbell and Garnett, *Life*, 221–22, 52–56, quote on 218.

72. *First Report of the Committee of the Aberdeen Mechanics' Institution* (Aberdeen: D. Chalmers, 1824).

73. *Twelfth Report of the Committee of the Aberdeen Mechanics' Institution* (Aberdeen: D. Chalmers, 1837).

74. Minute Book for School of Science and Art, no. 9 [Lo 3706 Ab 3.4], Aberdeen Library.

75. For examples, see James Clerk Maxwell to Richard Buckley Litchfield, October 15, 1857, in *SLP*, 1:540–41; James Clerk Maxwell to Henry Richmond Droop, November 14, 1857, in *SLP*, 1:557; and James Clerk Maxwell to Jane Cay, February 27, 1857, in *SLP*, 1:496.

76. James Clerk Maxwell to Richard Buckley Litchfield, May 29, 1857, in *SLP*, 1:508.

77. Campbell and Garnett, *Life*, 314.

78. James Clerk Maxwell to Thomas Andrews, July 15, 1875, in *SLP*, 3:236.

79. Campbell and Garnett, *Life*, 162; and Jordi Cat, *Land, Lines, Colours, and Toys: Becoming James Clerk Maxwell* (Oxford: Oxford University Press, forthcoming), chap. 8.

80. Campbell and Garnett, *Life*, 258–59; and Horace Lamb, "Clerk Maxwell as Lecturer," in *James Clerk Maxwell: A Commemoration Volume, 1831–1931*,

ed. Joseph John Thomson (Cambridge: Cambridge University Press, 1931), 142–46, on 144.

81. James Clerk Maxwell to James David Forbes, March 30, 1857, in *SLP*, 1:498.

82. James Clerk Maxwell to R. B. Litchfield, March 5, 1858, quoted in Campbell and Garnett, *Life*, 304–5.

83. James Clerk Maxwell to Charles Benjamin Tayler, July 8, 1853, in *SLP*, 1:220–21.

84. Campbell and Garnett, *Life*, 200.

85. James Clerk Maxwell, "Introductory Lecture, Aberdeen," in *SLP*, 1:544.

86. James Clerk Maxwell, "Inaugural Lecture at Marischal College, Aberdeen, November 3, 1856," in *SLP*, 1:431, 1:425.

87. James Clerk Maxwell, "Introductory Lecture on Experimental Physics," in *The Scientific Papers of James Clerk Maxwell*, ed. W. D. Niven, 2 vols. (Cambridge: Cambridge University Press), 2:252 (hereafter cited as *SP*).

88. James Clerk Maxwell, "Molecules (A Lecture)," in *SP*, 2:372.

89. Maxwell, "Introductory Lecture, Aberdeen," in *SLP*, 1:544.

90. James Clerk Maxwell, "Review of P. G. Tait, *Recent Advances in Physical Sciences*," in *SLP*, 3:310.

91. Campbell and Garnett, *Life*, 234.

92. Maxwell, "Inaugural Lecture at Marischal College, Aberdeen, November 3, 1856," 1:420.

93. James Clerk Maxwell, "Inaugural Lecture at King's College London, October 1860," in *SLP*, 1:668.

94. Ibid., 1:670.

95. James Clerk Maxwell, "Introductory Lecture, Aberdeen," in *SLP*, 1:547.

96. Maxwell, "Introductory Lecture on Experimental Physics," 2:248.

97. James Clerk Maxwell, "Review of Tait's *Thermodynamics*," in *SP*, 2:660.

98. Maxwell, "Introductory Lecture on Experimental Physics," 2:242.

99. Ibid., 2:251.

100. Maxwell, "Introductory Lecture, Aberdeen," 1:543.

101. Maxwell, "Inaugural Lecture at Marischal College, Aberdeen, November 3, 1856," 1:420.

102. Ibid., 1:426.

103. Ibid., 1:430.

104. Ibid., 1:427.

105. Maxwell, "Introductory Lecture, Aberdeen," 1:543.

106. Ibid.; and Campbell and Garnett, *Life*, 82–83.

107. Maxwell, "Introductory Lecture, Aberdeen," 1:542.

108. Maxwell, "Introductory Lecture on Experimental Physics," 2:250.

109. Adrian Desmond, *Huxley: From Devil's Disciple to Evolution's High Priest* (Reading, MA: Addison-Wesley, 1997), 361.

110. Thomas Henry Huxley, "Technical Education," in *Collected Essays of T. H. Huxley*, 9 vols. (Bristol, UK: Thoemmes Press, 2001), 3:407–8 (hereafter cited as *CE*).

111. Desmond, *Huxley*, 252.

112. Ibid., 362.

113. Thomas Henry Huxley, "Administrative Nihilism," in *Collected Essays*, vol. 1 (New York: D. Appleton, 1902), 257, 287.

114. Thomas Henry Huxley to George Howell, MP, January 2, 1880, quoted in Leonard Huxley, *Life and Letters of Thomas Henry Huxley*, 2 vols. (1900; repr., New York: AMS Press, 1979), 1:510 (hereafter cited as *LL*).

115. Thomas Henry Huxley, "Administrative Nihilism," 1:251–52.

116. Thomas Henry Huxley, "A Liberal Education; and Where to Find It," in *CE*, 3:90–91.

117. Ibid., 3:84.

118. Paul White, "Ministers of Culture: Arnold, Huxley, and Liberal Anglican Reform of Learning," *History of Science* 43 (2005): 115–38.

119. Thomas Henry Huxley, "A Lobster; or, the Study of Zoology," in *CE*, 9:225.

120. Ibid., 9:226.

121. Thomas Henry Huxley, "Liberal Education; and Where to Find It," 3:103; and Thomas Henry Huxley, "Scientific Education: Notes of an after-Dinner Speech," in *CE*, 3:117.

122. Thomas Henry Huxley, undated draft, quoted in *LL*, 2:329–30.

123. Thomas Henry Huxley, "Presidential Address to the Royal Society," *Nature* 33, no. 840 (1885): 118.

124. Paul White, *Thomas Huxley: Making the "Man of Science"* (Cambridge: Cambridge University Press, 2003), 150.

125. Thomas Henry Huxley to Frederick Dyster, February 27, 1855, quoted in *LL*, 1:149.

126. Desmond, *Huxley*, 640; and Adrian Desmond, *The Politics of Evolution: Morphology, Medicine, and Reform in Radical London* (Chicago: University of Chicago Press, 1989). Note that this relationship soured over the years, particularly as Huxley became more antisocialist. By the end of his life he was not particularly beloved by the working classes.

127. White, *"Man of Science,"* 81.

128. Thomas Henry Huxley to Charles Kingsley, September 23, 1860, quoted in *LL*, 1:238.

129. Thomas Henry Huxley to Charles Kingsley, May 22, 1863, quoted in *LL*, 1:261.

130. Desmond, *Huxley*, 361–62.

131. Thomas Henry Huxley, "Liberal Education; and Where to Find It," 3:77.

132. Ibid., 3:108.

133. Ibid., 3:83.

134. White, *"Man of Science,"* 80.

135. Thomas Henry Huxley, "On the Educational Value of the Natural History Sciences," in *CE*, 3:60.

136. Thomas Henry Huxley, "Liberal Education; and Where to Find It," 3:85–86.

137. Ibid., 3:79.

138. Thomas Henry Huxley, "On Natural History, as Knowledge, Discipline, and Power," in *The Scientific Memoirs of Thomas Henry Huxley*, ed. Michael Foster and E. Ray Lankester, 5 vols. (London: Macmillan, 1902), 1:313–14 (hereafter cited as *SM*).

139. Thomas Henry Huxley, "Scientific Education: Notes of an after-Dinner Speech," 3:125–26.

140. Thomas Henry Huxley, "Lobster; or, the Study of Zoology," 9:219–20.

141. White, *"Man of Science,"* 77.

142. Thomas Henry Huxley, "The Progress of Science," in *CE*, 1:53–54.

143. Thomas Henry Huxley, "On the Educational Value of the Natural History Sciences," 3:59.

144. Ibid., 3:45. Emphasis in the original.

145. White, *"Man of Science,"* 78, 126.

146. Thomas Henry Huxley to Charles Kingsley, September 23, 1860, quoted in *LL*, 1:237.

147. Thomas Henry Huxley, "Technical Education," 3:424.

148. Desmond, *Huxley*, 363.

149. Thomas Henry Huxley to Charles Kingsley, September 23, 1860, quoted in *LL*, 1:237.

150. Thomas Henry Huxley, "Scientific Education: Notes of an after-Dinner Speech," 3:133.

151. Ibid., 3:119.

152. Bernard Lightman, *The Origins of Agnosticism: Victorian Unbelief and the Limits of Knowledge* (Baltimore: Johns Hopkins University Press, 1987), 125–31.

153. Thomas Henry Huxley, "Scientific Education: Notes of an after-Dinner Speech," 3:123, 3:129–30.

154. Frederick Denison Maurice, *A Few Words on Secular and Denominational Education: In a Letter to the Members of the Working Men's College* (London: Macmillan, 1870).

155. "To Our Readers," *Christian Socialist*, July 5, 1851.

156. Matthew Stanley, *Practical Mystic: Religion, Science, and A. S. Eddington* (Chicago: University of Chicago Press, 2007), 6–7, 239–45.

157. Frank M. Turner, *Contesting Cultural Authority: Essays in Victorian Intellectual Life* (Cambridge: Cambridge University Press, 1993), 101–24.

158. Bernard Lightman, *Victorian Popularizers of Science: Designing Nature for New Audiences* (Chicago: University of Chicago Press, 2007), 372–77.

159. Henrietta Litchfield, *Litchfield: A Memoir*, 136–40.

160. R. B. Litchfield, *Beginnings of the Working Men's College*, 11.

161. S. J. D. Green, "Religion and the Rise of the Common Man: Mutual Improvement Societies, Religious Associations and Popular Education in Three Industrial Towns in the West Riding of Yorkshire, c. 1850–1900," in *Cities, Class and Communication: Essays in Honor of Asa Briggs*, ed. Derek Fraser, 25–43 (New York: Harvester Wheatsheaf, 1990); and Rose, *Intellectual Life of the British Working Classes*.

Chapter Five

1. The idea that science and freedom have a symbiotic relationship has a long history. For a critical discussion, see Frank M. Turner, "Science and Religious Freedom," in *Freedom and Religion in the Nineteenth Century*, ed. Richard J. Helmstadter, 54–86 (Stanford, CA: Stanford University Press, 1997).

2. Thomas Henry Huxley to Frederick Dyster, February 29, 1860, Huxley Papers, Imperial College London (hereafter HP), 15.110.

3. Thomas Henry Huxley to Joseph Hooker, May 22, 1889, quoted in Leonard Huxley, *Life and Letters of Thomas Henry Huxley*, 2 vols. (1900; repr., New York: AMS Press, 1979), 2:249 (hereafter cited as *LL*).

4. Thomas Henry Huxley to Edwin Lankester, May 25, 1889, quoted in *LL*, 2:250.

5. Edward Frankland to Thomas Henry Huxley, February 16, 1869, HP 16.251.

6. Thomas Henry Huxley, "Autobiography," in *Collected Essays of T. H. Huxley*, 9 vols., 1:1–17 (Bristol, UK: Thoemmes Press, 2001), 1:16–17 (hereafter cited as *CE*).

7. Thomas Henry Huxley, "Agnosticism and Christianity," in *CE*, 5:312–13.

8. Thomas Henry Huxley to Sir John Skelton, March 7, 1887, in *LL*, 2:195–96; and Thomas Henry Huxley to H. de Varigny, November 25, 1891, in *LL*, 2:309.

9. John Tyndall to Thomas Henry Huxley, April 22, 1869, HP 8.70.

10. Colin A. Russell, *Edward Frankland: Chemistry, Controversy and Conspiracy in Victorian England* (Cambridge: Cambridge University Press, 1996), 336–37.

11. Thomas Henry Huxley to John Donnelly, October 10, 1890, in *LL*, 2:171.

12. Thomas Henry Huxley, "Technical Education," in *CE*, 3:417.

13. "Truths," HP 31.

14. Thomas Henry Huxley to Henry Roscoe, June 29, 1887, in *LL*, 2:166.

15. Thomas Henry Huxley, "On the Hypothesis That Animals Are Automata, and Its History," in *CE*, 1:250.

16. Thomas Henry Huxley, "The Rede Lecture," in *The Scientific Memoirs of Thomas Henry Huxley*, ed. Michael Foster and E. Ray Lankester, 5 vols. (London: Macmillan, 1902), 5:73 (hereafter cited as *SM*).

17. Thomas Henry Huxley, prologue to *Essays upon Some Controverted Questions*, by Thomas Henry Huxley (New York: D. Appleton, 1892), 15.

18. Thomas Henry Huxley to Frederick Dyster, February 27, 1855, quoted in *LL*, 1:149.

19. Thomas Henry Huxley to Charles Lyell, March 17, 1860, quoted in *LL*, 1:228.

20. Thomas Henry Huxley to Charles Kingsley, May 5, 1863, quoted in *LL*, 1:348.

21. Thomas Henry Huxley, "Scientific and Pseudo-Scientific Realism," in *CE*, 5:79.

22. Thomas Henry Huxley, "Agnosticism and Christianity," 5:327.

23. Thomas Henry Huxley, prologue to *Essays upon Some Controverted Questions*, 17–18.

24. Ibid., 17.

25. Ibid., 21.

26. Ibid., 14–15.

27. Thomas Henry Huxley to Charles Kingsley, May 5, 1863, quoted in *LL*, 1:260.

28. John Tyndall to Thomas Henry Huxley, May 5, 1886, HP 8.245.

29. Thomas Henry Huxley to George John Romanes, November 3, 1892, quoted in *LL*, 2:361.

30. Thomas Henry Huxley, preface to *CE*, 5:viii–ix.

31. Thomas Henry Huxley to Robert Taylor, June 3, 1889, quoted in *LL*, 2:243.

32. Thomas Henry Huxley, "Mr. Balfour's Attack on Agnosticism," *Nineteenth Century* 37 (1895): 530.

33. Thomas Henry Huxley, "Agnosticism: A Rejoinder," in *CE*, 5:286.

34. Thomas Henry Huxley to Leonard Huxley, January 20, 1885, quoted in *LL*, 2:98; and Thomas Henry Huxley, "The Evolution of Theology: An Anthropological Study," in *CE*, 4:368–69.

35. Thomas Henry Huxley to Estlin Carpenter, October 11, 1890, quoted in *LL*, 2:283.

36. Thomas Henry Huxley, "Mr. Balfour's Attack on Agnosticism," 530.

37. Thomas Henry Huxley, "Evolution of Theology: An Anthropological Study," 4:371.

38. Thomas Henry Huxley, "A Modern Symposium: Influence on Morality of a Decline in Religious Belief," *Ninteenth Century* 1 (1877): 538.

39. Thomas Henry Huxley to Frederick Dyster, January 30, 1859, HP 15.106.

40. Thomas Henry Huxley, "Agnosticism," in *CE*, 5:216.

41. Thomas Henry Huxley, "Social Diseases and Worse Remedies," in *CE*, 9:192.

42. Thomas Henry Huxley to *Times*, December 9, 1890, and Thomas Henry Huxley to *Times*, December 27, 1890, in *CE*, 9:240–42.

43. Thomas Henry Huxley, "Agnosticism," 5:210–11.

44. Thomas Henry Huxley, preface to *CE*, 5:viii.

45. Thomas Henry Huxley, "Government: Anarchy or Regimentation," in *CE*, 1:384–85.

46. Ibid., 1:385.

47. Thomas Henry Huxley, "Agnosticism and Christianity," 5:313.

48. Ibid., 5:318.

49. Thomas Henry Huxley to Herbert Spencer, December 27, 1880, quoted in *LL*, 2:19.

50. Thomas Henry Huxley to unknown, March 12, 1883, in *LL*, 2:430.

51. Hugh McLeod, *Secularisation in Western Europe, 1848–1914* (London: Macmillan, 2000), 225–38; and Denis G. Paz, *Popular Anti-Catholicism in Mid-Victorian England* (Stanford, CA: Stanford University Press, 1992).

52. Thomas Henry Huxley to James Thomas Knowles, May 22, 1889, quoted in *LL*, 2:242.

53. Thomas Henry Huxley to Leonard Huxley, January 20, 1885, quoted in *LL*, 2:98.

54. Thomas Henry Huxley, "Scientific Education: Notes of an after-Dinner Speech," in *CE*, 3:120.

55. Ibid., 3:120–21.

56. Thomas Henry Huxley, "On Descartes' 'Discourse Touching the Method of Using One's Reason Rightly and of Seeking Scientific Truth,'" in *CE*, 1:196.

57. Ibid., 1:197.

58. Thomas Henry Huxley, "Mr. Darwin's Critics," in *CE*, 2:124.

59. Ibid., 2:125.

60. Ibid., 2:145.

61. Ibid., 2:147.

62. Ibid., 2:149.

63. Thomas Henry Huxley, "An Apologetic Irenicon," *Fortnightly Review* 52 (1892): 558.

64. Thomas Henry Huxley, "On the Physical Basis of Life," in *CE*, 1:156.

65. Thomas Henry Huxley, "The Scientific Aspects of Positivism," in *Lay Sermons, Addresses and Reviews* (New York: D. Appleton, 1872), 150.

66. Thomas Henry Huxley to M. Foster, July 18, 1883, quoted in *LL*, 2:61.

67. Thomas Henry Huxley, "Prologue to *Essays upon Some Controverted Questions*," in *CE*, 5:1–58, on 5:17.

68. Thomas Henry Huxley, "Administrative Nihilism," in *Collected Essays*, vol. 1 (New York: D. Appleton, 1902), 266.

69. John Tyndall to Thomas Henry Huxley, December 27, 1887, HP 8.260.

70. John Tyndall, "Scientific Use of the Imagination," in *Fragments of Science: A Series of Detached Essays, Addresses, and Reviews*, vol. 2 (New York: D. Appleton, 1897), 133.

71. Thomas Henry Huxley, "The Coming of Age of *The Origin of Species*," in *CE*, 2:227–43, on 2:229.

72. Thomas Henry Huxley, "Science and Pseudo-Science," in *CE*, 5:93.

73. Thomas Henry Huxley, prefatory note to *Freedom in Science and Teaching*, by Ernst Haeckel (New York: D. Appleton, 1879), xvii.

74. Thomas Henry Huxley to John Tyndall, July 22, 1874, quoted in *LL*, 2:442.

75. Thomas Henry Huxley, "Joseph Priestley," *Macmillan's Magazine* 30 (1874): 2–3.

76. Ibid., 3–4.

77. Ibid., 7, 11.

78. Ibid., 12–13.

79. Ibid., 22–23.

80. Ibid., 25–26.

81. Ibid., 31–32.

82. Henry Edward Armstrong, "Our Need to Honour Huxley's Will: Huxley Memorial Lecture," in *H. E. Armstrong and the Teaching of Science, 1880–1930*, ed. W. H. Brock (Cambridge: Cambridge University Press, 1973).

83. J. H. Rylance, "The Relation of Miracles to the Christian Faith," in *Christian Truth and Modern Opinion: Seven Sermons Preached in New-York by Clergymen of the Protestant Episcopal Church* (New York: Thomas Whittaker, 1874), 102–3.

84. St. George Jackson Mivart to Richard Owen, October 26, 1871, Richard Owen Papers, Mivart Correspondence, 19/258–271 (R), 267.

85. Duke of Argyll, "Professor Huxley and the Duke of Argyll," in *Nineteenth Century*, ed. James Knowles, vol. 29 (London: Kegan Paul, Trench, Trübner, 1891), 687.

86. Duke of Argyll, "Professor Huxley on the Warpath," *Nineteenth Century*, no. 167 (1891): 20, 21.

87. Thomas Henry Huxley, "Science and Pseudo-Science," 5:122.

88. Ibid., 5:124.

89. Thomas Henry Huxley, "The School Boards: What They Can Do, and What They May Do," in *CE*, 3:377.

90. Ibid., 3:380.

91. Ibid., 3:386.

92. Ibid., 3:395.

93. Ibid., 3:396.

94. Ibid., 3:397.

95. Ibid., 3:398–99.

96. Thomas Henry Huxley, "Agnosticism: A Rejoinder," 268–69.

97. Thomas Henry Huxley, "School Boards," 402.

98. Thomas Henry Huxley, "Prologue to *Essays upon Some Controverted Questions*," 5:57.

99. Ibid., 5:58.

100. Thomas Henry Huxley to Frederick William Farrar, November 6, 1894, quoted in *LL*, 2:406.

101. Thomas Henry Huxley to Charles Kingsley, September, 23, 1860, quoted in *LL*, 1:238.

102. James Clerk Maxwell to Lewis Campbell, September 15, 1853, quoted in Lewis Campbell and William Garnett, *The Life of James Clerk Maxwell* (London: Macmillan, 1882), 192 (hereafter cited as *Life*).

103. Campbell and Garnett, *Life*, 322.

104. James Clerk Maxwell to Lewis Campbell, December 22, 1857, quoted in Campbell and Garnett, *Life*, 293.

105. Ibid., 294.

106. Ibid.

107. Ibid.

108. See the letters between Thomson and Stokes regarding these issues from February 1849, found in David B. Wilson, ed., *The Correspondence between Sir George Gabriel Stokes and Sir William Thomson, Baron Kelvin of Largs*, vol. 1, *1846–1869* (Cambridge: Cambridge University Press, 1990), xv, 59–64.

109. Maxwell served as elder from 1863 to his death in 1879. The records for this period can be found at the National Archive of Scotland, CH2/446/1 Session Minutes 1838–1906.

110. Campbell and Garnett, *Life*, 233.

111. Ibid., 172.

112. Draft letter to Francis W. H. Petrie, ca. March 15, 1875, in *The Scientific Letters and Papers of James Clerk Maxwell*, ed. P. M. Harman, 3 vols. (Cambridge: Cambridge University Press, 1990), 3:194n4 (hereafter cited as *SLP*). Also see Jerrold McNatt, "James Clerk Maxwell's Refusal to Join the Victoria Institute," *Perspectives on Science and Christian Faith* 56 (2004): 204–15.

113. Draft letter to Francis W. H. Petrie, ca. March 15, 1875, in *SLP*, 3:194. See also Campbell and Garnett, *Life*, 404–5.

114. J. P. Ellens, "Which Freedom for Early Victorian Britain?," in *Freedom*

and Religion in the Nineteenth Century, ed. Richard Helmstadter (Stanford, CA: Stanford University Press, 1997), 92.

115. Richard Owen, "On Some Instances," Owen Collection, Library and Archives, The Natural History Museum, London, OC 59.1/63–98, 96–97.

116. Ellens, "Which Freedom for Early Victorian Britain?"

117. William Hodson Brock and Roy M. McLeod, "The Scientists' Declaration: Reflexions on Science and Belief in the Wake of Essays and Reviews, 1864–5," *British Journal for the History of Science* 9, no. 1 (1976): 39–66.

118. John F. W. Herschel, "Science and Scripture," *Athenaeum,* no. 1925 (1864).

119. William Stanley Jevons to John F. W. Herschel, September 27, 1864, John Herschel Letters, Royal Society, RS 10.320 12677.

120. James Clerk Maxwell to Lewis Campbell, March 7, 1852, quoted in Campbell and Garnett, *Life,* 179.

121. James Clerk Maxwell to R. B. Litchfield, March 5, 1858, in *SLP,* 1:587–90, on 1:589. Carpenter made similar points in William Benjamin Carpenter, *Principles of Mental Physiology: With Their Applications to the Training and Discipline of the Mind and the Study of Its Morbid Conditions,* 4th ed. (1876; repr., New York: D. Appleton, 1884), 404–6.

122. For example, see *SLP,* 3:823–26; or Wilson, *Correspondence,* 1:186–88, 1:190.

123. Melinda Baldwin, "Tyndall and Stokes: Correspondence, Peer Review, and the Physical Sciences in Victorian Britain," in *Tyndall and Nineteenth Century Science,* ed. Bernard Lightman, Michael Reidy, and Joshua Howe (Chicago: University of Chicago Press, forthcoming).

124. John Tyndall to Hermann Helmholtz, January 13, 1868, quoted in Baldwin, "Tyndall and Stokes."

125. James Clerk Maxwell, "Appendix: From a Letter of Reference for William Kingdon Clifford, *circa* July 1871," in *SLP,* 2:666.

126. Thomas Henry Huxley to John Tyndall, December 13, 1878, HP 8.217; and John Tyndall to Thomas Henry Huxley, July 16, 1868, HP 1.52.

127. James Clerk Maxwell to Herbert Spencer, December 5, 1873, in *SLP,* 2:956–61.

128. James Clerk Maxwell to Herbert Spencer, December 17, 1873, in *SLP,* 2:962–63; and James Clerk Maxwell to Herbert Spencer, December 17, 1874, in *SLP,* 3:152–54.

129. James Clerk Maxwell, *SLP,* 1:401n12.

130. Lord Kelvin to George Gabriel Stokes, July 29, 1897, in David B. Wilson, ed., *The Correspondence between Sir George Gabriel Stokes and Sir William Thomson, Baron Kelvin of Largs,* vol. 2, *1870–1901* (Cambridge: Cambridge University Press, 1990), 695–96.

131. James Clerk Maxwell to William Turner Thiselton-Dyer, February 12, 1876, quoted in *SLP*, 3:276–77. See Graeme Gooday, "Nature in the Laboratory: Domestication and Discipline with the Microscope in Victorian Life Science," *British Journal for the History of Science* 24, no. 3 (1991): 307–41.

132. P. M. Harman, introduction to *SLP*, 2:2–5.

133. James Clerk Maxwell to John Tyndall, April 20, 1864, in *SLP*, 2:147.

134. James Clerk Maxwell to Henry Richmond Droop, July 19, 1865, quoted in Campbell and Garnett, *Life*, 343, and in *SLP*, 2:226–27.

135. James Clerk Maxwell to Cecil James Monro, March 15, 1871, in *SLP*, 2:617–20, on 2:617.

136. James Clerk Maxwell to Peter Guthrie Tait, late August–early September 1873, in *SLP*, 2:921.

137. Adrian Desmond, *Huxley: From Devil's Disciple to Evolution's High Priest* (Reading, MA: Addison-Wesley, 1997), 406.

138. Edwin Ray Lankester, *The Kingdom of Man* (New York: Henry Holt, 1907), 86.

139. James Clerk Maxwell, quoted in Campbell and Garnett, *Life*, 420. Maxwell died before finishing the review.

140. P. M. Harman, introduction to *SLP*, 3:29.

141. Lord Kelvin, "Presidential Address," in *Proceedings of the Royal Society of London*, vol. 59, 107–24 (London: Harrison and Sons, 1896), 109–10.

142. George Gabriel Stokes to Lord Kelvin, December 1, 1895, quoted in Wilson, *Correspondence*, 2:630–31.

143. Charles Kingsley, *The Water Babies: A Fairy Tale for a Land-Baby* (New York: Macmillan, 1885), 141–42.

144. On the Belfast Address generally, see Frank Turner, "John Tyndall and Victorian Scientific Naturalism," in *John Tyndall: Essays on a Natural Philosopher*, ed. W. H. Brock, Norman D. McMillan, and R. Charles Mollan, 169–80 (Dublin: Royal Dublin Society, 1981); and Ruth Barton, "John Tyndall, Pantheist: A Rereading of the Belfast Address," *Osiris* 2, no. 3 (1987): 111–34.

145. Desmond, *Huxley*, 423.

146. Note that a poem incorrectly labeled as being written in response to Belfast (titled "Molecular Evolution") appears in Campbell and Garnett, *Life*, 637.

147. The letter appears in *SLP*, 3:99–102. The poem also appears (slightly altered) in Campbell and Garnett, *Life*, 638. The early history of the Red Lions is described in Brian Gardiner, "Edward Forbes, Richard Owen and the Red Lions," *Archives of Natural History* 20, no. 3 (1993): 349–72.

148. Campbell and Garnett, *Life*, 639–41.

149. Lord Kelvin to George Gabriel Stokes, April 29, 1859, in Wilson, *Correspondence*, 1:238–43.

150. Desmond, *Huxley*, 375–79.

Chapter Six

1. On the free will debates in Victorian culture and thought, see Roger Smith, *Free Will and the Human Sciences in Britain, 1870–1910* (London: Pickering and Chatto, 2013); John Robert Reed, *Victorian Will* (Athens: Ohio University Press, 1989); Leon Stephen Jacyna, "The Physiology of Mind, the Unity of Nature, and the Moral Order in Victorian Thought," *British Journal for the History of Science* 14, no. 2 (1981): 109–32; Rick Rylance, *Victorian Psychology and British Culture, 1850–1880* (Oxford: Oxford University Press, 2000), 25–45; Lorraine J. Daston, "The Theory of Will versus the Science of Mind," in *The Problematic Science: Psychology in Nineteenth-Century Thought*, ed. Mitchell G. Ash, 88–118 (New York: Praeger, 1982); and Kurt Danziger, "Mid-Nineteenth Century British Psycho-Physiology: A Neglected Chapter in the History of Psychology," in *Problematic Science*, 119–46.

2. For example, Henry Maudsley, *The Physiology and Pathology of Mind* (New York: D. Appleton, 1867); and Alexander Bain, "On the Correlation of Force and Its Bearing on the Mind," *Macmillan's Magazine* 16 (1867): 372–83. See also Danziger, "British Psycho-Physiology," 134–38; and Rick Rylance, *Victorian Psychology*, 164–75.

3. For the earlier history of the development of the mind sciences, see Christopher Fox, ed., *Inventing Human Science: Eighteenth-Century Domains* (Berkeley: University of California Press, 1995); Simon Schaffer, "States of Mind: Enlightenment and Natural Philosophy," in *The Languages of Psyche: Mind and Body in Enlightenment Thought*, ed. George Sebastian Rousseau, 233–90 (Berkeley: University of California Press, 1990); Fernando Vidal, *The Sciences of the Soul: The Early Modern Origins of Psychology* (Chicago: University of Chicago Press, 2011); Roy Porter, *Flesh in the Age of Reason: The Modern Foundations of Body and Soul* (New York: W. W. Norton, 2003); and John Yolton, *Thinking Matter: Materialism in Eighteenth-Century Britain* (Minneapolis: University of Minnesota Press, 1983).

4. On the concept of humans as automata, see Minsoo Kang, *Sublime Dreams of Living Machines: The Automaton in the European Imagination* (Cambridge, MA.: Harvard University Press, 2011). On the construction of automata and their cultural meaning, see Jessica Riskin, ed., *Genesis Redux: Essays in the History and Philosophy of Artificial Life* (Chicago: University of Chicago Press, 2007); and Adelheid Voskuhl, "The Mechanics of Sentiment: Music-Playing Women Automata and the Culture of Affect in Late Eighteenth-Century Europe" (PhD diss., Cornell University, 2007).

5. Kang, *Sublime Dreams of Living Machines*, 8, 148.

6. Roger Smith, *Free Will*, passim, particularly 1, 8, 100, 161. On the political stakes, see Schaffer, "States of Mind," 233–35; and Stefan Collini, "The Idea of

'Character' in Victorian Political Thought," *Transactions of the Royal Historical Society*, 5th ser., 35 (1985): 29–50. On the theological and moral stakes, see Leon Stephen Jacyna, "Immanence or Transcendence: Theories of Life and Organization in Britain, 1790–1835," *Isis* 74, no. 3 (1983): 318–19.

7. Robert M. Young, *Mind, Brain, and Adaptation in the Nineteenth Century: Cerebral Localization and Its Biological Context from Gall to Ferrier* (Oxford: Oxford University Press, 1970); Leon Stephen Jacyna, "Scientific Naturalism in Victorian Britain: An Essay in the Social History of Ideas" (unpublished PhD thesis, University of Edinburgh, 1980); and John Van Whye, *Phrenology and the Origins of Victorian Scientific Naturalism* (Burlington, VT: Ashgate, 2004).

8. Young, *Mind, Brain, and Adaptation in the Nineteenth Century*, 88–89.

9. Anson Rabinbach, *The Human Motor: Energy, Fatigue, and the Origins of Modernity* (Berkeley: University of California Press, 1992), 65–66; and Emil Du Bois-Reymond, "Ueber die Grenzen des Naturerkennens," in *Reden* (Leipzig: Veit, 1886). Ernst Cassirer, *Determinism and Indeterminism in Modern Physics: Historical and Systematic Studies in the Problem of Causality* (New Haven, CT: Yale University Press, 1956), points to this speech as framing the determinism problem for the late nineteenth century. See also Keith Anderton's PhD diss., "The Limits of Science: A Social, Political, and Moral Agenda for Epistemology in Nineteenth Century Germany" (Harvard University, Department of the History of Science, 1993); and Gabriel Finkelstein's PhD diss., "Emil Du Bois-Reymond: The Making of a German Liberal Scientist" (Princeton University, Department of History, 1996).

10. Alison Winter, *Mesmerized: Powers of Mind in Victorian Britain* (Chicago: University of Chicago Press, 1998), 7.

11. Jacyna, "Physiology of Mind, the Unity of Nature, and the Moral Order in Victorian Thought," 110.

12. Edwin Clarke and L. S. Jacyna, *Nineteenth-Century Origins of Neuroscientific Concepts* (Berkeley: University of California Press, 1987), 114–24; and Ruth Leys, "Background to the Reflex Controversy: William Alison and the Doctrine of Sympathy before Hall," *Studies in History of Biology* 4 (1980): 1–66.

13. Edwin Clarke and L. S. Jacyna, *Nineteenth-Century Origins of Neuroscientific Concepts*, 127, 39.

14. Winter, *Mesmerized: Powers of Mind in Victorian Britain*, 288–90.

15. Edwin Clarke and L. S. Jacyna, *Nineteenth-Century Origins of Neuroscientific Concepts*, 140–41.

16. Young, *Mind, Brain, and Adaptation in the Nineteenth Century*, 100–150.

17. Alan Richardson, *British Romanticism and the Science of the Mind* (Cambridge: Cambridge University Press, 2004).

18. Thomas Henry Huxley, "Thoughts and Doings," HP 31.

19. M. Norton Wise, "The Gender of Automata in Victorian Britain," in *Gen-*

esis Redux: Essays in the History and Philosophy of Artificial Life, edited by Jessica Riskin, 163–95 (Chicago: University of Chicago Press, 2007), 184.

20. [Thomas Henry Huxley], "Contemporary Literature—Science," *Westminister Review* 63 (1855): 239–53.

21. Ibid., 241–42.

22. William Benjamin Carpenter, *Nature and Man: Essays Scientific and Philosophical* (London: Kegan Paul, Trench, 1888), 66–67.

23. William Benjamin Carpenter, *Principles of Human Physiology, with Their Chief Applications to Pathology, Hygiene, and Forensic Medicine*, 4th ed. (London: Samuel Bentley, 1853), 19–88.

24. Ibid., 95–97, 122–23.

25. Ibid., 345.

26. Thomas Henry Huxley, "Evidence as to the Miracle of the Resurrection," Metaphysical Society, January 11, 1876.

27. Thomas Henry Huxley, "On the Educational Value of the Natural History Sciences," in *Collected Essays of T. H. Huxley*, 9 vols. (Bristol, UK: Thoemmes Press, 2001), 3:41–42 (hereafter cited as *CE*).

28. Thomas Henry Huxley to John Tyndall, June 14, 1870, HP 8.79.

29. Thomas Henry Huxley, "On Our Knowledge of the Causes of the Phenomena of Organic Nature," in *CE*, 2:311.

30. Ibid., 2:315.

31. Ibid., 2:316.

32. John Tyndall, "Scientific Materialism," in *Fragments of Science: A Series of Detached Essays, Addresses, and Reviews*, vol. 2 (New York: D. Appleton, 1897), 80–83.

33. The social and political consequences of applying energy physics to humans is documented in Rabinbach, *Human Motor*. Note that the term *energy* was still in flux during this period, and the historical actors often refer to *force*.

34. Richard L. Kremer, *The Thermodynamics of Life and Experimental Physiology, 1770–1880* (New York: Garland Publishing, 1990), 23–25.

35. Ibid., 215–31, 37–55. On Helmholtz, see David Cahan, ed., *Hermann von Helmholtz and the Foundations of Nineteenth-Century Science* (Berkeley: University of California Press, 1993); and Robert Brain and M. Norton Wise, "Muscles and Engines: Indicator Diagrams and Helmholtz's Graphical Methods," in *The Science Studies Reader*, ed. Mario Biagioli (New York: Routledge, Chapman and Hall, 1999). On his relationships with British scientists, see David Cahan, "Helmholtz and the British Scientific Elite: From Force Conservation to Energy Conservation," *Notes and Records of the Royal Society* 66, no. 1 (2012): 55–68. On German thinking on materialism and determinism, see Frederick Gregory, "Scientific versus Dialectical Materialism: A Clash of Ideologies in Nineteenth-Century German Radicalism," *Isis* 68 (1977): 206–23.

36. Kremer, *Thermodynamics of Life and Experimental Physiology*, 370–75.

37. Edwin Lankester, *On Food* (London: Robert Hardwicke, 1861), 3.

38. Ibid., 125–26.

39. Colin A. Russell, *Edward Frankland: Chemistry, Controversy and Conspiracy in Victorian England* (Cambridge: Cambridge University Press, 1996), 421–25. Also see Edward Frankland, "On the Origin of Muscular Power," *Philosophical Magazine* 32, no. 215 (1866): 182–89.

40. Samuel Houghton, *On the Natural Constants of the Healthy Urine of Man, and a Theory of Work Founded Thereon* (Dublin: University of Dublin Press, 1860), 6.

41. Hermann von Helmholtz, "The Application of the Law of the Conservation of Force to Organic Nature," in *Selected Writings of Hermann von Helmholtz*, ed. Russell Kahl (Middletown, CT: Wesleyan University Press, 1971), 119. The relevant experiments are described in Edward Smith, "The Influence of the Labour of the Treadwheel over Respiration and Pulsation," *Medical Times and Gazette* 14 (1857): 601–3; and Edward Smith and William Ralph Milner, "Report on the Action of Prison Diet and Discipline on the Bodily Functions of Prisoners," *British Association for the Advancement of Science* 31 (1861): 44–81.

42. Thomas Henry Huxley, "The Progress of Science," in *CE*, 1:94.

43. Ibid., 1:95.

44. Maurice Mandelbaum, *History, Man, and Reason: A Study in Nineteenth-Century Thought* (Baltimore: Johns Hopkins University Press, 1971), 298–304.

45. Thomas Henry Huxley, "Hume: With Helps to the Study of Berkeley," in *CE*, 6:90.

46. Thomas Henry Huxley, "On the Present State of Knowledge as to the Structure and Functions of Nerve," in *The Scientific Memoirs of Thomas Henry Huxley*, ed. Michael Foster and E. Ray Lankester, 5 vols. (London: Macmillan, 1902), 1:320 (hereafter cited as *SM*).

47. Thomas Henry Huxley, "Mr. Darwin's Critics," in *CE*, 2:158.

48. Thomas Henry Huxley, "Science and Morals," in *CE*, 9:135–36; and Thomas Henry Huxley, "Hume," 6:95.

49. Thomas Henry Huxley, "Hume," 6:123–24.

50. Thomas Henry Huxley, "Has a Frog a Soul and of What Nature Is That Soul, Supposing It to Exist?" Papers of the Metaphysical Society, 1 (1869–74), November 8, 1870, 1–7, Manchester College Library, Oxford, 2–3.

51. Ibid., 7.

52. Carpenter, *Principles of Human Physiology*, 359.

53. Ibid., 374.

54. Quoted in Jacyna, "Scientific Naturalism in Victorian Britain," 205.

55. Thomas Henry Huxley, "On the Hypothesis That Animals Are Automata, and Its History," in *CE*, 1:203. Emphasis in the original.

56. Ibid., 1:218.

57. Ibid., 1:226–27.

58. Ibid., 1:228–29.

59. Ibid., 1:238.

60. Ibid., 1:240.

61. Ibid., 1:242–43.

62. Ibid., 1:243–44.

63. Thomas Henry Huxley, "Hume," 6:214.

64. Thomas Henry Huxley, *Introductory Science Primer* (London: Appleton, 1887), 94 (hereafter cited as *ISP*).

65. Thomas Henry Huxley, "Hume," 6:61.

66. On the role of free will in the formation of psychology as a discipline, see Roger Smith, *Free Will*, chap. 3.

67. Tyndall, "Scientific Materialism," 87–88.

68. Thomas Henry Huxley, "Hume," 6:200.

69. Ibid., 6:192–93.

70. Thomas Henry Huxley, "Hume," 6:212. Emphasis in the original.

71. Ibid., 6:213.

72. Thomas Henry Huxley, "On the Hypothesis That Animals Are Automata, and Its History," 1:240–42.

73. Thomas Henry Huxley, "On Descartes' 'Discourse Touching the Method of Using One's Reason Rightly and of Seeking Scientific Truth,'" in *CE*, 1:187.

74. Pierre Simon de Laplace, *A Philosophical Essay on Probabilities*, trans. F. W. Truscott (New York: John Wiley and Sons, 1902), 4.

75. John Tyndall, "The Belfast Address," in *Fragments of Science: A Series of Detached Essays, Addresses, and Reviews*, vol. 2 (New York: D. Appleton, 1897), 180–81.

76. Thomas Henry Huxley, "On the Physical Basis of Life," in *CE*, 1:158.

77. Thomas Henry Huxley, "Draft of Mr. Balfour's Attack on Agnosticism, Part II," in *Huxley, Prophet of Science*, by Houston Peterson (New York: Longmans, Green, 1932), 324.

78. Katharine Anderson, *Predicting the Weather: Victorians and the Science of Meteorology* (Chicago: University of Chicago Press, 2005), 39, and chap. 1 more generally on the question of scientific prediction in Victorian science.

79. Thomas Henry Huxley, "On the Method of Zadig," in *CE*, 4:1–23; and Matthew Stanley, "Predicting the Past: Ancient Eclipses and Airy, Newcomb, and Huxley on the Authority of Science," *Isis* 103, no. 2 (2012): 254–77.

80. Tyndall, "Scientific Materialism," 2:85.

81. Thomas Henry Huxley to Herbert Spencer, August 3, 1861, quoted in Leonard Huxley, *Life and Letters of Thomas Henry Huxley*, 2 vols. (1900; repr., New York: AMS Press, 1979), 1:249 (hereafter cited as *LL*).

82. Arthur James Balfour, *The Foundations of Belief: Being Notes Introductory to the Study of Theology* (London: Longmans, Green, 1895), 21.

83. Thomas Henry Huxley, "Hume," 6:220–21.

84. William Kingdon Clifford, *Lectures and Essays*, ed. Leslie Stephen and Frederick Pollock, 2 vols. (London: Macmillan, 1879), 2:53.

85. Thomas Henry Huxley, "On the Reception of the *Origin of Species*," in *The Life and Letters of Charles Darwin*, ed. Francis Darwin, 2 vols. (New York: D. Appleton, 1911), 1:556–57. Huxley was largely correct that the basic concepts of free will and determinism were originally creations of religious debate. See Roger Smith, *Free Will*, 58.

86. Thomas Henry Huxley, "Draft of Mr. Balfour's Attack on Agnosticism, Part II," 324.

87. Thomas Henry Huxley, "On the Hypothesis That Animals Are Automata, and Its History," 1:249.

88. Thomas Henry Huxley, "Joseph Priestley," *Macmillan's Magazine* 30 (1874): 480.

89. Roger Smith, *Free Will*, chap. 1; and Jacyna, "Physiology of Mind, the Unity of Nature, and the Moral Order in Victorian Thought," 120–24.

90. William Samuel Lilly, "Materialism and Morality," *Fortnightly Review* 40 (October 1886): 576.

91. Ibid., 584–85. Emphasis in the original.

92. Ibid., 589.

93. Ibid., 591.

94. Thomas Henry Huxley, "Mr. Darwin's Critics," in *CE*, 2:168.

95. Thomas Henry Huxley, "Hume," 6:221.

96. Ibid., 6:222.

97. Thomas Henry Huxley, "A Modern Symposium: The Soul and Future Life," in *The Nineteenth Century: A Monthly Review*, ed. James Knowles, vol. 2 (London: Henry S. King, 1877), 339.

98. Thomas Henry Huxley, "On Descartes' 'Discourse Touching the Method of Using One's Reason Rightly and of Seeking Scientific Truth,'" 1:192–93. Roger Smith, *Free Will*, chap. 6, provides a close reading of Huxley's *Evolution and Ethics* that shows the importance of moral issues relating to the will even within a naturalistic worldview.

99. Jacyna, "Physiology of Mind, the Unity of Nature, and the Moral Order in Victorian Thought," 118.

100. Carpenter, *Principles of Human Physiology*, 18.

101. Ibid., 371.

102. Ibid., 793–94.

103. Ibid., 796–97. Emphasis in the original. See also Roger Smith, *Free Will*, 84–86.

104. Carpenter, *Principles of Human Physiology*, 800. Emphasis in the original.

105. Ibid., 813. Emphasis in the original.

106. Ibid., 1104.

107. Ruth Barton, "Sunday Lecture Societies: Naturalistic Scientists, Unitarians, and Secularists Unite against Sabbatarian Legislation," in *Victorian Scientific Naturalism: Community, Identity, Continuity*, ed. Bernard Lightman and Gowan Dawson (Chicago: University of Chicago Press, forthcoming).

108. William Benjamin Carpenter, *The Doctrine of Human Automatism* (London: Sunday Lecture Society, 1875), 4.

109. Ibid., 27–28.

110. Ibid., 32.

111. Carpenter, *Principles of Mental Physiology*, viii–ix. The separation of volition and automatic functions also had significance for gender. See Wise, "Gender of Automata in Victorian Britain," 185–86.

112. Carpenter, *Principles of Mental Physiology*, xii–xiii.

113. Ibid., xiv–xv.

114. James Clerk Maxwell to Charles Benjamin Tayler, July 8, 1853, in *The Scientific Letters and Papers of James Clerk Maxwell*, ed. P. M. Harman, 3 vols. (Cambridge: Cambridge University Press, 1990), 2:220–21 (hereafter cited as *SLP*).

115. Ibid.

116. Frederick Denison Maurice, *Theological Essays* (Cambridge: Macmillan, 1853), 424–25.

117. Reed, *Victorian Will*, 11–16, 38–39.

118. Carpenter, *Principles of Mental Physiology*, 428. Emphasis in the original.

119. Lewis Campbell and William Garnett, *The Life of James Clerk Maxwell* (London: Macmillan, 1882), 240 (hereafter cited as *Life*).

120. On Bain, see Rick Rylance, *Victorian Psychology*, chap. 5.

121. James Clerk Maxwell, "On Colour Vision," in *The Scientific Papers of James Clerk Maxwell*, ed. W. D. Niven, 2 vols. (Cambridge: Cambridge University Press, 1890), 2:267–79 (hereafter cited as *SP*).

122. Henry Thomas Buckle, *History of Civilization in England*, vol. 1 (London: J. W. Parker and Sons, 1857), 17–18; and Reed, *Victorian Will*, 97–102.

123. Campbell and Garnett, *Life*, 240.

124. Ibid., 243.

125. James Clerk Maxwell to Lewis Campbell, December 22, 1857, in Campbell and Garnett, *Life*, 294.

126. James Clerk Maxwell to Lewis Campbell, April 21, 1862, in *SLP*, 1:711–12, on 1:712.

127. This concern was widespread. See John Ruskin, "Unto This Last," in *The Works of John Ruskin*, ed. E. T. Cook and Alexander Wedderburn (London: George Allen, 1903), 1:29–30.

128. James Thomson, *Collected Papers in Physics and Engineering* (Cam-

bridge: Cambridge University Press, 1912), lvii–lviii. On William Thomson's views on free will, see Crosbie Smith and M. Norton Wise, *Energy and Empire: A Biographical Study of Lord Kelvin* (Cambridge: Cambridge University Press, 1989), 612–33.

129. James Clerk Maxwell to Lewis Campbell, April 21, 1862, in *SLP*, 1:712.

130. Carpenter, *Principles of Mental Physiology*, xxxv–xxxvi. Emphasis in the original.

131. David B. Wilson, "A Physicist's Alternative to Materialism: The Religious Thought of George Gabriel Stokes," in *Energy and Entropy: Science and Culture in Victorian Britain*, ed. Patrick Brantlinger (Bloomington: Indiana University Press, 1989), 201. Wilson also notes that P. G. Tait used a similar description for the will, but without the train metaphor, in an 1862 encyclopedia article.

132. James Clerk Maxwell, "Essay for the Eranus Club on Science and Free Will, 11 February 1873," in *SLP*, 2:814–23.

133. Ibid., 2:815.

134. Ibid., 2:817.

135. Balfour Stewart and J. Norman Lockyer, "The Sun as a Type of the Material Universe," *Macmillan's Magazine* 18 (1868): 246–57, 319–27.

136. Balfour Stewart, *The Conservation of Energy* (New York: D. Appleton, 1876), 159–60.

137. Ibid., 160–62.

138. William James, "Are We Automata?," *Mind* 4, no. 13 (1879): 15–16. See also Roger Smith, *Free Will*, 29–30.

139. [James Clerk Maxwell], "The Conservation of Energy," in *Nature: A Weekly Illustrated Journal of Science*, vol. 9, 198–200 (London: Macmillan, 1874), 199. Maxwell's authorship of this review is argued for in Philip Marston, "Maxwell and Creation: Acceptance, Criticism, and His Anonymous Publication," *American Journal of Physics* 75, no. 8 (August 2007): 731–40.

140. Maxwell, "Essay for the Eranus Club on Science and Free Will," 2:823.

141. James Clerk Maxwell, "Molecules (A Lecture)," in *SP*, 2:361–77, on 2:373.

142. James Clerk Maxwell, "Atom," in *SP*, 2:445–84, particularly 2:461.

143. Mary Jo Nye, "The Moral Freedom of Man and the Determinism of Nature: The Catholic Synthesis of Science and History in the *Revue des questions scientifiques*," *British Journal for the History of Science* 9, no. 3 (1976): 274–92. See also Ian Hacking, "Nineteenth Century Cracks in the Concept of Determinism," *Journal of the History of Ideas* 44, no. 3 (1983): 455–75, on 464–65.

144. James Clerk Maxwell to Francis Galton, February 26, 1879, in *SLP*, 3:756–58, on 3:757.

145. James Clerk Maxwell, "A Review of *Paradoxical Philosophy*," in *SP*, 2:760.

146. Ibid. Emphasis in the original.

147. Maxwell to Galton, February 26, 1879, 3:757–58.

148. Maxwell, *"Paradoxical,"* 2:760. See also James Clerk Maxwell, "Hermann Ludwig Ferdinand Helmholtz," in *SP*, 2:592–98.

149. Ibid., 2:593.

150. Ibid., 2:595–96.

151. James Clerk Maxwell, "Essay for the Eranus Club on Psychophysik," in *SLP*, 3:604.

152. Ibid., 3:601–2.

153. Maxwell, *"Paradoxical,"* 2:760, 2:759.

154. Maxwell, "Psychophysik," 3:606–7.

155. Ibid., 3:598–99.

156. Ibid., 3:607.

157. Maxwell, *"Paradoxical,"* 2:760–61.

158. Ibid., 2:762.

159. Ibid., 2:761; James Clerk Maxwell, "'Analogies in Nature,'" in *SLP*, 1:380; and James Clerk Maxwell to C. J. Ellicott, November 22, 1876, in Campbell and Garnett, *Life*, 393–95, on 394.

160. P. M. Harman, *The Natural Philosophy of James Clerk Maxwell* (Cambridge: Cambridge University Press, 1998), 124–29; and Theodore M. Porter, "A Statistical Survey of Gases: Maxwell's Social Physics," *Historical Studies in the Physical Sciences* 12, no. 1 (1981): 77–116. For an overview of Maxwell's approach to kinetic theory, see Theodore M. Porter, *The Rise of Statistical Thinking: 1820–1900* (Princeton, NJ: Princeton University Press, 1986), 111–26; Stephen Brush, *The Kind of Motion We Call Heat: A History of the Kinetic Theory of Gases in the 19th Century* (New York: Elsevier, 1976); and Stephen Brush, C. W. F. Everitt, and Elizabeth Garber, eds., *Maxwell on Heat and Statistical Mechanics: On "Avoiding All Personal Enquiries" of Molecules* (Bethlehem, PA: Lehigh University Press, 1995).

161. For instance, see James Clerk Maxwell to John William Strutt, December 6, 1870, in *SLP*, 2:582–83. William Thomson was particularly interested in the temporal directionality of thermodynamics. See Crosbie Smith and M. Norton Wise, *Energy and Empire*, 612–33. The social and cultural significance of thermodynamics is discussed in Stephen Brush, *The Temperature of History: Phases of Science and Culture in the Nineteenth Century* (New York: Burt Franklin, 1978); Bruce Clarke, "Allegories of Victorian Thermodynamics," *Configurations* 4, no. 1 (1996): 67–90; and Greg Myers, "Nineteenth-Century Popularizations of Thermodynamics and the Rhetoric of Social Prophecy," in *Energy and Entropy: Science and Culture in Victorian Britain*, ed. Patrick Brantlinger, 67–90 (Bloomington: Indiana University Press, 1989).

162. James Clerk Maxwell to Peter Guthrie Tait, December 11, 1867, in *SLP*, 2:328–33, on 2:332.

163. The name *demon* came from Thomson, not Maxwell. See James Clerk

Maxwell, "Note to Tait 'Concerning Demons,'" in *SLP*, 3:185–86. *Demon* has become the default title for Maxwell's neat-fingered being, however, and I will use it here. Further significance of the name *demon* is discussed in Silvan S. Schweber, "Demons, Angels and Probability: Some Aspects of British Science in the Nineteenth Century," in *Physics as Natural Philosophy: Essays in Honor of Laszlo Tisza on His Seventy-Fifth Birthday*, ed. Abner Shimony and Herman Feshbach, 319–63 (Cambridge, MA: MIT Press, 1982).3.

164. James Clerk Maxwell to John William Strutt, December 6, 1870, in *SLP*, 2:582–83.

165. On Maxwell's metaphors see Jordi Cat, "On Understanding: Maxwell on the Methods of Illustration and Scientific Metaphor," *Studies in History and Philosophy of Modern Physics* 32, no. 3 (2001): 395–441; Mary Hesse, *Models and Analogies in Science* (Notre Dame, IN: University of Notre Dame Press, 1966); and Mary Hesse, *Revolutions and Reconstructions in the Philosophy of Science* (Bloomington: Indiana University Press, 1980).

166. Cat, "On Understanding," 424.

167. See Norris Pope, "Dickens's 'The Signalman' and Information Problems in the Railway Age," *Technology and Culture* 42, no. 3 (July 2001): 436–61. The assertion that the demon's effectiveness relies on information becomes key to many of the twentieth-century attempts to refute the demon. These various attempts are discussed in John Earman and John D. Norton, "Exorcist XIV: The Wrath of Maxwell's Demon. Part I: From Maxwell to Szilard," *Studies in History and Philosophy of Modern Physics* 29, no. 4 (1998): 435–71; John Earman and John D. Norton, "Exorcist XIV: The Wrath of Maxwell's Demon. Part II: From Szilard to Landauer and Beyond," *Studies in History and Philosophy of Modern Physics* 30, no. 1 (1999): 1–40. Some of these attempts note that the demon only functions if free will is real: Orly Shenker, "Maxwell's Demon and Baron Munchausen: Free Will as a Perpetuum Mobile," *Studies in History and Philosophy of Modern Physics* 30, no. 3 (1999): 347–72.

168. Cat, "On Understanding," 425.

169. Edward Daub, "Maxwell's Demon," *Studies in History and Philosophy of Science* 1 (1970): 213–27, on 224.

170. Maxwell to Strutt, December 6, 1870, 2:582–83; and James Clerk Maxwell, "'Concerning Demons,'" in *SLP*, 3:185–86. Exactly how anthropomorphic the demon needs to be in order to function remains a matter of contention. See N. Katherine Hayles, *Chaos Bound: Orderly Disorder in Contemporary Literature and Science* (Ithaca, NY: Cornell University Press, 1990), 43.

171. P. M. Harman, *Energy, Force, and Matter: The Conceptual Development of Nineteenth-Century Physics* (Cambridge: Cambridge University Press, 1982), 140.

172. Crosbie Smith and M. Norton Wise, *Energy and Empire*, 623.

173. Cat, "On Understanding," 424, 428–29.

174. James Clerk Maxwell, "Address to the Mathematical and Physical Sections of the British Association (1870)," in *SP*, 2:215–29.

175. Cat, "On Understanding," 430.

176. Crosbie Smith and M. Norton Wise, *Energy and Empire*, 625, argue that the demon was intended to show what was *distinctive* about conscious creatures, which is certainly an important part of Maxwell's reasoning.

177. Maxwell, "Eranus Club on Science and Free Will," 3:818.

178. Ibid., 3:819.

179. Maxwell, "'Analogies in Nature,'" 1:381.

180. Maxwell, "Molecules," 2:361–78.

181. James Clerk Maxwell, *Theory of Heat* (London: Longmans, Green, 1872), 308–9.

182. James Clerk Maxwell, "Diffusion," in *SP*, 2:625–46, on 2:646.

183. James Clerk Maxwell to R. B. Litchfield, September 23, 1857, in Campbell and Garnett, *Life*, 281.

184. Quoted in Roger Smith, *Free Will*, 166.

185. Roger Smith, *Free Will*, 27.

186. James H. Leuba, *The Belief in God and Immortality: A Psychological, Anthropological and Statistical Study* (1916; repr., Chicago: Open Court, 1921), 278. On Leuba, see George Marsden, *The Soul of the American University: From Protestant Establishment to Established Nonbelief* (Oxford: Oxford University Press, 1994), 292–96. Edward J. Larson and Larry Witham replicated Leuba's study in 1996: "Scientists Are Still Keeping the Faith," *Nature* 386, no. 6624 (1997): 435–36; and "Leading Scientists Still Reject God," *Nature* 394, no. 6691 (1998): 313. Interestingly, they found that physical scientists were no longer more religious than biologists.

Chapter Seven

1. Charles Taylor, *A Secular Age* (Cambridge, MA: Harvard University Press, 2007), 571–72.

2. Frank M. Turner, "The Victorian Conflict between Science and Religion: A Professional Dimension," *Isis* 69, no. 3 (1978): 356–76.

3. Frank M. Turner, *Contesting Cultural Authority: Essays in Victorian Intellectual Life* (Cambridge: Cambridge University Press, 1993), 204–11.

4. D. S. L. Cardwell, *The Organisation of Science in England: A Retrospect* (London: William Heinemann, 1972), 120–31.

5. For example, see Leonard Huxley, *Life and Letters of Thomas Henry Huxley*, 2 vols. (1900; repr., New York: AMS Press, 1979), 2:33 (hereafter cited as *LL*).

6. Adrian Desmond, *Huxley: From Devil's Disciple to Evolution's High Priest* (Reading, MA: Addison-Wesley, 1997), 419.

7. Peter J. Bowler, *Reconciling Science and Religion: The Debate in Early-Twentieth-Century Britain* (Chicago: University of Chicago Press, 2001), 62–66.

8. Gerald L. Geison, *Michael Foster and the Cambridge School of Physiology: The Scientific Enterprise in Late Victorian Society* (Princeton, NJ: Princeton University Press, 1978), 130–47.

9. Roy M. MacLeod, "The 'Naturals' and Victorian Cambridge: Reflections on the Anatomy of an Elite, 1851–1914," *Oxford Review of Education* 6, no. 2 (1980): 185.

10. Philip J. Pauly, "The Appearance of Academic Biology in Late Nineteenth-Century America," *Journal of the History of Biology* 17, no. 3 (1984): 378.

11. Henry Fairfield Osborn, "Enduring Recollections," *Nature* 115 (1925): 726–28; and Ronald Rainger, "Vertebrate Paleontology as Biology: Henry Fairfield Osborn and the American Museum of Natural History," in *American Development of Biology*, ed. Ronald Rainger, Keith Benson, and Jane Maienschein, 219–56 (Rutgers, NJ: Rutgers University Press, 1988).

12. Adrian Desmond, "Redefining the X Axis: 'Professionals,' 'Amateurs' and the Making of Mid-Victorian Biology: A Progress Report," *Journal of the History of Biology* 34, no. 1 (2001): 33.

13. Janet Howarth, "Science Education in Late-Victorian Oxford: A Curious Case of Failure?," *English Historical Review* 102, no. 403 (1987): 339; and Desmond, *Huxley*, 539.

14. Desmond, "Redefining," 28. Desmond points out that many of these new teachers came from industrial, Dissenting regions of Britain and were thus particularly receptive to Huxley's outlook. Huxley's efforts toward reforming science education are described in detail in Cyril Bibby, *T. H. Huxley: Scientist, Humanist, and Educator* (London: Watts, 1959), 123–93. For an overview of the further development of science education in this period, see David Layton, "The Schooling of Science in England, 1854–1939," in *The Parliament of Science: The British Association for the Advancement of Science, 1831–1981*, ed. Roy MacLeod and Peter Collins, 188–210 (Northwood, UK: Science Reviews, 1981).

15. Thomas Henry Huxley to Anton Dohrn, July 7, 1871, quoted in *LL*, 1:389.

16. Graeme Gooday, "Nature in the Laboratory: Domestication and Discipline with the Microscope in Victorian Life Science," *British Journal for the History of Science* 24, no. 3 (1991): 307–41.

17. Desmond, "Redefining," 28.

18. Thomas Henry Huxley, "Address on Behalf of the National Association for the Promotion of Technical Education," in *Collected Essays of T. H. Huxley*, 9 vols. (Bristol, UK: Thoemmes Press, 2001), 3:444 (hereafter cited as *CE*).

19. James Elwick, "Economies of Scales: Evolutionary Naturalists and the

Victorian Examination System," in *Victorian Scientific Naturalism: Community, Identity, Continuity*, ed. Bernard Lightman and Gowan Dawson (Chicago: University of Chicago Press, forthcoming).

20. Quoted in Elwick, "Economies of Scales."

21. J. D. Hooker to Thomas Henry Huxley, 23 Friday (before summer 1856), John Hooker Papers, Letters to T. H. Huxley, ca. 1851–1894 (JDH/2/13).

22. Elwick, "Economies of Scales."

23. Quoted in Elwick, "Economies of Scales."

24. Bernard Lightman, *Victorian Popularizers of Science: Designing Nature for New Audiences* (Chicago: University of Chicago Press, 2007), 369–71.

25. Ibid., 378–81.

26. Ibid., 395.

27. Bibby, *T. H. Huxley*, 111.

28. Desmond, "Redefining," 32.

29. On how scientists rewrite their discipline's history for current purposes, see Richard Staley, *Einstein's Generation* (Chicago: University of Chicago Press, 2008); Peter Galison, "Re-reading the Past from the End of Physics: Maxwell's Equations in Retrospect," in *Functions and Uses of Disciplinary Histories*, ed. Loren Graham, Wolf Lepenies, and Peter Weingart (Dordrecht, Netherlands: D. Reidel, 1983); and G. Nigel Gilbert and Michael Mulkay, "Experiments Are the Key: Participants' Histories and Historians' Histories of Science," *Isis* 75, no. 1 (1984): 105–25.

30. Turner, *Contesting Cultural Authority*, chap. 4; and James R. Moore, "Theodicy and Society: The Crisis of the Intelligentsia," in *Victorian Faith in Crisis: Essays on Continuity and Change in Nineteenth-Century Religious Belief*, ed. Richard J. Helmstadter and Bernard Lightman, 153–86 (Stanford, CA: Stanford University Press, 1990).

31. Lightman, *Victorian Popularizers of Science*, 372–77.

32. Desmond, "Redefining," 32.

33. William Benjamin Carpenter, *Nature and Man: Essays Scientific and Philosophical* (London: Kegan Paul, Trench, 1888), 66–67.

34. Thomas Henry Huxley to John Tyndall, October 27, 1896, HP 8.196.

35. On Faraday's religion, see Geoffrey Cantor, *Michael Faraday: Sandemanian and Scientist: A Study of Science and Religion in the Nineteenth Century* (London: Macmillan, 1991); and Geoffrey Cantor, "Reading the Book of Nature: The Relation between Faraday's Religion and His Science," in *Faraday Rediscovered: Essays on the Life and Work of Michael Faraday, 1791–1867*, ed. David Gooding and Frank A. J. L. James, 69–81 (London: Macmillan, 1985).

36. C. Tomlinson, "Michael Faraday," *Graphic* 20, no. 508 (August 23, 1879): 183.

37. James Clerk Maxwell, "Faraday," in *The Scientific Papers of James Clerk*

Maxwell, ed. W. D. Niven, 2 vols. (Cambridge: Cambridge University Press, 1890), 2:792 (hereafter cited as *SP*).

38. Ibid., 2:792–93.

39. Bence Jones, *The Life and Letters of Faraday*, 2 vols. (London: Longmans, Green, 1870). For example, 1:265, and 2:230, 2:244, 2:325–26, 2:428–35.

40. John Tyndall, "Review of *Life and Letters of Faraday*," in *Fragments of Science: A Series of Detached Essays, Addresses, and Reviews*, vol. 1 (New York: D. Appleton, 1915), 418–19.

41. Ibid., 1:419. Emphasis in the original.

42. John Tyndall, "Science and Man," in *Fragments of Science: A Series of Detached Essays, Addresses, and Reviews*, vol. 2 (New York: D. Appleton, 1915), 367–68.

43. Joseph Dalton Hooker, "Presidential Address," *Report of the Thirty-Eighth Meeting of the British Association for the Advancement of Science* (London: John Murray, 1869), lviii–lxxv, on lxviii and lxxii.

44. John Tyndall, *Faraday as a Discoverer* (London: Longmans, Green, 1868), 39.

45. Ibid., 148–49.

46. Ibid., 151.

47. Ibid., 152.

48. Silvanus P. Thompson, *Michael Faraday: His Life and Work* (London: Cassell, 1898).

49. Ibid., 290.

50. Ibid., 292.

51. Ibid., 298.

52. J. G. Crowther, *British Scientists of the Nineteenth Century* (London: K. Paul, Trench, Trubner, 1935), xii.

53. Thomas Henry Huxley, "Owen's Position in the History of Anatomical Science," in *The Scientific Memoirs of Thomas Henry Huxley*, ed. Michael Foster and E. Ray Lankester, 5 vols. (London: Macmillan, 1902), 4:660.

54. Ibid., 4:677.

55. Ibid., 4:681.

56. Ibid., 4:682–83.

57. Thomas Henry Huxley, *Evolution and Ethics and Other Essays* (New York: D. Appleton, 1899), 104–5.

58. John Hedley Brooke, *Science and Religion: Some Historical Perspectives* (Cambridge: Cambridge University Press, 1991), 223.

59. Taylor, *Secular Age*, 95.

60. Ibid., 562–66.

61. Examples of the secularization thesis can be found in Alan D. Gilbert, *Religion and Society in Industrial England: Church, Chapel and Social Change* (London: Longman, 1976); Steve Bruce, *Religion in the Modern World: From*

Cathedrals to Cults (Oxford: Oxford University Press, 1996); and Owen Chadwick, *The Secularization of the European Mind in the Nineteenth Century: The Gifford Lectures in the University of Edinburgh for 1973–74* (Cambridge: Cambridge University Press, 1975). The particularly influential exchange theory of secularization appeared in Rodney Stark and William Sims Bainbridge, *A Theory of Religion* (New Brunswick, NJ: Rutgers University Press, 1987). The role of science in secularization is generally overstated. See John Hedley Brooke, "That Modern Science Has Secularized Western Culture," in *Galileo Goes to Jail and Other Myths about Science and Religion*, ed. Ronald L. Numbers, 224–34 (Cambridge, MA: Harvard University Press, 2009).

62. S. J. D. Green, *Religion in the Age of Decline: Organisation and Experience in Industrial Yorkshire, 1870–1920* (Cambridge: Cambridge University Press, 1996), chap. 9.

63. For examples, see Steve Bruce, ed., *Religion and Modernization: Sociologists and Historians Debate the Secularization Thesis* (Oxford: Clarendon Press, 1992); S. C. Williams, *Religious Belief and Popular Culture in Southwark, c. 1880–1939* (Oxford: Oxford University Press, 1999); and Jeffrey Cox, *The English Churches in a Secular Society: Lambeth, 1870–1930* (Oxford: Oxford University Press, 1982). Moderate examinations of secularization can be found in Hugh McLeod, *Secularisation in Western Europe, 1848–1914* (London: Macmillan, 2000); and Hugh McLeod, *Religion and Society in England, 1850–1914* (New York: St. Martin's, 1996).

64. Jose Harris, *Private Lives, Public Spirit: A Social History of Britain, 1870–1914* (Oxford: Oxford University Press, 1993).

65. Ronald L. Numbers, *Science and Christianity in Pulpit and Pew* (Oxford: Oxford University Press, 2007), 132–33.

66. George Marsden, *The Soul of the American University: From Protestant Establishment to Established Nonbelief* (Oxford: Oxford University Press, 1994), 155–59.

67. Lightman, *Victorian Popularizers of Science*, 41–42.

68. Ibid., 187–89.

69. Bowler, *Reconciling Science and Religion*, 31–41.

70. Numbers, *Science and Christianity in Pulpit and Pew*, 135.

71. Bowler, *Reconciling Science and Religion*, 18–19.

72. Ibid., 407, 20.

73. Ibid., 60, 196.

74. John Burnham, *How Superstition Won and Science Lost* (New Brunswick, NJ: Rutgers University Press, 1987), 159.

75. Jon H. Roberts and James Turner, *The Sacred and the Secular University* (Princeton, NJ: Princeton University Press, 2000).

76. R. Clinton Ohlers, "The End of Miracles: Scientific Naturalism in America, 1839–1934" (PhD diss., University of Pennsylvania Press, 2007), 95.

77. Daniel J. Kevles, *The Physicists: The History of a Scientific Community in Modern America* (Cambridge, MA: Harvard University Press, 1995), 15; and Burnham, *How Superstition Won and Science Lost*, 160–69.

78. John L. Rudolph, "Epistemology for the Masses: The Origins of 'the Scientific Method' in American Schools," *History of Education Quarterly* 45, no. 3 (2005): 346.

79. Kevles, *Physicists*, 23–24. There is some confusion about whether Huxley's address was part of the actual opening ceremonies: Marsden, *Soul of the American University*, 152.

80. Albert E. Moyer, *A Scientist's Voice in American Culture: Simon Newcomb and the Rhetoric of Scientific Method* (Berkeley: University of California Press, 1992).

81. See Ohlers, "End of Miracles," 161–76. On philosophical naturalism in America more generally, including relevant primary sources, see John Ryder, ed., *American Philosophic Naturalism in the Twentieth Century* (New York: Prometheus Books, 1994).

82. Jay Martin, *The Education of John Dewey: A Biography* (New York: Columbia University Press, 2002), 39–40; and Thomas C. Dalton, *Becoming John Dewey* (Bloomington: Indiana University Press, 2002), 9–10, 29. As with the British naturalists, Dewey's precise relationship to religion was quite complex, and cannot be fully discussed here.

83. Martin, *Education of John Dewey*, 41.

84. Dalton, *Becoming John Dewey*, 34–36.

85. Ibid., 45–47, 125–46. On Johns Hopkins as a pioneer in teaching science in a nonsectarian, nonreligious environment, see Marsden, *Soul of the American University*, chap. 9.

86. Dalton, *Becoming John Dewey*, 53–57.

87. Martin, *Education of John Dewey*, 72–88, on 87.

88. Adam R. Shapiro, "Between Training and Popularization: Regulating Science Textbooks in Secondary Education," *Isis* 103, no. 1 (2012): 99–110; Adam R. Shapiro, "Civic Biology and the Origin of the School Antievolution Movement," *Journal of the History of Biology* 41, no. 3 (2008): 409–33; Rudolph, "Epistemology for the Masses"; and John L. Rudolph, "Turning Science to Account: Chicago and the General Science Movement in Science Education, 1905–1920," *Isis* 96, no. 3 (2005): 353–89.

89. Rudolph, "Turning Science to Account," 367–71.

90. Shapiro, "Between Training and Popularization," 102; and Rudolph, "Epistemology for the Masses," 343–44, 366.

91. John Dewey, *"The Influence of Darwin on Philosophy," and Other Essays in Contemporary Thought* (New York: Henry Holt, 1910), 2–3.

92. John Dewey, *How We Think* (1910; repr., New York: Prometheus 1991), 148, 27. Emphasis in the original.

93. Rudolph, "Epistemology for the Masses," 366–72.

94. A similar phenomenon occurred with the widespread acceptance of the "warfare thesis." See James R. Moore, *The Post-Darwinian Controversies: A Study of the Protestant Struggle to Come to Terms with Darwin in Great Britain and America, 1870–1900* (Cambridge: Cambridge University Press, 1979), chap. 1.

95. Lightman, *Victorian Popularizers of Science*, 163–64.

Conclusion

1. Jim Endersby, *Imperial Nature: Joseph Hooker and the Practices of Victorian Science* (Chicago: University of Chicago Press, 2008), 317–23.

2. Michael Behe, *Darwin's Black Box: The Biochemical Challenge to Evolution* (New York: Free Press, 1996). These processes are described as "irreducibly complex" (39).

3. Philip Kitcher, "Born-Again Creationism," in *Intelligent Design Creationism and Its Critics*, ed. Robert Pennock (Cambridge, MA: MIT Press, 2001), 262–63 (hereafter cited as *IDC*).

4. Alvin Plantinga, "Methodological Naturalism?," in *IDC*, 355–57.

5. Jeffrey Koperski, "Two Bad Ways to Attack Intelligent Design and Two Good Ones," *Zygon* 43, no. 2 (2008): 442–43.

6. *Tammy Kitzmiller, et al. v. Dover Area School District, et al.*, 400 F. Supp. 2d 707 (M.D. Pa. 2005), 64.

7. Barbara Forrest, "The Wedge at Work: How Intelligent Design Creationism Is Wedging Its Way into the Cultural and Academic Mainstream," in *IDC*, 18–22.

8. George Marsden, *The Soul of the American University: From Protestant Establishment to Established Nonbelief* (Oxford: Oxford University Press, 1994), 267–70.

9. Koperski, "Two Bad Ways to Attack Intelligent Design and Two Good Ones," 439–40; and Larry Laudan, "The Demise of the Demarcation Problem," in *But Is It Science? The Philosophical Question in the Creation/Evolution Controversy*, ed. Michael Ruse, 337–50 (Buffalo, NY: Prometheus Books, 1988).

10. Laudan, "Demise of the Demarcation Problem," 349. On the origin of the category of "pseudoscience," see Michael D. Gordin, *The Pseudoscience Wars: Immanuel Velikovsky and the Birth of the Modern Fringe* (Chicago: University of Chicago Press, 2012).

11. Larry Laudan, "Science at the Bar—Causes for Concern," in *But Is It Science? The Philosophical Question in the Creation/Evolution Controversy*, ed. Michael Ruse (Buffalo, NY: Prometheus Books, 1988), 354.

12. For example, Phillip E. Johnson, *Reason in the Balance: The Case against*

Naturalism in Science, Law, and Education (Downers Grove, IL: InterVarsity Press, 1993), 90.

13. Plantinga, "Methodological Naturalism?," 344–45.

14. J. A. Stone, "Religious Naturalism and the Religion-Science Dialogue: A Minimalist View," *Zygon* 37, no. 2 (2002): 381–94; J. A. Stone, "Varieties of Religious Naturalism," *Zygon* 38, no. 1 (2003): 89–95; Mikael Stenmark, *Scientism: Science, Ethics, and Religion* (Aldershot, UK: Ashgate, 2001); Willem B. Drees, *Religion, Science, and Naturalism* (Cambridge: Cambridge University Press, 1996); and David Ray Griffin, *Religion and Scientific Naturalism: Overcoming the Conflicts* (Albany: SUNY Press, 2000).

15. Griffin, *Religion and Scientific Naturalism*, 45–51.

16. Ernan McMullin, "Evolution and Special Creation," in *The Philosophy of Biology*, ed. David Hull and Michael Ruse (Oxford: Oxford University Press, 1998), 701–2.

17. Koperski, "Two Bad Ways to Attack Intelligent Design and Two Good Ones," 436.

18. Michael Ruse, "Methodological Naturalism under Attack," *South African Journal of Philosophy* 24, no. 1 (2005): 45–46.

19. William James, *The Varieties of Religious Experience* (New York: Penguin, 1958), 37.

Bibliography

Anderson, Katharine. *Predicting the Weather: Victorians and the Science of Meteorology.* Chicago: University of Chicago Press, 2005.

Anderton, Keith. "The Limits of Science: A Social, Political, and Moral Agenda for Epistemology in Nineteenth Century Germany." PhD diss., Harvard University, Department of the History of Science, 1993.

"Anniversary of the Working Men's College." *Cambridge Chronicle,* May 24, 1856.

Argyll, George Douglas Campbell (Duke of). "Letter to the Editor." *Times.* February 8, 1892.

———. "Professor Huxley and the Duke of Argyll." In *Nineteenth Century,* edited by James Knowles, vol. 29. London: Kegan Paul, Trench, Trübner, 1891.

———. "Professor Huxley on the Warpath." *Nineteenth Century,* no. 167 (1891): 1–34.

———. *The Reign of Law.* 5th ed. 1867. Reprint, London: Strahan, 1870.

Armstrong, Henry Edward. "Our Need to Honour Huxley's Will: Huxley Memorial Lecture." In *H. E. Armstrong and the Teaching of Science, 1880–1930,* edited by W. H. Brock. Cambridge: Cambridge University Press, 1973.

Astore, William J. *Observing God: Thomas Dick, Evangelicalism, and Popular Science in Victorian Britain and America.* London: Ashgate, 2001.

Atkinson, Thomas Dinham. *Cambridge Described and Illustrated: Being a Short History of the Town and University.* Cambridge: Macmillan, 1897.

Bain, Alexander. "On the Correlation of Force and Its Bearing on the Mind." *Macmillan's Magazine* 16 (1867): 372–83.

Baldwin, Melinda. "Tyndall and Stokes: Correspondence, Peer Review, and the Physical Sciences in Victorian Britain." In *Tyndall and Nineteenth Century Science,* edited by Bernard Lightman, Michael Reidy, and Joshua Howe. Chicago: University of Chicago Press, forthcoming.

Balfour, Arthur James. *The Foundations of Belief: Being Notes Introductory to the Study of Theology.* London: Longmans, Green, 1895.

Barton, Ruth. "Evolution: The Whitworth Gun in Huxley's War for the Liberation of Science from Theology." In *The Wider Domain of Evolutionary Thought*, edited by D. Oldroyd, 261–87. Dordrecht, Netherlands: Springer, 1983.

———. "'An Influential Set of Chaps': The X-Club and Royal Society Politics, 1864–85." *British Journal for the History of Science* 23, no. 1 (1990): 53–82.

———. "John Tyndall, Pantheist: A Rereading of the Belfast Address." *Osiris* 2, no. 3 (1987): 111–34.

———. "'Men of Science': Language, Identity and Professionalization in the Mid-Victorian Scientific Community." *History of Science* 41 (2003): 73–119.

———. "Sunday Lecture Societies: Naturalistic Scientists, Unitarians, and Secularists Unite against Sabbatarian Legislation." In *Victorian Scientific Naturalism: Community, Identity, Continuity*, edited by Bernard Lightman and Gowan Dawson. Chicago: University of Chicago Press, forthcoming.

Bebbington, David W. *Evangelicalism in Modern Britain: A History from the 1730s to the 1980s*. Abingdon, UK: Routledge, 1989.

———. "Science and Evangelical Theology in Britain from Wesley to Orr." In *Evangelicals and Science in Historical Perspective*, edited by David N. Livingstone, D. G. Hart, and Mark A. Noll, 120–41. Oxford: Oxford University Press, 1999.

Behe, Michael. *Darwin's Black Box: The Biochemical Challenge to Evolution*. New York: Free Press, 1996.

Bibby, Cyril. *T. H. Huxley: Scientist, Humanist, and Educator*. London: Watts, 1959.

Black, Max. *Models and Metaphors: Studies in Language and Philosophy*. Ithaca, NY: Cornell University Press, 1962.

Bowler, Peter J. *Evolution: The History of an Idea*. 3rd ed. Berkeley: University of California Press, 2003.

———. *Fossils and Progress: Paleontology and the Idea of Progressive Evolution in the Nineteenth Century*. New York: Science History Publications, 1976.

———. *Reconciling Science and Religion: The Debate in Early-Twentieth-Century Britain*. Chicago: University of Chicago Press, 2001.

Brain, Robert, and M. Norton Wise. "Muscles and Engines: Indicator Diagrams and Helmholtz's Graphical Methods." In *The Science Studies Reader*, edited by Mario Biagioli. New York: Routledge, Chapman and Hall, 1999.

Brantlinger, Patrick, ed. *Energy and Entropy: Science and Culture in Victorian Britain*. Bloomington: Indiana University Press, 1989.

Brock, William Hodson, and Roy M. McLeod. "The Scientists' Declaration: Reflexions on Science and Belief in the Wake of Essays and Reviews, 1864–5." *British Journal for the History of Science* 9, no. 1 (1976): 39–66.

Brock, Willam Hodson, Norman D. McMillan, and R. Charles Mollan, eds. *John Tyndall: Essays on a Natural Philosopher*. Dublin: Royal Dublin Society, 1981.

Brooke, John Hedley. Introduction to *Evangelicals and Science in Historical Perspective*, edited by David N. Livingstone, D. G. Hart, and Mark A. Noll. Oxford: Oxford University Press, 1999.

———. *Science and Religion: Some Historical Perspectives*. Cambridge: Cambridge University Press, 1991.

———. "That Modern Science Has Secularized Western Culture." In *Galileo Goes to Jail and Other Myths about Science and Religion*, edited by Ronald L. Numbers, 224–34. Cambridge, MA: Harvard University Press, 2009.

Brooke, John Hedley, and Geoffrey Cantor. *Reconstructing Nature: The Engagement of Science and Religion*. Oxford: Oxford University Press, 1998.

Brooke, John Hedley, and Reijer Hooykaas. *New Interactions between Theology and Natural Science*. Milton Keynes, UK: Open University Press, 1974.

Brose, Olive J. *Frederick Denison Maurice: Rebellious Conformist, 1805–1872*. Athens: Ohio University Press, 1971.

Brown, Alan Willard. *The Metaphysical Society: Victorian Minds in Crisis, 1869–1880*. New York: Columbia University Press, 1947.

Brown, Callum G. *The Social History of Religion in Scotland since 1730*. London: Methuen, 1987.

Brown, Stewart, and Michael Fry, eds. *Scotland in the Age of Disruption*. Edinburgh: Edinburgh University Press, 1993.

Brown, William, and Andrew Fabian, eds. *Darwin (The Darwin College Lectures)*. Cambridge: Cambridge University Press, 2010.

Browne, Janet. *Charles Darwin: The Power of Place*. Vol. 2. New York: Knopf, 2002.

———. *Charles Darwin: Voyaging*. Vol. 1. New York: Knopf, 1995.

Bruce, Steve, ed. *Religion and Modernization: Sociologists and Historians Debate the Secularization Thesis*. Oxford: Clarendon Press, 1992.

———. *Religion in the Modern World: From Cathedrals to Cults*. Oxford: Oxford University Press, 1996.

Brush, Stephen. *The Kind of Motion We Call Heat: A History of the Kinetic Theory of Gases in the 19th Century*. New York: Elsevier, 1976.

———. *The Temperature of History: Phases of Science and Culture in the Nineteenth Century*. New York: Burt Franklin, 1978.

Brush, Stephen, C. W. F. Everitt, and Elizabeth Garber, eds. *Maxwell on Heat and Statistical Mechanics: On "Avoiding All Personal Enquries" of Molecules*. Bethlehem, PA: Lehigh University Press, 1995.

———. *Maxwell on Saturn's Rings*. Cambridge, MA: MIT Press, 1983.

Buchwald, Jed Z. "Modifying the Continuum." In *Wranglers and Physicists: Studies on Cambridge Mathematical Physics in the Nineteenth Century*, edited by P. M. Harman. Manchester, UK: Manchester University Press, 1985.

Buckle, Henry Thomas. *History of Civilization in England*. Vol. 1. London: J. W. Parker and Sons, 1857.

Burchfield, Joe D. *Lord Kelvin and the Age of the Earth*. New York: Science History Publications, 1975.

Burnham, John. *How Superstition Won and Science Lost*. New Brunswick, NJ: Rutgers University Press, 1987.

Burns, R. M. *The Great Debate on Miracles: From Joseph Glanvill to David Hume*. Lewisburg, PA: Bucknell University Press, 1981.

Cahan, David. "Helmholtz and the British Scientific Elite: From Force Conservation to Energy Conservation." *Notes and Records of the Royal Society* 66, no. 1 (2012): 55–68.

———, ed. *Hermann von Helmholtz and the Foundations of Nineteenth-Century Science*. Berkeley: University of California Press, 1993.

Campbell, Lewis, and William Garnett. *The Life of James Clerk Maxwell*. London: Macmillan, 1882.

Cannon, Susan Faye. *Science in Culture: The Early Victorian Period*. New York: Dawson, 1978.

Cantor, Geoffrey. *Michael Faraday: Sandemanian and Scientist: A Study of Science and Religion in the Nineteenth Century*. London: Macmillan, 1991.

———. "Reading the Book of Nature: The Relation between Faraday's Religion and His Science." In *Faraday Rediscovered: Essays on the Life and Work of Michael Faraday, 1791–1867*, edited by David Gooding and Frank A. J. L. James, 69–81. London: Macmillan, 1985.

Cardwell, D. S. L. *The Organisation of Science in England: A Retrospect*. London: William Heinemann, 1972.

Carpenter, William Benjamin. *The Doctrine of Human Automatism*. London: Sunday Lecture Society, 1875.

———. "Nature and Law." In *Nature and Man: Essays Scientific and Philosophical*, 365–85. London: Kegan Paul, Trench, 1888.

———. *Nature and Man: Essays Scientific and Philosophical* London: Kegan Paul, Trench, 1888.

———. *Principles of Human Physiology, with Their Chief Applications to Psychology, Pathology, Therapeutics, Hygiene, and Forensic Medicine*. 4th ed. London: Samuel Bentley, 1853.

———. *Principles of Mental Physiology: With Their Applications to the Training and Discipline of the Mind and the Study of Its Morbid Conditions*. 4th ed. 1876. Reprint, New York: D. Appleton, 1884.

Carter, Stephen L. *The Culture of Disbelief: How American Law and Politics Trivialize Religious Devotion*. New York: Anchor Books, 1994.

Cassirer, Ernst. *Determinism and Indeterminism in Modern Physics: Historical and Systematic Studies in the Problem of Causality*. New Haven, CT: Yale University Press, 1956.

Cat, Jordi. *Land, Lines, Colours and Toys: Becoming James Clerk Maxwell* (Oxford: Oxford University Press, forthcoming).

————. "On Understanding: Maxwell on the Methods of Illustration and Scientific Metaphor." *Studies in the History and Philosophy of Modern Physics* 32, no. 3 (2001): 395–441.

Chadwick, Owen. *The Secularization of the European Mind in the Nineteenth Century: The Gifford Lectures in the University of Edinburgh for 1973–74.* Cambridge: Cambridge University Press, 1975.

Cheyne, A .C. *Studies in Scottish Church History.* Edinburgh: T&T Clark, 1999.

Clark, J. F. M. "'The Ants Were Duly Visited': Making Sense of John Lubbock, Scientific Naturalism and the Senses of Social Insects." *British Journal for the History of Science* 30, no. 2 (1997): 151–76.

Clarke, Bruce. "Allegories of Victorian Thermodynamics." *Configurations* 4, no. 1 (1996): 67–90.

Clarke, Edwin, and L. S. Jacyna. *Nineteenth-Century Origins of Neuroscientific Concepts.* Berkeley: University of California Press, 1987.

Clifford, William Kingdon. *Lectures and Essays.* Edited by Leslie Stephen and Frederick Pollock. 2 vols. London: Macmillan, 1879.

Collini, Stefan. "The Idea of 'Character' in Victorian Political Thought." *Transactions of the Royal Historical Society,* 5th series, 35 (1985): 29–50.

Corsi, Pietro. *Science and Religion: Baden Powell and the Anglican Debate, 1800–1860.* Cambridge: Cambridge University Press, 1988.

Cox, Jeffrey. *The English Churches in a Secular Society: Lambeth, 1870–1930.* Oxford: Oxford University Press, 1982.

Crowther, J. G. *British Scientists of the Nineteenth Century.* London: K. Paul, Trench, Trubner, 1935.

Dalton, Thomas C. *Becoming John Dewey.* Bloomington: Indiana University Press, 2002.

Danziger, Kurt. "Mid-Nineteenth Century British Psycho-Physiology: A Neglected Chapter in the History of Psychology." In *The Problematic Science: Psychology in Nineteenth-Century Thought,* edited by Mitchell G. Ash, 119–46. New York: Praeger, 1982.

Daston, Lorraine J. "The Theory of Will versus the Science of Mind." In *The Problematic Science: Psychology in Nineteenth-Century Thought,* edited by Mitchell G. Ash, 88–118. New York: Praeger, 1982.

Daub, Edward. "Maxwell's Demon." *Studies in History and Philosophy of Science* 1 (1970): 213–27.

Dawson, Gowan. "First among Equals." *Times Literary Supplement,* January 9, 2009, 7–8.

Dawson, Gowan, and Bernard Lightman, eds. *Victorian Scientific Naturalism: Community, Identity, Continuity.* Chicago: University of Chicago Press, forthcoming.

Dear, Peter. "Miracles, Experiments, and the Ordinary Course of Nature." *Isis* 81, no. 4 (December 1990): 663–83.

"Defining Darwinism: One Hundred and Fifty Years of Debate." *Studies in the History of Philosophy of Biology and Biomedical Science* 42 (2011): 2–4.

Depew, David J. "Darwinian Controversies: An Historiographical Recounting." *Science and Education* 19 (2010): 323–66.

Desmond, Adrian. *Archetypes and Ancestors: Palaeontology in Victorian London, 1850–1875*. Chicago: University of Chicago Press, 1984.

———. *Huxley: From Devil's Disciple to Evolution's High Priest*. Reading, MA: Addison-Wesley, 1997.

———. *The Politics of Evolution: Morphology, Medicine, and Reform in Radical London*. Chicago: University of Chicago Press, 1989.

———. "Redefining the X Axis: 'Professionals,' 'Amateurs' and the Making of Mid-Victorian Biology: A Progress Report." *Journal of the History of Biology* 34, no. 1 (2001): 3–50.

Dewey, John. *How We Think*. 1910. Reprint, New York: Prometheus, 1991.

———. *"The Influence of Darwin on Philosophy," and Other Essays in Contemporary Thought*. New York: Henry Holt, 1910.

Di Gregorio, Mario A. *T. H. Huxley's Place in Natural Science*. New Haven, CT: Yale University Press, 1984.

Drees, Willem B. *Religion, Science, and Naturalism*. Cambridge: Cambridge University Press, 1996.

Du Bois-Reymond, Emil. "Ueber die Grenzen des Naturerkennens," in *Reden*. Leipzig: Veit, 1886.

Earman, John, and John D. Norton. "Exorcist XIV: The Wrath of Maxwell's Demon. Part I: From Maxwell to Szilard." *Studies in the History and Philosophy of Modern Physics* 29, no. 4 (1998): 435–71.

———. "Exorcist XIV: The Wrath of Maxwell's Demon. Part II: From Szilard to Landauer and Beyond." *Studies in the History and Philosophy of Modern Physics* 30, no. 1 (1999): 1–40.

Ellens, J. P. "Which Freedom for Early Victorian Britain?" In *Freedom and Religion in the Nineteenth Century*, edited by Richard Helmstadter, 87–119. Stanford, CA: Stanford University Press, 1997.

Elwick, James. "Economies of Scales: Evolutionary Naturalists and the Victorian Examination System." In *Victorian Scientific Naturalism: Community, Identity, Continuity*, edited by Bernard Lightman and Gowan Dawson. Chicago: University of Chicago Press, forthcoming.

Endersby, Jim. *Imperial Nature: Joseph Hooker and the Practices of Victorian Science*. Chicago: University of Chicago Press, 2008.

Everitt, C. W. F. "Maxwell's Scientific Creativity." In *Springs of Scientific Creativity*, edited by Rutherford Aris, Horace B. Davis, and Roger H. Stuewer, 71–141. Minneapolis: University of Minnesota Press, 1983.

Fichman, Martin. "Biology and Politics: Defining the Boundaries." In *Victorian*

Science in Context, edited by Bernard Lightman, 94–118. Chicago: University of Chicago Press, 1997.

Finkelstein, Gabriel. "Emil Du Bois-Reymond: The Making of a German Liberal Scientist." PhD diss., Princeton University, Department of History, 1996.

First Report of the Committee of the Aberdeen Mechanics' Institution. Aberdeen: D. Chalmers, 1824.

Forrest, Barbara. "The Wedge at Work: How Intelligent Design Creationism Is Wedging Its Way into the Cultural and Academic Mainstream." In *Intelligent Design Creationism and Its Critics: Philosophical, Theological, and Scientific Perspectives*, edited by Robert Pennock, 5–53. Cambridge, MA: MIT Press, 2001.

Fox, Christopher, ed. *Inventing Human Science: Eighteenth-Century Domains.* Berkeley: University of California Press, 1995.

Frankland, Edward. "On the Origin of Muscular Power." *Philosophical Magazine* 31 (1866): 485–503.

Fyfe, Aileen. *Science and Salvation: Evangelical Popular Science Publishing in Victorian Britain.* Chicago: University of Chicago Press, 2004.

Galison, Peter. *How Experiments End.* Chicago: University of Chicago Press, 1987.

———. *Image and Logic: A Material Culture of Microphysics.* Chicago: University of Chicago Press, 1997.

———. "Re-reading the Past from the End of Physics: Maxwell's Equations in Retrospect." In *Functions and Uses of Disciplinary Histories*, edited by Loren Graham, Wolf Lepenies, and Peter Weingart. Dordrecht, Netherlands: D. Reidel, 1983.

Gardiner, Brian. "Edward Forbes, Richard Owen and the Red Lions." *Archives of Natural History* 20, no. 3 (1993): 349–72.

Geison, Gerald L. *Michael Foster and the Cambridge School of Physiology: The Scientific Enterprise in Late Victorian Society.* Princeton, NJ: Princeton University Press, 1978.

Gilbert, Alan D. *Religion and Society in Industrial England: Church, Chapel and Social Change.* London: Longman, 1976.

Gilbert, G. Nigel, and Michael Mulkay. "Experiements Are the Key: Participants' Histories and Historians' Histories of Science." *Isis* 75, no. 1 (1984): 105–25.

Gooday, Graeme. "Nature in the Laboratory: Domestication and Discipline with the Microscope in Victorian Life Science." *British Journal for the History of Science* 24, no. 3 (1991): 307–41.

Goodwin, Harvey. *Education for Working Men: An Address Delivered in the Town-Hall of Cambridge.* Cambridge: Deighton, Bell, 1855.

Gordin, Michael D. *The Pseudoscience Wars: Immanuel Velikovsky and the Birth of the Modern Fringe*. Chicago: University of Chicago Press, 2012.

Grant, Robert. *Miracle and Natural Law in Graeco-Roman and Early Christian Thought*. Eugene, OR: Wipf and Stock, 1952.

Green, S. J. D. "Religion and the Rise of the Common Man: Mutual Improvement Societies, Religious Associations and Popular Education in Three Industrial Towns in the West Riding of Yorkshire, c. 1850–1900." In *Cities, Class and Communication: Essays in Honor of Asa Briggs*, edited by Derek Fraser, 25–43. New York: Harvester Wheatsheaf, 1990.

———. *Religion in the Age of Decline: Organisation and Experience in Industrial Yorkshire, 1870–1920*. Cambridge: Cambridge University Press, 1996.

Gregory, Frederick. "Scientific versus Dialectical Materialism: A Clash of Ideologies in Nineteenth-Century German Radicalism." *Isis* 68 (1977): 206–23.

Griffin, David Ray. *Religion and Scientific Naturalism: Overcoming the Conflicts*. Albany: SUNY Press, 2000.

"The Guardian and Christian Socialism." *Christian Socialist*, March 21, 1851.

Hacking, Ian. "Nineteenth Century Cracks in the Concept of Determinism." *Journal of the History of Ideas* 44, no. 3 (1983): 455–75.

Harman, P. M. "Edinburgh Philosophy and Cambridge Physics: The Natural Philosophy of James Clerk Maxwell." In *Wranglers and Physicists: Studies on Cambridge Mathematical Physics in the Nineteenth Century*, edited by P. M. Harman, 202–24. Manchester, UK: Manchester University Press, 1985.

———. *Energy, Force, and Matter: The Conceptual Development of Nineteenth-Century Physics*. Cambridge: Cambridge University Press, 1982.

———. *The Natural Philosophy of James Clerk Maxwell*. Cambridge: Cambridge University Press, 1998.

———, ed. *The Scientific Letters and Papers of James Clerk Maxwell*. 3 vols. Cambridge: Cambridge University Press, 1990.

———, ed. *Wranglers and Physicists: Studies on Cambridge Mathematical Physics in the Nineteenth Century*. Manchester, UK: Manchester University Press, 1985.

Harris, Jose. *Private Lives, Public Spirit: A Social History of Britain, 1870–1914*. Oxford: Oxford University Press, 1993.

Harrison, J. F. C. *Learning and Living, 1790–1960: A Study in the History of the English Adult Education Movement*. Toronto: University of Toronto Press, 1961.

———. *Robert Owen and the Owenites in Britain and America: The Quest for the New Moral World*. London: Routledge and Kegan Paul, 1969.

Harrison, Peter. "The Development of the Concept of the Laws of Nature." In *Creation: Law and Probability*, edited by Fraser Watts. Aldershot, UK: Ashgate, 2008.

Hayles, N. Katherine. *Chaos Bound: Orderly Disorder in Contemporary Literature and Science*. Ithaca, NY: Cornell University Press, 1990.

Hearnshaw, F. J. C. *The Centenary History of King's College London, 1828–1928*. London: G. G. Harrap, 1929.

Heathcote Hamilton, Godfrey. *Queen Square: Its Neighbourhood & Its Institutions*. London: L. Parsons, 1926.

Helmholtz, Hermann von. "The Application of the Law of the Conservation of Force to Organic Nature." In *Selected Writings of Hermann von Helmholtz*, edited by Russell Kahl, 109–21. Middletown, CT: Wesleyan University Press, 1971.

Hendry, John. *James Clerk Maxwell and the Theory of the Electromagnetic Field*. Bristol, UK, and Boston: A. Hilger, 1986.

Herrnstein Smith, Barbara. *Natural Reflections: Human Cognition at the Nexus of Science and Religion*. New Haven, CT: Yale University Press, 2009.

Herschel, John F. W. *A Preliminary Discourse on the Study of Natural Philosophy*. London: Longman, Rees, Orme, Brown, and Green, and J. Taylor, 1830.

———. "Science and Scripture." *Athenaeum*, no. 1925 (1864).

Hesse, Mary. *Models and Analogies in Science*. Notre Dame, IN: University of Notre Dame Press, 1966.

———. *Revolutions and Reconstructions in the Philosophy of Science*. Bloomington: Indiana University Press, 1980.

Hilton, Boyd. *The Age of Atonement: The Influence of Evangelicalism on Social and Economic Thought, 1785–1865*. Oxford: Oxford University Press, 1991.

Hirst, Thomas. *Hirst Journals*, November 6, 1864, vol. 4, fol. 1702. Quoted in Ruth Barton, "'An Influential Set of Chaps': The X-Club and Royal Society Politics, 1864–85." *British Journal for the History of Science* 23, no. 1 (1990): 57.

"History of the Working Men's College." *Working Men's College Magazine* 1, no. 24 (1860): 188–92.

Hooker, Joseph Dalton. "Presidential Address." In *Report of the Thirty-Eighth Meeting of the British Association for the Advancement of Science*. London: John Murray, 1869.

Hooykaas, Reijer. *Principle of Uniformity in Geology, Biology and Theology*. Leiden, Netherlands: Brill, 1963.

Houghton, Samuel. *On the Natural Constants of the Healthy Urine of Man, and a Theory of Work Founded Thereon*. Dublin: University of Dublin Press, 1860.

Howarth, Janet. "Science Education in Late-Victorian Oxford: A Curious Case of Failure?" *English Historical Review* 102, no. 403 (1987): 334–71.

Huxley, Leonard. *Life and Letters of Thomas Henry Huxley*. 2 vols. 1900. Reprint, New York: AMS Press, 1979.

[Huxley, Thomas Henry]. "Contemporary Literature—Science." *Westminister Review* 63 (1855): 239–53.

Huxley, Thomas Henry. "Address on Behalf of the National Association for the Promotion of Technical Education." In *Collected Essays of T. H. Huxley*, vol. 3, 427–51. Bristol, UK: Thoemmes Press, 2001.

——. "Administrative Nihilism." In *Collected Essays*, vol. 1, 251–87. New York: D. Appleton, 1902.

——. "Agnosticism." In *Collected Essays of T. H. Huxley*, vol. 5, 202–62. Bristol, UK: Thoemmes Press, 2001.

——. "Agnosticism: A Rejoinder." In *Collected Essays of T. H. Huxley*, vol. 5, 263–308. Bristol, UK: Thoemmes Press, 2001.

——. "Agnosticism and Christianity." In *Collected Essays of T. H. Huxley*, vol. 5, 309–65. Bristol, UK: Thoemmes Press, 2001.

——. "Agnosticism, a Symposium." *Agnostic Annual* 1 (1884): 5–6.

——. "An Apologetic Irenicon." *Fortnightly Review* 52 (1892): 557–71.

——. "Autobiography." In *Collected Essays of T. H. Huxley*, vol. 1, 1–17. Bristol, UK: Thoemmes Press, 2001.

——. "The Coming of Age of *The Origin of Species*." In *Collected Essays of T. H. Huxley*, vol. 2, 227–43. Bristol, UK: Thoemmes Press, 2001.

——. "The Darwinian Hypothesis." In *Collected Essays of T. H. Huxley*, vol. 2. Bristol, UK: Thoemmes Press, 2001.

——. "Draft of Mr. Balfour's Attack on Agnosticism, Part II." In *Huxley, Prophet of Science*, by Houston Peterson. New York: Longmans, Green, 1932.

——. "An Episcopal Trilogy." In *Collected Essays of T. H. Huxley*, vol. 5, 126–59. Bristol, UK: Thoemmes Press, 2001.

——. "Evidence as to the Miracle of the Resurrection." Metaphysical Society, January 11, 1876.

——. *Evolution and Ethics and Other Essays*. New York: D. Appleton, 1899.

——. "Evolution in Biology." In *Collected Essays of T. H. Huxley*, vol. 2. Bristol, UK: Thoemmes Press, 2001.

——. "The Evolution of Theology: An Anthropological Study." In *Collected Essays of T. H. Huxley*, vol. 4, 287–372. Bristol, UK: Thoemmes Press, 2001.

——. "Geological Contemporaneity and Persistent Types of Life." In *Collected Essays of T. H. Huxley*, vol. 8. Bristol, UK: Thoemmes Press, 2001.

——. "Geological Reform." In *Collected Essays of T. H. Huxley*, vol. 8, 305–39. Bristol, UK: Thoemmes Press, 2001.

——. "Government: Anarchy or Regimentation." In *Collected Essays of T. H. Huxley*, vol. 1, 383–430. Bristol, UK: Thoemmes Press, 2001.

——. "Has a Frog a Soul and of What Nature Is That Soul, Supposing It to Exist?" Papers of the Metaphysical Society 1 (1869–74), November 8, 1870, 1–7, Manchester College Library, Oxford.

——. "Hume: With Helps to the Study of Berkeley." In *Collected Essays of T. H. Huxley*, vol. 6, 3–240. Bristol, UK: Thoemmes Press, 2001.

——. *Introductory Science Primer*. London: Appleton, 1887.

———. "Joseph Priestley." *Macmillan's Magazine* 30 (1874): 1–37.

———. "A Liberal Education; and Where to Find It." In *Collected Essays of T. H. Huxley*, vol. 3. Bristol, UK: Thoemmes Press, 2001.

———. "A Lobster; or, the Study of Zoology." In *Collected Essays of T. H. Huxley*, vol. 9, 196–228. Bristol, UK: Thoemmes Press, 2001.

———. "A Modern Symposium: Influence on Morality of a Decline in Religious Belief." *Nineteenth Century* 1 (1877): 536–9.

———. "A Modern Symposium: The Soul and Future Life." In *The Nineteenth Century: A Monthly Review*, edited by James Knowles, vol. 2. London: Henry S. King, 1877.

———. "Mr. Balfour's Attack on Agnosticism." *Nineteenth Century* 37 (1895): 527–40.

———. "Mr. Darwin's Critics." In *Collected Essays of T. H. Huxley*, vol. 2, 120–86. Bristol, UK: Thoemmes Press, 2001.

———. "On Certain Zoological Arguments Commonly Adduced in Favour of the Hypothesis of the Progressive Development of Animal Life in Time." In *The Scientific Memoirs of Thomas Henry Huxley*, edited by Michael Foster and E. Ray Lankester, vol. 1, 300–304. London: Macmillan, 1902.

———. "On Descartes' 'Discourse Touching the Method of Using One's Reason Rightly and of Seeking Scientific Truth.'" In *Collected Essays of T. H. Huxley*, vol. 1, 166–98. Bristol, UK: Thoemmes Press, 2001.

———. "On Natural History, as Knowledge, Discipline, and Power." In *The Scientific Memoirs of Thomas Henry Huxley*, edited by Michael Foster and E. Ray Lankester, vol. 1, 305–14. London: Macmillan, 1902.

———. "On Our Knowledge of the Causes of the Phenomena of Organic Nature." In *Collected Essays of T. H. Huxley*, vol. 2, 303–475. Bristol, UK: Thoemmes Press, 2001.

———. "On the Animals Which Are Most Nearly Intermediate between Birds and Reptiles." In *Notices of the Proceedings at the Meetings of Members of the Royal Institution of Great Britain with Abstracts of the Discourses Delivered at Evening Meetings*, vol. 5, 278–86. London: William Clowes and Sons, 1869.

———. "On the Educational Value of the Natural History Sciences." In *Collected Essays of T. H. Huxley*, vol. 3. Bristol, UK: Thoemmes Press, 2001.

———. "On the Hypothesis That Animals Are Automata, and Its History." In *Collected Essays of T. H. Huxley*, vol. 1, 199–250. Bristol, UK: Thoemmes Press 1874.

———. "On the Method of Paleontology." In *The Scientific Memoirs of Thomas Henry Huxley*, edited by Michael Foster and E. Ray Lankester, vol. 1, 432–44. London: Macmillan, 1902.

———. "On the Method of Zadig." In *Collected Essays of T. H. Huxley*, vol. 4, 1–23. Bristol, UK: Thoemmes Press, 2001.

———. "On the Persistent Types of Animal Life." In *The Scientific Memoirs of Thomas Henry Huxley*, edited by Michael Foster and E. Ray Lankester, vol. 2, 90–93. London: Macmillan, 1902.

———. "On the Physical Basis of Life." In *Collected Essays of T. H. Huxley*, vol. 1, 130–65. Bristol, UK: Thoemmes Press, 2001.

———. "On the Present State of Knowledge as to the Structure and Functions of Nerve." In *The Scientific Memoirs of Thomas Henry Huxley*, edited by Michael Foster and E. Ray Lankester, vol. 1, 315–20. London: Macmillan, 1902.

———. "On the Reception of the *Origin of Species*." In *The Life and Letters of Charles Darwin*, edited by Francis Darwin, vol. 1. New York: D. Appleton, 1911.

———. "On the Theory of the Vertebrate Skull." In *The Scientific Memoirs of Thomas Henry Huxley*, edited by Michael Foster and E. Ray Lankester, vol. 1, 538–606. London: Macmillan, 1902.

———. "The Origin of Species." In *Collected Essays of T. H. Huxley*, vol. 2, 22–79. Bristol, UK: Thoemmes Press, 2001.

———. "Owen's Position in the History of Anatomical Science." In *The Scientific Memoirs of Thomas Henry Huxley*, edited by Michael Foster and E. Ray Lankester, vol. 4, 658–89. London: Macmillan, 1902.

———. "Possibilities and Impossibilites." In *Collected Essays of T. H. Huxley*, vol. 5. Bristol, UK: Thoemmes Press, 2001.

———. Preface to *Collected Essays of T. H. Huxley*, vol. 5. Bristol, UK: Thoemmes Press, 2001.

———. Prefatory note to *Freedom in Science and Teaching*, by Ernst Haeckel. New York: D. Appleton, 1879.

———. "Presidential Address to the Royal Society." *Nature* 33, no. 840 (1885): 112–19.

———. "The Progress of Science." In *Collected Essays of T. H. Huxley*, vol. 1, 42–129. Bristol, UK: Thoemmes Press, 2001.

———. Prologue to *Essays upon Some Controverted Questions*, by Thomas Henry Huxley, 1–40. New York: D. Appleton, 1892.

———. "Prologue to *Essays upon Some Controverted Questions*." In *Collected Essays of T. H. Huxley*, vol. 5, 1–58. Bristol, UK: Thoemmes Press, 2001.

———. "The Rede Lecture." In *The Scientific Memoirs of Thomas Henry Huxley*, edited by Michael Foster and E. Ray Lankester, vol. 5, 70–79. London: Macmillan, 1902.

———. "Review of Haeckel's *Natural History of Creation*." *Academy* 1 (1869).

———. "Review of *Vestiges of the Natural History of Creation*." In *The Scientific Memoirs of Thomas Henry Huxley*, edited by Michael Foster and E. Ray Lankester, vol. 5, 1–19. London: Macmillan, 1902.

———. "The School Boards: What They Can Do, and What They May Do." In

Collected Essays of T. H. Huxley, vol. 3, 373–403. Bristol, UK: Thoemmes Press, 2001.

———. "Science and 'Church Policy.'" *Reader* 4 (1864): 821.

———. "Science and Morals." In *Collected Essays of T. H. Huxley*, vol. 9. Bristol, UK: Thoemmes Press, 2001.

———. "Science and Pseudo-Science." In *Collected Essays of T. H. Huxley*, vol. 5, 90–125. Bristol, UK: Thoemmes Press, 2001.

———. "Science and Religion." *Builder* 17 (1859): 35–36.

———. "Scientific and Pseudo-Scientific Realism." In *Collected Essays of T. H. Huxley*, vol. 5, 59–89. Bristol, UK: Thoemmes Press, 2001.

———. "The Scientific Aspects of Positivism." In *Lay Sermons, Addresses and Reviews*, 128–51. New York: D. Appleton, 1872.

———. "Scientific Education: Notes of an after-Dinner Speech." In *Collected Essays of T. H. Huxley*, vol. 3, 111–33. Bristol, UK: Thoemmes Press, 2001.

———. "Social Diseases and Worse Remedies." In *Collected Essays of T. H. Huxley*, vol. 9. Bristol, UK: Thoemmes Press, 2001.

———. "Technical Education." In *Collected Essays of T. H. Huxley*, vol. 3. Bristol, UK: Thoemmes Press, 2001.

———. "Time and Life: Mr. Darwin's *Origin of Species*." *Macmillan's Magazine* 1 (1859): 7.

———. "Universities: Actual and Ideal." In *Collected Essays of T. H. Huxley*, vol. 3, 189–234. Bristol, UK: Thoemmes Press, 2001.

Jacyna, Leon Stephen. "Immanence or Transcendence: Theories of Life and Organization in Britain, 1790–1835." *Isis* 74, no. 3 (1983): 310–29.

———. "The Physiology of Mind, the Unity of Nature, and the Moral Order in Victorian Thought." *British Journal for the History of Science* 14, no. 2 (1981): 109–32.

———. "Science and Social Order in the Thought of A. J. Balfour." *Isis* 71, no. 1 (1980): 11–34.

———. "Scientific Naturalism in Victorian Britain: An Essay in the Social History of Ideas." Unpublished PhD thesis, University of Edinburgh, 1980.

James, William. "Are We Automata?" *Mind* 4, no. 13 (1879): 1–22.

———. *The Varieties of Religious Experience*. New York: Penguin, 1958.

Jensen, J. Vernon. "The X Club: Fraternity of Victorian Scientists." *British Journal for the History of Science* 5, no. 1 (1970): 63–72.

Johnson, Phillip E. *Darwin on Trial*. Crowborough, UK: Monarch, 1991.

———. "Evolution as Dogma: The Establishment of Naturalism." In *Intelligent Design Creationism and Its Critics: Philosophical, Theological, and Scientific Perspectives*, edited by Robert Pennock, 59–76. Cambridge, MA: MIT Press, 2001.

———. *Reason in the Balance: The Case against Naturalism in Science, Law, and Education*. Downers Grove, IL: InterVarsity Press, 1993.

Jones, Bence. *The Life and Letters of Faraday*. 2 vols. London: Longmans, Green, 1870.

Jungnickel, Christa, and Russell McCormmach. *Intellectual Mastery of Nature: Theoretical Physics from Ohm to Einstein*. 2 vols. Vol. 1. Chicago: University of Chicago Press, 1986.

Kang, Minsoo. *Sublime Dreams of Living Machines: The Automaton in the European Imagination*. Cambridge, MA: Harvard University Press, 2011.

Kargon, Robert. "Model and Analogy in Victorian Science: Maxwell's Critique of the French Physicists." *Journal of the History of Ideas* 30, no. 3 (1969): 423–36.

Kelvin, Lord. "Presidential Address." In *Proceedings of the Royal Society of London*, vol. 59, 107–24. London: Harrison and Sons, 1896.

Kevles, Daniel J. *The Physicists: The History of a Scientific Community in Modern America*. Cambridge, MA: Harvard University Press, 1995.

Kingsley, Charles. *The Water Babies: A Fairy Tale for a Land-Baby*. New York: Macmillan, 1885.

Kingsley, Frances Eliza Grenfell, ed. *Charles Kingsley: His Letters and Memories of His Life*. 2 vols. Vol. 1. London: Henry S. King, 1877.

Kitcher, Philip. "Born-Again Creationism." In *Intelligent Design Creationism and Its Critics: Philosophical, Theological, and Scientific Perspectives*, edited by Robert Pennock, 257–87. Cambridge, MA: MIT Press, 2001.

Knight, David. "Thomas Henry Huxley and Philsophy of Science." In *Thomas Henry Huxley's Place in Science and Letters: Centenary Essays*, edited by Alan P. Barr, 51–66. Athens: University of Georgia Press, 1997.

Knudsen, Ole. "Mathematics and Physical Reality in William Thomson's Electromagnetic Theory." In *Wranglers and Physicists: Studies on Cambridge Mathematical Physics in the Nineteenth Century*, edited by P. M. Harman, 149–79. Manchester, UK: Manchester University Press, 1985.

Koperski, Jeffrey. "Two Bad Ways to Attack Intelligent Design and Two Good Ones." *Zygon* 43, no. 2 (2008): 433–49.

Kremer, Richard L. *The Thermodynamics of Life and Experimental Physiology, 1770–1880*. New York: Garland Publishing, 1990.

LaFollette, Marcel C. *Creationism, Science, and the Law: The Arkansas Case*. Cambridge, MA: MIT Press, 1983.

Lamb, Horace. "Clerk Maxwell as Lecturer." In *James Clerk Maxwell: A Commemoration Volume, 1831–1931*, edited by Joseph John Thomson. Cambridge: Cambridge University Press, 1931.

Lambert, Kevin. "Mind over Matter: Language, Mathematics, and Electromagnetism in Nineteenth Century Britain." PhD diss., UCLA, 2005.

———. "The Uses of Analogy: James Clerk Maxwell's 'On Faraday's Lines of Force' and Early Victorian Analogical Argument." *British Journal for the History of Science* 44, no. 1 (2011): 61–88.

Lang, Andrew. "Science and Demonology." *Illustrated London News*, January 30, 1894, 822.

Langley, J. N. "A Few Thoughts on the Franchise." *Working Men's College Magazine* 1, no. 28 (1861): 45–48.

Lankester, Edwin. *On Food*. London: Robert Hardwicke, 1861.

Lankester, Edwin Ray. *The Kingdom of Man*. New York: Henry Holt, 1907.

Laplace, Pierre Simon de. *A Philosophical Essay on Probabilities*. Translated by F. W. Truscott. New York: John Wiley and Sons, 1902.

Larson, Edward J. *Trial and Error: The American Controversy over Creation and Evolution*. Oxford: Oxford University Press, 2003.

Larson, Edward J., and Larry Witham. "Leading Scientists Still Reject God." *Nature* 394, no. 6691 (1998): 313.

———. "Scientists Are Still Keeping the Faith," *Nature* 386, no. 6624 (1997): 435–36.

Laudan, Larry. "The Demise of the Demarcation Problem." In *But Is It Science? The Philosophical Question in the Creation/Evolution Controversy*, edited by Michael Ruse, 337–50. Buffalo, NY: Prometheus Books, 1988.

———. "Science at the Bar—Causes for Concern." In *But Is It Science? The Philosophical Question in the Creation/Evolution Controversy*, edited by Michael Ruse, 351–55. Buffalo, NY: Prometheus Books, 1988.

Layton, David. "The Schooling of Science in England, 1854–1939." In *The Parliament of Science: The British Association for the Advancement of Science, 1831–1981*, edited by Roy MacLeod and Peter Collins, 188–210. Northwood, UK: Science Reviews, 1981.

"Letters on the Public Health." *Christian Socialist*, December 7, 1850.

Leuba, James H. *The Belief in God and Immortality: A Psychological, Anthropological and Statistical Study*. 1916. Reprint, Chicago: Open Court, 1921.

Leys, Ruth. "Background to the Reflex Controversy: William Alison and the Doctrine of Sympathy before Hall." *Studies in History of Biology* 4 (1980): 1–66.

Lightman, Bernard. "Fighting Even with Death: Balfour, Scientific Naturalism, and Thomas Henry Huxley's Final Battle." In *Thomas Henry Huxley's Place in Science and Letters*, edited by Alan P. Barr, 323–50. Atlanta: University of Georgia Press, 1997.

———. "Huxley and Scientific Agnosticism: The Strange History of a Failed Rhetorical Strategy." *British Journal for the History of Science* 35, no. 3 (2002): 271–89.

———. *The Origins of Agnosticism: Victorian Unbelief and the Limits of Knowledge*. Baltimore: Johns Hopkins University Press, 1987.

———. "Pope Huxley and the Church Agnostic: The Religion of Science." *Historical Papers* (1983): 150–63.

——. *Victorian Popularizers of Science: Designing Nature for New Audiences.*
Chicago: University of Chicago Press, 2007.

——. "Victorian Sciences and Religions: Discordant Harmonies." In *Osiris.*
Vol. 16, *Science in Theistic Contexts (Cognitive Dimensions)*, edited by John
Hedley Brooke, Margaret J. Osler, and Jitse van der Meer, 343–66. Chicago:
University of Chicago Press, 2001.

Lilly, William Samuel. "Materialism and Morality." *Fortnightly Review* 40 (October 1886): 575–94.

Lindberg, David, and Ronald L. Numbers, eds. *God and Nature: Historical Essays on the Encounter between Christianity and Science.* Berkeley: University
of California Press, 1986.

Litchfield, Henrietta. *Richard Buckley Litchfield: A Memoir Written for His
Friends by His Wife.* Cambridge: Cambridge University Press, 1910.

Litchfield, R. B. *The Beginnings of the Working Men's College.* London: Working Men's College, 1902.

Livingstone, David N. *Darwin's Forgotten Defenders: The Encounter between
Evangelical Theology and Evolutionary Thought.* Grand Rapids, MI: Eerdmans Publishing, 1987.

——. "That Huxley Defeated Wilberforce in Their Debate over Evolution and
Religion." In *Galileo Goes to Jail and Other Myths about Science and Religion*, edited by Ronald L. Numbers, 152–60. Cambridge, MA: Harvard University Press, 2009.

Livingstone, David N., D. G. Hart, and Mark A. Noll, eds. *Evangelicals and Science in Historical Perspective.* Oxford: Oxford University Press, 1999.

Lot, Parson. "Bible Politics; or, God Justified to the People." *Christian Socialist*,
November 9, 1850.

——. "Bible Politics; or, God Justified to the People II." *Christian Socialist*, November 23, 1850, 25–26.

——. "My Political Creed." *Christian Socialist*, December 14, 1850, 50.

Lubenow, W. C. *The Cambridge Apostles, 1820–1914: Liberalism, Imagination, and Friendship in British Intellectual and Professional Life.* Cambridge:
Cambridge University Press, 1998.

Ludlow, John. "A New Idea." In *Church and State in the Modern Age: A Documentary History*, edited by J. F. Maclear. Oxford: Oxford University Press,
1995.

Lushington, Godfrey. "Shall We Learn Latin?" *Working Men's College Magazine* 1, no. 17 (1860): 71–78.

Lyons, Sherrie. "Thomas Huxley: Fossils, Persistence, and the Argument from
Design." *Journal of the History of Biology* 26, no. 3 (1993): 545–69.

MacLeod, Roy M. "The 'Naturals' and Victorian Cambridge: Reflections on the
Anatomy of an Elite, 1851–1914." *Oxford Review of Education* 6, no. 2 (1980):
177–96.

——. "The X-Club: A Social Network of Science in Late-Victorian England." *Notes and Records of the Royal Society of London* 24, no. 2 (1970): 305–22.

Mandelbaum, Maurice. *History, Man, and Reason: A Study in Nineteenth-Century Thought.* Baltimore: Johns Hopkins University Press, 1971.

Marsden, George. *The Soul of the American University: From Protestant Establishment to Established Nonbelief.* Oxford: Oxford University Press, 1994.

Marston, Philip. "Maxwell and Creation: Acceptance, Criticism, and His Anonymous Publication." *American Journal of Physics* 75, no. 8 (August 2007): 731–40.

Martin, Jay. *The Education of John Dewey: A Biography.* New York: Columbia University Press, 2002.

Maudsley, Henry. *The Physiology and Pathology of Mind.* New York: D. Appleton, 1867.

Maurice, Frederick Denison. *A Few Words on Secular and Denominational Education: In a Letter to the Members of the Working Men's College.* London: Macmillan, 1870.

——. "Introductory Lecture on the Studies of the (London) Working Men's College." *Working Men's College Magazine* 1, no. 1 (1859): 1–8.

——. "Notebook." King's College London Archives, box 5199.M3 (1872).

——. "Personal Explanation." *Working Men's College Magazine* 1, no. 26 (1861): 13–15.

——. *Theological Essays.* Cambridge: Macmillan, 1853.

Maurice, John Frederick. *The Life of Frederick Denison Maurice: Chiefly Told in His Own Letters.* New York: Charles Scribner, 1884.

[Maxwell, James Clerk]. "The Conservation of Energy." In *Nature: A Weekly Illustrated Journal of Science*, vol. 9. London: Macmillan, 1874.

Maxwell, James Clerk. "Address to the Mathematical and Physical Sections of the British Association (1870)." In *The Scientific Papers of James Clerk Maxwell*, edited by W. D. Niven, vol. 2, 215–29. Cambridge: Cambridge University Press, 1890.

——. "Appendix: From a Letter of Reference for William Kingdon Clifford, *circa* July 1871." In *The Scientific Letters and Papers of James Clerk Maxwell*, edited by P. M. Harman, vol. 2. Cambridge: Cambridge University Press, 1990.

——. "Appendix II: Fragments of an Apostles Essay 'What Is the Nature of Evidence of Design? 1853." In *The Scientific Letters and Papers of James Clerk Maxwell*, edited by P. M. Harman, vol. 1. Cambridge: Cambridge University Press, 1990.

——. "Atom." In *The Scientific Papers of James Clerk Maxwell*, edited by W. D. Niven, vol. 2, 445–84. Cambridge: Cambridge University Press, 1890.

——. "Attraction." In *The Scientific Papers of James Clerk Maxwell*, edited by W. D. Niven, vol. 2, 485–91. Cambridge: Cambridge University Press, 1890.

———. "Concerning Demons." In *The Scientific Letters and Papers of James Clerk Maxwell*, edited by P. M. Harman, vol. 3. Cambridge: Cambridge University Press, 1990.

———. "Diffusion." In *The Scientific Papers of James Clerk Maxwell*, edited by W. D. Niven, vol. 2, 625–46. Cambridge: Cambridge University Press, 1890.

———. "Draft on the Methods of Physical Science." In *The Scientific Letters and Papers of James Clerk Maxwell*, edited by P. M. Harman, vol. 3. Cambridge: Cambridge University Press, 1990.

———. "Essay for the Apostles on 'Analogies in Nature,' February 1856." In *The Scientific Letters and Papers of James Clerk Maxwell*, edited by P. M. Harman, vol. 1. Cambridge: Cambridge University Press, 1990.

———. "Essay for the Eranus Club on Psychophysik." In *The Scientific Letters and Papers of James Clerk Maxwell*, edited by P. M. Harman, vol. 3. Cambridge: Cambridge University Press, 1990.

———. "Essay for the Eranus Club on Science and Free Will, 11 February 1873." In *The Scientific Letters and Papers of James Clerk Maxwell*, edited by P. M. Harman, vol. 2, 814–23. Cambridge: Cambridge University Press, 1990.

———. "Ether." In *The Scientific Papers of James Clerk Maxwell*, edited by W. D. Niven, vol. 2, 763–75. Cambridge: Cambridge University Press, 1890.

———. "Faraday." In *The Scientific Papers of James Clerk Maxwell*, edited by W. D. Niven, vol. 2, 786–93. Cambridge: Cambridge University Press, 1890.

———. "Hermann Ludwig Ferdinand Helmholtz." In *The Scientific Papers of James Clerk Maxwell*, edited by W. D. Niven, vol. 2, 592–98. Cambridge: Cambridge University Press, 1890.

———. "Inaugural Lecture at Aberdeen." Quoted in Reginald Victor Jones, "James Clerk Maxwell at Aberdeen, 1856–1860." *Notes and Records of the Royal Society of London* 28, no. 1 (1973).

———. "Inaugural Lecture at King's College London, October 1860." In *The Scientific Letters and Papers of James Clerk Maxwell*, edited by P. M. Harman, vol. 1, 662–74. Cambridge: Cambridge University Press, 1990.

———. "Inaugural Lecture at Marischal College, Aberdeen, November 3, 1856." In *The Scientific Letters and Papers of James Clerk Maxwell*, edited by P. M. Harman, vol. 1, 419–32. Cambridge: Cambridge University Press, 1990.

———. "Inaugural Lecture at Marischal College, Aberdeen, November 1857." In *The Scientific Letters and Papers of James Clerk Maxwell*, edited by P. M. Harman, vol. 1. Cambridge: Cambridge University Press, 1990.

———. "Introductory Lecture, Aberdeen." In *The Scientific Letters and Papers of James Clerk Maxwell*, edited by P. M. Harman, vol. 1, 542–47. Cambridge: Cambridge University Press, 1990.

———. "Introductory Lecture on Experimental Physics." In *The Scientific Papers of James Clerk Maxwell*, edited by W. D. Niven, vol. 2, 241–55. Cambridge: Cambridge University Press, 1890.

———. "Molecules (A Lecture)." In *The Scientific Papers of James Clerk Maxwell*, edited by W. D. Niven, vol. 2, 361–77. Cambridge: Cambridge University Press, 1890.

———. "Note to Tait 'Concerning Demons.'" In *The Scientific Letters and Papers of James Clerk Maxwell*, edited by P. M. Harman, vol. 3. Cambridge: Cambridge University Press, 1990.

———. "On Colour Vision." In *The Scientific Papers of James Clerk Maxwell*, edited by W. D. Niven, vol. 2. Cambridge: Cambridge University Press, 1890.

———. "On Faraday's Lines of Force." In *The Scientific Papers of James Clerk Maxwell*, edited by W. D. Niven, vol. 1, 155–229. Cambridge: Cambridge University Press, 1890.

———. "On Physical Lines of Force." In *The Scientific Papers of James Clerk Maxwell*, edited by W. D. Niven, vol. 1. Cambridge: Cambridge University Press, 1890.

———. "On the Dynamical Evidence of the Molecular Constitution of Bodies." In *The Scientific Papers of James Clerk Maxwell*, edited by W. D. Niven, vol. 2, 418–38. Cambridge: Cambridge University Press, 1890.

———. "On the (Physical) Dynamical Explanation of Electric Phenomena." In *The Scientific Letters and Papers of James Clerk Maxwell*, edited by P. M. Harman, vol. 3. Cambridge: Cambridge University Press, 1990.

———. "On the Theory of Colours in Relation to Colour-Blindness." In *The Scientific Papers of James Clerk Maxwell*, edited by W. D. Niven, vol. 1, 119–25. Cambridge: Cambridge University Press, 1890.

———. "Plateau on Soap-Bubbles." In *The Scientific Papers of James Clerk Maxwell*, edited by W. D. Niven, vol. 2, 393–99. Cambridge: Cambridge University Press, 1890.

———. "Review of *An Essay on the Mathematical Principles of Physics*." In *The Scientific Papers of James Clerk Maxwell*, edited by W. D. Niven, vol. 2, 338–42. Cambridge: Cambridge University Press, 1890.

———. "A Review of *Paradoxical Philosophy*." In *The Scientific Papers of James Clerk Maxwell*, edited by W. D. Niven, vol. 2, 756–62. Cambridge: Cambridge University Press, 1890.

———. "Review of P. G. Tait, *Recent Advances in Physical Sciences*." In *The Scientific Letters and Papers*, edited by P. M. Harman, vol. 3, 308–12. Cambridge: Cambridge University Press, 1990.

———. "Review of Tait's *Thermodynamics*." In *The Scientific Papers of James Clerk Maxwell*, edited by W. D. Niven, vol. 2, 660–71. Cambridge: Cambridge University Press, 1890.

———. "Review of Thomson and Tait's *Natural Philosophy*." In *The Scientific Papers of James Clerk Maxwell*, edited by W. D. Niven, vol. 2, 776–84. Cambridge: Cambridge University Press, 1890.

———. "Review of Thomson, *Papers on Electrostatics and Magnetism*." In *The*

Scientific Letters and Papers of James Clerk Maxwell, edited by P. M. Harman, vol. 2, 301–7. Cambridge: Cambridge University Press, 1990.

———. *Theory of Heat.* London: Longmans, Green, 1872.

———. *A Treatise on Electricity and Magnetism.* 2 vols. Oxford: Clarendon Press, 1873.

McLeod, Hugh. *Religion and Society in England, 1850–1914.* New York: St. Martin's, 1996.

———. *Secularisation in Western Europe, 1848–1914.* London: Macmillan, 2000.

McMullin, Ernan. "Evolution and Special Creation." In *The Philosophy of Biology*, edited by David Hull and Michael Ruse, 698–733. Oxford: Oxford University Press, 1998.

McNatt, Jerrold. "James Clerk Maxwell's Refusal to Join the Victoria Institute." *Perspectives on Science and Christian Faith* 56 (2004): 204–15.

"Meeting of the Cambridge Working Men's College." *Working Men's College Magazine* 1, no. 3 (1859): 53–57.

Moore, James R. *The Post-Darwinian Controversies: A Study of the Protestant Struggle to Come to Terms with Darwin in Great Britain and America, 1870–1900.* Cambridge: Cambridge University Press, 1979.

———. "Theodicy and Society: The Crisis of the Intelligentsia." In *Victorian Faith in Crisis: Essays on Continuity and Change in Nineteenth-Century Religious Belief*, edited by Richard J. Helmstadter and Bernard Lightman, 153–86. Stanford, CA: Stanford University Press, 1990.

Morgan, Mary S., and Margaret Morrison, eds. *Models as Mediators: Perspectives on Natural and Social Science.* Cambridge: Cambridge University Press, 1999.

Moyer, Albert E. *A Scientist's Voice in American Culture: Simon Newcomb and the Rhetoric of Scientific Method.* Berkeley: University of California Press, 1992.

Mullin, Robert Bruce. *Miracles and the Modern Religious Imagination.* New Haven, CT: Yale University Press, 1996.

———. "Science, Miracles, and the Prayer-Gauge Debate." In *When Science and Christianity Meet*, edited by David C. Lindberg and Ronald L. Numbers, 203–24. Chicago: University of Chicago Press, 2003.

Myers, Greg. "Nineteenth-Century Popularizations of Thermodynamics and the Rhetoric of Social Prophecy." In *Energy and Entropy: Science and Culture in Victorian Britain*, edited by Patrick Brantlinger, 67–90. Bloomington: Indiana University Press, 1989.

Niven, W. D., ed. *The Scientific Papers of James Clerk Maxwell.* 2 vols. Cambridge: Cambridge University Press, 1890.

Noakes, Richard. "Spiritualism, Science, and the Supernatural in Mid-Victorian Britain." In *The Victorian Supernatural*, edited by Nicola Bown, Carolyn

Burdett, and Pamela Thurschwell, 23–43. Cambridge: Cambridge University Press, 2004.

Noll, Mark A., David W. Bebbington, and George A. Rawlyk, eds. *Evangelicalism: Comparative Studies of Popular Protestantism in North America, the British Isles, and Beyond, 1700–1990*. New York and Oxford: Oxford University Press, 1994.

Norman, Edward R. *The Victorian Christian Socialists*. Cambridge: Cambridge University Press, 1987.

Numbers, Ronald L. *The Creationists: From Scientific Creationism to Intelligent Design*. Cambridge, MA: Harvard University Press, 2006.

———. *Science and Christianity in Pulpit and Pew*. Oxford: Oxford University Press, 2007.

———. "Science without God: Natural Laws and Christian Beliefs." In *When Science and Christianity Meet*, edited by David C. Lindberg and Ronald L. Numbers, 265–85. Chicago: University of Chicago Press, 2003.

Nye, Mary Jo. "The Moral Freedom of Man and the Determinism of Nature: The Catholic Synthesis of Science and History in the *Revue des questions scientifiques*." *British Journal for the History of Science* 9, no. 3 (1976): 274–92.

O'Connor, Ralph. "Reflections on Popular Science in Britain: Genres, Categories, Historians." *Isis* 100 (2009): 333–45.

Ohlers, R. Clinton. "The End of Miracles: Scientific Naturalism in America, 1839–1934." PhD diss., University of Pennsylvania Press, 2007.

Olson, Richard. *Scottish Philosophy and British Physics, 1750–1880: A Study in the Foundations of the Victorian Scientific Style*. Princeton, NJ: Princeton University Press, 1975.

Oppenheim, Janet. *The Other World: Spiritualism and Psychical Research in England, 1850–1914*. Cambridge: Cambridge University Press, 1985.

Osborn, Henry Fairfield. "Enduring Recollections." *Nature* 115 (1925): 726–28.

"Our Principles." *Christian Socialist*, November 2, 1850.

Paley, William. *Evidences of Christianity*. 1794. Reprint, London: W. Clowes and Sons, 1851.

———. *Natural Theology*. 1802. Reprint, Boston: Gould and Lincoln, 1872.

Pauly, Philip J. "The Appearance of Academic Biology in Late Nineteenth-Century America." *Journal of the History of Biology* 17, no. 3 (1984): 369–97.

Paz, Denis G. *Popular Anti-Catholicism in Mid-Victorian England*. Stanford, CA: Stanford University Press, 1992.

Pennock, Robert T., ed. *Intelligent Design Creationism and Its Critics: Philosophical, Theological, and Scientific Perspectives*. Cambridge, MA: MIT Press, 2001.

———. "Naturalism, Evidence and Creationism: The Case of Philip Johnson." *Biology and Philosophy* 11, no. 4 (1996): 543–99.

——. *Tower of Babel: The Evidence against the New Creationism*. Cambridge, MA MIT Press, 1999.

Peterson, Houston. *Huxley, Prophet of Science*. New York: Longmans, Green, 1932.

Plantinga, Alvin. "Methodological Naturalism?" In *Intelligent Design Creationism and Its Critics: Philosophical, Theological, and Scientific Perspectives*, edited by Robert Pennock, 339–62. Cambridge, MA: MIT Press, 2001.

——. "When Faith and Reason Clash: Evolution and the Bible." In *Intelligent Design Creationism and Its Critics: Philosophical, Theological, and Scientific Perspectives*, edited by Robert Pennock, 113–45. Cambridge, MA: MIT Press, 2001.

Pope, Norris. "Dickens's 'The Signalman' and Information Problems in the Railway Age." *Technology and Culture* 42, no. 3 (July 2001): 436–61.

Porter, Enid. *Victorian Cambridge: Josiah Chater's Diaries, 1844–1884*. London: Phillimore, 1975.

Porter, Roy. *Flesh in the Age of Reason: The Modern Foundations of Body and Soul*. New York: W. W. Norton, 2003.

Porter, Theodore M. *The Rise of Statistical Thinking: 1820–1900*. Princeton, NJ: Princeton University Press, 1986.

——. "A Statistical Survey of Gases: Maxwell's Social Physics." *Historical Studies in the Physical Sciences* 12, no. 1 (1981): 77–116.

Powell, Baden. *The Connexion of Natural and Divine Truth; or, the Study of the Inductive Philosophy, Considered as Subservient to Theology*. London: J. W. Parker, 1838.

——. *Essays on the Spirit of Inductive Philosophy, the Unity of Worlds, and the Philosophy of Creation*. London: Longman, Brown, Green, and Longmans, 1855.

"Proceedings at a Meeting of the Working Men's College." *Cambridge Chronicle*, March 26, 1858.

Rabinbach, Anson. *The Human Motor: Energy, Fatigue, and the Origins of Modernity*. Berkeley: University of California Press, 1992.

Raia, Courtenay Grean. "The Substance of Things Hoped For: Faith, Science and Psychical Research in the Victorian Fin de Siècle." PhD diss., University of California, Los Angeles, 2005.

Rainger, Ronald. "Vertebrate Paleontology as Biology: Henry Fairfield Osborn and the American Museum of Natural History." In *American Development of Biology*, edited by Ronald Rainger, Keith Benson, and Jane Maienschein, 219–56. Rutgers, NJ: Rutgers University Press, 1988.

Reed, John Robert. *Victorian Will*. Athens: Ohio University Press, 1989.

"Report of the Manchester Working Men's College Meeting." *Working Men's College Magazine* 1, no. 2 (1859): 30–38.

Richards, Joan L. "God, Truth, and Mathematics in Nineteenth-Century England." In *The Invention of Physical Science: Intersections of Mathematics, Theology, and Natural Philosophy since the Seventeenth Century: Essays in Honor of Erwin N. Hiebert*, edited by Mary Jo Nye, Joan L. Richards, and R. H. Stuewer, 51–80. Dordrecht, Netherlands: Kluwer Academic Publishers, 1992.

——. *Mathematical Visions: The Pursuit of Geometry in Victorian England*. San Diego: Academic Press, 1988.

Richardson, Alan. *British Romanticism and the Science of the Mind*. Cambridge: Cambridge University Press, 2004.

Riskin, Jessica, ed. *Genesis Redux: Essays in the History and Philosophy of Artificial Life*. Chicago: University of Chicago Press, 2007.

Robbins, Keith. "Religion and Community in Scotland and Wales since 1800." In *A History of Religion in Britain*, edited by Sheridan Gilley and W. J. Sheils, 363–80. Cambridge, MA: Blackwell, 1994.

Roberts, Jon H., and James Turner. *The Sacred and the Secular University*. Princeton, NJ: Princeton University Press, 2000.

Roderick, Gordon W., and Michael D. Stephens. *Scientific and Technical Education in Nineteenth-Century England: A Symposium*. Newton Abbot, UK: David and Charles, 1972.

Rose, Jonathan. *The Intellectual Life of the British Working Classes*. New Haven, CT: Yale University Press, 2001.

Rouse, Joseph. *How Scientific Practices Matter: Reclaiming Philosophical Naturalism*. Chicago: University of Chicago Press, 2003.

Rudolph, John L. "Epistemology for the Masses: The Origins of 'the Scientific Method' in American Schools." *History of Education Quarterly* 45, no. 3 (2005): 341–76.

——. "Turning Science to Account: Chicago and the General Science Movement in Science Educatia–1920." *Isis* 96, no. 3 (2005): 353–89.

Rupke, Nicolaas A. *Richard Owen: Biology without Darwin*. Chicago: University of Chicago Press, 2009.

Ruse, Michael. *Can a Darwinian Be a Christian? The Relationship between Science and Religion*. Cambridge: Cambridge University Press, 2001.

——. *Evolutionary Naturalism: Selected Essays*. London: Routledge, 1995.

——. "Methodological Naturalism under Attack." *South African Journal of Philosophy* 24, no. 1 (2005): 44–60.

——. "A Philosopher's Day in Court." In *But Is It Science? The Philosophical Question in the Creation/Evolution Controversy*, edited by Michael Ruse, 13–35. Buffalo, NY: Prometheus Books, 1988.

——. "The Relationship between Science and Religion in Britain, 1830–1870." *Church History* 44, no. 4 (1975): 505–22.

Ruskin, John. "Unto This Last." In *The Works of John Ruskin*, edited by E. T. Cook and Alexander Wedderburn. London: George Allen, 1903.

Russell, Colin A. *Edward Frankland: Chemistry, Controversy and Conspiracy in Victorian England*. Cambridge: Cambridge University Press, 1996.

Ryder, John, ed. *American Philosophic Naturalism in the Twentieth Century*. New York: Prometheus Books, 1994.

Rylance, J. H. "The Relation of Miracles to the Christian Faith." In *Christian Truth and Modern Opinion: Seven Sermons Preached in New-York by Clergymen of the Protestant Episcopal Church*, 101–36. New York: Thomas Whittaker, 1874.

Rylance, Rick. *Victorian Psychology and British Culture, 1850–1880*. Oxford: Oxford University Press, 2000.

Schaffer, Simon. "Accurate Measurement Is an English Science." In *The Values of Precision*, edited by M. Norton Wise, 135–72. Princeton, NJ: Princeton University Press, 1995.

———. "Metrology, Metrication, and Victorian Values." In *Victorian Science in Context*, edited by Bernard Lightman, 438–74. Chicago: University of Chicago Press, 1997.

———. "States of Mind: Enlightenment and Natural Philosophy." In *The Languages of Psyche: Mind and Body in Enlightenment Thought*, edited by George Sebastian Rousseau, 233–90. Berkeley: University of California Press, 1990.

"Scheme of Mathematical Study." *Working Men's College Magazine* 1, no. 1 (1859): 9–10.

Schweber, Silvan S. "Demons, Angels and Probability: Some Aspects of British Science in the Nineteenth Century." In *Physics as Natural Philosophy: Essays in Honor of Laszlo Tisza on His Seventy-Fifth Birthday*, edited by Abner Shimony and Herman Feshbach, 319–63. Cambridge, MA: MIT Press, 1982.

Secord, James. *Victorian Sensation: The Extraordinary Publication, Reception, and Secret Authorship of "Vestiges of the Natural History of Creation."* Chicago: University of Chicago Press, 2000.

Shapiro, Adam R. "Between Training and Popularization: Regulating Science Textbooks in Secondary Education." *Isis* 103, no. 1 (2012): 99–110.

———. "Civic Biology and the Origin of the School Antievolution Movement." *Journal of the History of Biology* 41, no. 3 (2008): 409–33.

Shenker, Orly. "Maxwell's Demon and Baron Munchausen: Free Will as a Perpetuum Mobile." *Studies in History and Philosophy of Modern Physics* 30, no. 3 (1999): 347–72.

Siegel, Daniel M. *Innovation in Maxwell's Electromagnetic Theory: Molecular Vortices, Displacement Current, and Light*. Cambridge: Cambridge University Press, 1991.

———. "Mechanical Image and Reality in Maxwell's Electromagnetic Theory." In *Wranglers and Physicists: Studies on Cambridge Mathematical Physics in the Nineteenth Century*, edited by P. M. Harman. Manchester, UK: Manchester University Press, 1985.

———. "Thomson, Maxwell, and the Universal Ether in Victorian Physics." In *Conceptions of Ether: Studies in the History of Ether Theories, 1740–1900*, edited by G. N. Cantor and Michael Jonathan Sessions Hodge. Cambridge: Cambridge University Press, 1981.

Simpson, Thomas K. *Figures of Thought: A Literary Appreciation of Maxwell's "Treatise on Electricity and Magnetism."* Santa Fe: Green Lion Press, 2005.

———. *Maxwell on the Electromagnetic Field: A Guided Study.* New Brunswick, NJ: Rutgers University Press, 1997.

Smith, Crosbie. *The Science of Energy: A Cultural History of Energy Physics in Victorian Britain.* Chicago: University of Chicago Press, 1998.

Smith, Crosbie, and M. Norton Wise. *Energy and Empire: A Biographical Study of Lord Kelvin.* Cambridge: Cambridge University Press, 1989.

Smith, Edward. "The Influence of the Labour of the Treadwheel over Respiration and Pulsation." *Medical Times and Gazette* 14 (1857): 601–3.

Smith, Edward, and William Ralph Milner. "Report on the Action of Prison Diet and Discipline on the Bodily Functions of Prisoners." *British Association for the Advancement of Science* 31 (1861): 44–81.

Smith, Roger. *Free Will and the Human Sciences in Britain, 1870–1910.* London: Pickering and Chatto, 2013.

Snyder, Laura J. *The Philosophical Breakfast Club: Four Remarkable Friends Who Transformed Science and Changed the World.* New York: Broadway, 2011.

———. *Reforming Philosophy: A Victorian Debate on Science and Society.* Chicago: University of Chicago Press, 2006.

Solly, Henry. "Reasons for a Working Men's College." *Working Men's College Magazine* 1, no. 18 (1860): 107–10.

Staley, Richard. *Einstein's Generation.* Chicago: University of Chicago Press, 2008.

Stanley, Matthew. "By Design: James Clerk Maxwell and the Evangelical Unification of Science." *British Journal for the History of Science* 45 (March 2012): 57–73.

———. "A Modern Natural Theology?" *Journal of Faith and Science Exchange* 3 (1999): 105–12.

———. "The Pointsman: Maxwell's Demon, Victorian Free Will, and the Boundaries of Science." *Journal of the History of Ideas* 69, no. 3 (July 2008): 467–91.

———. *Practical Mystic: Religion, Science, and A. S. Eddington.* Chicago: University of Chicago Press, 2007.

———. "Predicting the Past: Ancient Eclipses and Airy, Newcomb, and Huxley on the Authority of Science." *Isis* 103, no. 2 (2012): 254–77.

———. "The Uniformity of Natural Laws in Victorian Britain: Naturalism, Theism, and Scientific Practice." *Zygon* 46, no. 3 (September 2011): 536–60.

Stark, Rodney, and William Sims Bainbridge. *A Theory of Religion*. New Brunswick, NJ: Rutgers University Press, 1987.

Stenmark, Mikael. *Scientism: Science, Ethics, and Religion*. Aldershot, UK: Ashgate, 2001.

Stewart, Balfour. *The Conservation of Energy*. New York: D. Appleton, 1876.

Stewart, Balfour, and J. Norman Lockyer. "The Sun as a Type of the Material Universe." *Macmillan's Magazine* 18 (1868).

Stone, J. A. "Religious Naturalism and the Religion-Science Dialogue: A Minimalist View." *Zygon* 37, no. 2 (2002): 381–94.

———. "Varieties of Religious Naturalism." *Zygon* 38, no. 1 (2003): 89–95.

Tammy Kitzmiller, et al. v. Dover Area School District, et al., 400 F. Supp. 2d 707 (M.D. Pa. 2005).

Taylor, Charles. *A Secular Age*. Cambridge, MA: Harvard University Press, 2007.

Temple, Frederick. *The Relations between Religion and Science: Eight Lectures Preached before the University of Oxford in the Year 1884*. London: Macmillan, 1884.

Theerman, Paul. "James Clerk Maxwell and Religion." *American Journal of Physics* 54 (1986): 312–17.

Thompson, Silvanus P. *Michael Faraday: His Life and Work*. London: Cassell, 1898.

Thomson, James. *Collected Papers in Physics and Engineering*. Cambridge: Cambridge University Press, 1912.

Thomson, William. "On Geological Time." In *Popular Lectures and Addresses*, vol. 2. London: Macmillan, 1894.

———. "On Mechanical Antecedents of Motion, Heat, and Light." In *Report of the Annual Meeting of the British Association for the Advancement of Science*, vol. 24, 59–63. London: John Murray, 1854.

———. *Popular Lectures and Addresses*. 3 vols. London: Macmillan, 1894.

Tomlinson, C. "Michael Faraday." *Graphic* 20, no. 508 (August 23, 1879): 183.

"To Our Readers." *Christian Socialist*, July 5, 1851.

Topham, Jonathan R. "An Infinite Variety of Arguments: The Bridgewater Treatises and British Natural Theology in the 1830's." PhD diss., Lancester University, 1993.

———. "Science, Natural Theology, and Evangelicalism in Early Nineteenth-Century Scotland: Thomas Chalmers and the *Evidence* Controversy." In *Evangelicals and Science in Historical Perspective*, edited by David N. Livingstone, D. G. Hart, and Mark A. Noll, 142–74. Oxford: Oxford University Press, 1999.

Turner, Frank M. *Between Science and Religion: The Reaction to Scientific Naturalism in Late Victorian England*. New Haven, CT: Yale University Press, 1974.

———. *Contesting Cultural Authority: Essays in Victorian Intellectual Life*. Cambridge: Cambridge University Press, 1993.

———. "John Tyndall and Victorian Scientific Naturalism." In *John Tyndall: Essays on a Natural Philosopher*, edited by W. H. Brock, Norman D. McMillan, and R. Charles Mollan, 169–80. Dublin: Royal Dublin Society, 1981.

———. "Rainfall, Plagues, and the Prince of Wales: A Chapter in the Conflict of Religion and Science." *Journal of British Studies* 13, no. 2 (1974): 46–65.

———. "Science and Religious Freedom." In *Freedom and Religion in the Nineteenth Century*, edited by Richard J. Helmstadter, 54–86. Stanford, CA: Stanford University Press, 1997.

———. "The Victorian Conflict between Science and Religion: A Professional Dimension." *Isis* 69, no. 3 (1978): 356–76.

———. "The Victorian Crisis of Faith and the Faith That Was Lost." In *Victorian Faith in Crisis: Essays on Continuity and Change in Nineteenth-Century Religious Belief*, edited by Richard Helmstadter and Bernard Lightman, 9–38. Stanford, CA: Stanford University Press, 1990.

———. "Victorian Scientific Naturalism and Thomas Carlyle." *Victorian Studies* 18, no. 3 (1975): 325–43.

Twelfth Report of the Committee of the Aberdeen Mechanics' Institution. Aberdeen: D. Chalmers, 1837.

Tyndall, John. "Address." In *Report of the Forty-Fourth Meeting of the British Association for the Advancement of Science Held at Belfast in August 1874*. London: John Murray, 1875.

———. "Apology for the Belfast Address." In *Fragments of Science: A Series of Detached Essays, Addresses, and Reviews*, vol. 2, 202–23. New York: D. Appleton, 1915.

———. "The Belfast Address." In *Fragments of Science: A Series of Detached Essays, Addresses, and Reviews*, vol. 2, 135–201. New York: D. Appleton, 1897.

———. "The Constitution of Nature." In *Fragments of Science: A Series of Detached Essays, Addresses, and Reviews*, vol. 1, 3–27. New York: D. Appleton, 1915.

———. *Faraday as a Discoverer*. London: Longmans, Green, 1868.

———. *Fragments of Science: A Series of Detached Essays, Addresses, and Reviews*. 2 vols. New York: D. Appleton, 1915.

———. *Fragments of Science for Unscientific People: A Series of Detached Essays, Addresses, and Reviews*. 2 vols. New York: D. Appleton, 1871–97.

———. "Matter and Force." In *Fragments of Science: A Series of Detached Essays, Addresses, and Reviews*, vol. 2, 53–74. New York: D. Appleton, 1915.

――. "On the Study of Physics." In *Fragments of Science: A Series of Detached Essays, Addresses, and Reviews*, vol. 1, 281–304. New York: D. Appleton, 1915.

――. "The 'Prayer for the Sick': Hints towards a Serious Attempt to Estimate Its Value." *Contemporary Review* 20 (1872): 205.

――. "Professor Virchow and Evolution." In *Fragments of Science: A Series of Detached Essays, Addresses, and Reviews*, vol. 2, 373–418. New York: D. Appleton, 1915.

――. "Review of *Life and Letters of Faraday*." In *Fragments of Science: A Series of Detached Essays, Addresses, and Reviews*, vol. 1. New York: D. Appleton, 1915.

――. "The Rev. James Martineau and the Belfast Address." In *Fragments of Science: A Series of Detached Essays, Addresses, and Reviews*, vol. 2, 224–50. New York: D. Appleton, 1915.

――. "Science and Man." In *Fragments of Science: A Series of Detached Essays, Addresses, and Reviews*, vol. 2, 335–73. New York: D. Appleton, 1915.

――. "Scientific Materialism." In *Fragments of Science: A Series of Detached Essays, Addresses, and Reviews*, vol. 2, 75–90. New York: D. Appleton, 1897.

――. "Scientific Use of the Imagination." In *Fragments of Science: A Series of Detached Essays, Addresses, and Reviews*, vol. 2, 101–34. New York: D. Appleton, 1897.

"Usurpation and Slavery." *Christian Socialist*, November 16, 1850.

Van Whye, John. *Phrenology and the Origins of Victorian Scientific Naturalism*. Burlington, VT: Ashgate, 2004.

Vidal, Fernando. *The Sciences of the Soul: The Early Modern Origins of Psychology*. Chicago: University of Chicago Press, 2011.

Vincent, David. *Bread, Knowledge and Freedom: A Study of Nineteenth-Century Working Class Autobiography*. London: Europa Publications, 1981.

Voskuhl, Adelheid. "The Mechanics of Sentiment: Music-Playing Women Automata and the Culture of Affect in Late Eighteenth-Century Europe." PhD diss., Cornell University, 2007.

Warwick, Andrew. *Masters of Theory: Cambridge and the Rise of Mathematical Physics*. Chicago: University of Chicago Press, 2003.

"What Use Is It?" *Working Men's College Magazine*, no. 9, September 1, 1859, 138.

White, Paul. "Ministers of Culture: Arnold, Huxley, and Liberal Anglican Reform of Learning." *History of Science* 43 (2005): 115–38.

――. *Thomas Huxley: Making the "Man of Science."* Cambridge: Cambridge University Press, 2003.

Williams, S. C. *Religious Belief and Popular Culture in Southwark, c. 1880–1939*. Oxford: Oxford University Press, 1999.

Wilson, David B., ed. *The Correspondence between Sir George Gabriel Stokes and Sir William Thomson, Baron Kelvin of Largs.* 2 vols. Vol. 1, *1846–1869*. Cambridge: Cambridge University Press, 1990.

———. *The Correspondence between Sir George Gabriel Stokes and Sir William Thomson, Baron Kelvin of Largs.* 2 vols. Vol. 2, *1870–1901*. Cambridge: Cambridge University Press, 1990.

———. "The Educational Matrix: Physics Education at Early-Victorian Cambridge, Edinburgh and Glasgow Universities." In *Wranglers and Physicists: Studies on Cambridge Mathematical Physics in the Nineteenth Century*, edited by P. M. Harman, 12–48. Manchester, UK: Machester University Press, 1985.

———. *Kelvin and Stokes: A Comparative Study in Victorian Physics.* Bristol, UK: A. Hilger, 1987.

———. "A Physicist's Alternative to Materialism: The Religious Thought of George Gabriel Stokes." In *Energy and Entropy: Science and Culture in Victorian Britain*, edited by Patrick Brantlinger, 177–203. Bloomington: Indiana University Press, 1989.

———. *Seeking Nature's Logic: Natural Philosophy in the Scottish Enlightenment.* University Park: Pennsylvania State University Press, 2009.

Winter, Alison. *Mesmerized: Powers of Mind in Victorian Britain.* Chicago: University of Chicago Press, 1998.

Wise, M. Norton. "The Gender of Automata in Victorian Britain." In *Genesis Redux: Essays in the History and Philosophy of Artificial Life*, edited by Jessica Riskin, 163–95. Chicago: University of Chicago Press, 2007.

———. "The Maxwell Literature and British Dynamical Theory." *Historical Studies in the Physical Sciences* 13, no. 1 (1982): 175–205.

Yeo, Richard. *Defining Science: William Whewell, Natural Knowledge and Public Debate in Early Victorian Britain.* Cambridge: Cambridge University Press, 2003.

———. "The Principle of Plenitude and Natural Theology in Nineteenth Century Britain." *British Journal for the History of Science* 19 (1986): 263–82.

Yolton, John. *Thinking Matter: Materialism in Eighteenth-Century Britain.* Minneapolis: University of Minnesota Press, 1983.

Young, Robert M. *Darwin's Metaphor: Nature's Place in Victorian Culture.* Cambridge: Cambridge University Press, 1985.

———. *Mind, Brain, and Adaptation in the Nineteenth Century: Cerebral Localization and Its Biological Context from Gall to Ferrier.* Oxford: Oxford University Press, 1970.

Index